KU-261-689

THE INTERNATIONAL SYSTEM OF UNITS (SI)

QUANTITY	BASIC UNIT	SYMBOL
Length	meter	m
Mass	kilogram	kg
Time	second	s
Electric current	ampere	A
Thermodynamic temperature	degree kelvin	K
Luminous intensity	candela	cd

STANDARDIZED PREFIXES TO SIGNIFY POWERS OF 10

PREFIX	SYMBOL	POWER
atto	a	10^{-18}
femto	f	10^{-15}
pico	p	10^{-12}
nano	n	10^{-9}
micro	μ	10^{-6}
milli	m	10^{-3}
centi	c	10^{-2}
deci	d	10^{-1}
deka	da	10
hecto	h	10^{2}
kilo	k	10^{3}
mega	M	10^{6}
giga	G	10^{9}
tera	T	10^{12}

Introductory Circuits for Electrical and Computer Engineering

James W. Nilsson

PROFESSOR EMERITUS
IOWA STATE UNIVERSITY

Susan A. Riedel

MARQUETTE UNIVERSITY

Prentice Hall
Upper Saddle River, New Jersey 07458

Library of Congress Cataloging-in-Publication Data

Nilsson, James W.
    Introductory circuits for electrical and computer
        engineering / James W. Nilsson and Susan A. Riedel
    p.   cm.
    Includes index.
    ISBN 0-13-019855-2
1. Electric circuits.   I. Nilsson, James W.   II. Riedel, Susan A.   III. Title.

Vice President and Editorial Director, ECS: *Marcia J. Horton*
Publisher: *Tom Robbins*
Associate Editor: *Alice Dworkin*
Editorial Assistant: *Jody McDonnell*
Vice President and Director of Production and Manufacturing, ESM: *David W. Riccardi*
Executive Managing Editor: *Vince O'Brien*
Managing Editor: *David A. George*
Production Editor: *Irwin Zucker*
Composition: *PreTEX, Inc.*
Director of Creative Services: *Paul Belfanti*
Creative Director: *Carole Anson*
Art Director: *Maureen Eide*
Assistant Art Director: *John Christiana*
Cover and Interior Design: *Daniel Conte*
Cover Art: *Vanessa Piñeiro*
Audio/Visual Editor: *Xiaohong Zhu*
Manufacturing Manager: *Trudy Pisciotti*
Manufacturing Buyer: *Lisa McDowell*
Marketing Manager: *Holly Stark*
Marketing Assistant: *Karen Moon*

© 2002 by Prentice Hall
Prentice-Hall, Inc.
Upper Saddle River, New Jersey 07458

All rights reserved. No part of this book may be reproduced, in any format or by any means, without permission in writing from the publisher.

The author and publisher of this book have used their best efforts in preparing this book. These efforts include the development, research, and testing of the theories and programs to determine their effectiveness. The author and publisher make no warranty of any kind, expressed or implied, with regard to these programs or the documentation contained in this book. The author and publisher shall not be liable in any event for incidental or consequential damages in connection with, or arising out of, the furnishing, performance, or use of these programs.

LabView is a trademark of National Instruments, Inc., 11500 N Mopac Expressway, Austin, TX 78759-3504. Orcad and PSpice are registered trademarks of Cadence Design Systems, 13221 S.W 68th Parkway, Suite 200, Portland, OR 97223. MATLAB is a registered trademark of The MathWorks, Inc., 3 Apple Hill Drive, Natick, MA, 01760-2098.

Printed in the United States of America

10 9 8 7 6 5 4 3 2 1

ISBN 0-13-019855-2

Pearson Education Ltd., *London*
Pearson Education Australia Pty. Limited, *Sydney*
Pearson Education Singapore, Pte. Ltd.
Pearson Education North Asia Ltd., *Hong Kong*
Pearson Education Canada Inc., *Toronto*
Pearson Education Hispanoamericana, S.A., *Mexico*
Pearson Educatíon de Mexico, S.A. de C.V.
Pearson Education—Japan, *Tokyo*
Pearson Education Malaysia, Pte. Ltd.

To Anna

Contents

Preface

Introductory Circuits for Electrical and Computer Engineering is a one-semester version of the most widely used introductory circuits text of the past 15 years. Importantly, the underlying teaching approaches and philosophies remain unchanged. The goals are:

- *To build an understanding of concepts and ideas explicitly in terms of previous learning.* The learning challenges faced by students of engineering circuit analysis are prodigious; each new concept is built on a foundation of many other concepts. In *Electric Circuits*, much attention is paid to helping students recognize how new concepts and ideas fit together with those previously learned.

- *To emphasize the relationship between conceptual understanding and problem-solving approaches.* Developing the students' problem-solving skills continues to be the central challenge in this course. To address this challenge, examples and simple drill exercises are used to demonstrate problem-solving approaches and to offer students practice opportunities. We do so not with the primary aim of giving students procedural models for solving problems; rather, we emphasize problem solving as a thought process in which one applies conceptual understanding to the solution of a practical problem. As such, in both the textual development and in the worked-out examples, we place great emphasis on a problem-solving process based on concepts rather than the use of rote procedures. Students are encouraged to think through problems before attacking them, and we often pause to consider the broader implications of a specific problem-solving situation.

- *To provide students with a strong foundation of engineering practices.* There are limited opportunities in a sophomore-year circuit analysis course to introduce students to real-world engineering experiences. We continue to emphasize the opportunities that do exist by making a strong effort to develop problems and exercises that use realistic values and represent realizable physical situations. We have included many application-type problems and exercises to help stimulate students' interest in engineering. Many of these problems require the kind of insight an engineer is expected to display when solving problems.

WHAT'S NEW IN *INTRODUCTORY CIRCUITS FOR ELECTRICAL AND COMPUTER ENGINEERING*

The condensation of *Electric Circuits* to a one-semester textbook on introductory circuits for electrical and computer engineers has been accomplished by showing how the basic techniques of circuit analysis are used to analyze circuits of particular interest in the world of digital computation. Hence, after introducing circuit variables and basic circuit elements in Chapter 1 some circuit simplification techniques are introduced in Chapter 2 that are then used to facilitate the analysis of a digital-to-analog resistive ladder circuit.

The digital-to-analog resistive ladder is the first of a series of Practical Perspectives that supports the orientation of the textbook toward digital systems. The others are:

Chapter 4	The Operational Amplifier
Practical Perspective	The Flash Converter
Chapter 5	The Natural and Step Response of RL and RC Circuits
Practical Perspective	Dual Slope Analog-to-Digital Converter
Chapter 6	Natural and Step Response of RLC Circuits
Practical Perspective	Parasitic Inductance
Chapter 8	Introduction to the Laplace Transform
Practical Perspective	Two-Stage RC Ladder
Chapter 9	The Laplace Transform in Circuit Analysis
Practical Perspective	Creation of a Voltage Surge

Integration of Computer Tools

Computer tools cannot replace the traditional methods for mastering the study of electric circuits. They can, however, assist students in the learning process by providing a visual representation of a circuit's behavior, validating a calculated solution, reducing the computational burden of more complex circuits, and iterating toward a desired solution using parameter variation. This computational support is often invaluable in the design process.

Introductory Circuits for Electrical and Computer Engineering continues the support for two popular computer tools, PSpice and MATLAB, into the main text with the addition of icons identifying chapter problems suited for exploration with one or both of these tools. The icon [P] identifies those problems to investigate with PSpice, while the icon [M] identifies problems to investigate with MATLAB. Instructors are provided with computer files containing the PSpice or MATLAB simulation of the problems so marked.

DESIGN EMPHASIS

We continue to support the emphasis on design of circuits in two ways. First, design oriented chapter problems have been explicitly labeled with the icon ❖, enabling students and instructors to identify those problems with a design focus. Second, the identification of problems specifically suited to exploration with PSpice or MATLAB suggests design opportunities using one or both of these computer tools.

Text Design and Pedagogical Features

Introductory Circuits for Electrical and Computer Engineering continues the successful design introduced in the sixth edition of *Electric Circuits*, including the following features:

- *Practical Perspective introductions* are located opposite eight chapter opening pages and are highlighted with a second-color background.
- *Practical Perspective examples* at the end of these eight chapters are set apart in an easy-to-identify separate section.
- *Practical Perspective problems* in the Chapter Problem sets are indicated with an icon ◆ for easy reference.
- *Key terms* are set in boldface when they are first defined. They also appear in boldface in the chapter summaries. This makes it easier for students to find the definitions of important terms.
- *Design problems* in the Chapter Problem sets are indicated with an icon ❖ for easy reference.
- *PSpice problems* in the Chapter Problem sets are indicated with an icon [P] for easy reference.
- *MATLAB problems* in the Chapter Problem sets are indicated with an icon [M] for easy reference.

EXAMPLES, DRILL EXERCISES, AND HOMEWORK PROBLEMS

Solved Numerical Examples

Solved numerical examples are used extensively throughout the text to help students understand how theory is applied to circuit analysis. Because many students value worked examples more than any other aspect of the text, these examples represent an important opportunity to influence the development of student's problem-solving behavior. The nature and format of the examples in *Introductory Circuits for Electrical and Computer Engineering* are a reflection of the overall teaching approach of the text. When presenting a solution, we place great emphasis on the importance of problem solving as a thought process that applies underlying concepts, as we discussed earlier. By emphasizing this idea—even in the solution of simple problems—we hope to communicate that this approach to problem solving can help students handle the more complex problems they will encounter later on. Some characteristics of the examples include:

- encouraging the student to study the problem or the circuit and to make initial observations before diving into a solution pathway;

- emphasizing the individual stages of the solution as part of solving the problem systematically, without suggesting that there are rote procedures for problem solving;

- exploring decision making, that is, the idea that we are often faced with choosing among many different solution approaches; and

- suggesting that students challenge their results by emphasizing the importance of checking and testing answers based on their knowledge of circuit theory and the real world.

Drill Exercises

Drill exercises are included in the text to give students an opportunity to test their understanding of the material they have just read. The drill exercises are presented in a double-column format as a way of signaling to students that they should stop and solve the exercises before proceeding to the next section.

Homework Problems

The homework problems are one of the book's most attractive features. The problems are designed around the following objectives (in parentheses are the corresponding problem categories identified in the *Instructor's Manual* and an illustrative problem number):

- To give students practice in using the analytical techniques developed in the text (Practice; see Problem 3.7)

- To show students that analytical techniques are tools, not objectives (Analytical Tool; see Problem 3.2)

- To give students practice in choosing the analytical method to be used in obtaining a solution (Open Method; see Problem 3.48)

- To show students how the results from one solution can be used to find other information about a circuit's operation (Additional Information; see Problem 3.65)

- To encourage students to challenge the solution either by using an alternate method or by testing the solution to see if it makes sense in terms of known circuit behavior (Solution Check; see Problem 5.12)

- To introduce students to design oriented problems (Design; see Problem 4.30)

- To give students practice in deriving and manipulating equations where quantities of interest are expressed as functions of circuit variables such as R, L, C, ω, and so forth; this type of problem also supports the design process (Derivation; see Problem 7.27)

- To challenge students with problems that will stimulate their interest in both electrical and computer engineering (Practical; see Problem 9.76)

PREREQUISITES

In writing the first seven chapters of the text, we have assumed that the reader has taken a course in elementary differential and integral calculus. We have also assumed that the reader has had an introductory physics course, at either the high school or university level, that introduces the concepts of energy, power, electric charge, electric current, electric potential, and electromagnetic fields. In writing the final two chapters, we have assumed the student has had, or is enrolled in, an introductory course in differential equations.

SUPPLEMENTS

Students and professors are constantly challenged in terms of time and energy by the confines of the classroom and the importance of integrating new information and technologies into an electric circuits course. Through the following supplements, we believe we have succeeded in making some of these challenges more manageable.

PSpice for Introductory Circuits for Electrical and Computer Engineering

This supplement is published as a separate booklet to facilitate its use at a computer. This supplement presents topics in PSpice in the same order as those presented in the text, and expressly supports the use of OrCad PSpice Release 9.2.

Student Workbook

This new supplement is provided for those students who might benefit from some additional "coaching" in their problem solving skills. Each solution technique is presented as a recipe, or a series of solution steps, and illustrated for several example problems. Then problems are presented for the students to solve, and each step in the solution is prompted individually. Finally, students are directed to additional Chapter Problems from the text to which the technique may be applied. The workbook is available as a PDF document on the companion web site so students can print and use whatever sections of the workbook they need.

Instructor's Manual

The Instructor's Manual enables professors to orient themselves quickly to this text and the supplement package. This supplement can be found on the book's web site

http://www.prenhall.com/nilsson.

For easy reference, the following information is organized for each chapter:

- a chapter overview
- problem categorizations
- problem references by chapter section
- a list of examples

Solutions Manual

The solutions manual is available on CD, it contains solutions with supporting figures to each of the nearly 650 end-of-chapter problems. These solutions are presented on the CD in both PDF and LaTEX™ format. This supplement is available free to all adopting faculty, it is not available to students. Files for the PSpice solutions and MATLAB solutions for all indicated problems are included on the CD.

Companion Web Site

The companion web site to accompany the text is located at *http://www.prenhall.com/nilsson*. The following materials are available on the web site:

- Power Point slides and key figures from the text
- student workbook
- instructor's manual
- Syllabus Manager™
- sample chapters
- dynamic message board

Acknowledgments

We continue to express our appreciation for the contributions of Norman Wittels of Worcester Polytechnic Institute. His contributions to the Practical Perspectives greatly enhanced both this edition and the previous one.

There were many hard-working people behind the scenes of our publisher who deserve our thanks and gratitude for their efforts on behalf of this book. We thank Tom Robbins, Irwin Zucker, Jody McDonnell, and Alice Dworkin for all their hard work.

We are deeply indebted to the many instructors and students who have offered positive feedback, suggestions for improvement. Gary E. Ford and James McNames have been very helpful in shaping the contents of this book. We are especially thankful to Ken Kruempel who spent considerable time and effort proofreading and verifying the accuracy of content in the text and the solutions manual.

1 Circuit Variables and Circuit Elements

Chapter Contents

Electrical and computer engineering are exciting and challenging professions for anyone who has a genuine interest in, and aptitude for, applied science and mathematics. Over the past century and a half, electrical and computer engineers have played a dominant role in the development of systems that have changed the way people live and work. Satellite communication links, telephones, digital computers, televisions, diagnostic and surgical medical equipment, assembly-line robots, and electrical power tools are representative components of systems that define a modern technological society. As an electrical or computer engineer, you can participate in this ongoing technological revolution by improving and refining these existing systems and by discovering and developing new systems to meet the needs of our ever-changing society.

We begin our study with an overview of circuit analysis. This is followed by an introduction to the concepts of voltage, current, the basic circuit elements, power, and energy. Next we introduce both independent and dependent voltage and current sources. We conclude this introductory chapter with a discussion of electrical resistance, Ohm's law, and Kirchhoff's laws.

1.1 CIRCUIT ANALYSIS: AN OVERVIEW

Before becoming involved in the details of circuit analysis, we need to take a broad look at engineering design, specifically the design of electric circuits. The purpose of this overview is to provide you with a perspective on where circuit analysis fits within the whole of circuit design. Even though this book focuses on circuit analysis, we try to provide opportunities for circuit design where appropriate.

All engineering designs begin with a need, as shown in Fig. 1.1. This need may come from the desire to improve on an existing design, or it may be something brand-new. A careful assessment of the need results in design specifications, which are measurable characteristics of a proposed design. Once a design is proposed, the design specifications allow us to assess whether or not the design actually meets the need.

A concept for the design comes next. The concept derives from a complete understanding of the design specifications coupled with an insight into the need, which comes from education and experience. The concept may be realized as a sketch, as a written description, or in some other form. Often the next step is to translate the concept into a mathematical model. A commonly used mathematical model for electrical and computer systems is a **circuit model**.

The elements that comprise the circuit model are called **ideal circuit components**. An ideal circuit component is a mathematical model of an actual electrical component, like a battery or a light bulb. It is important for the ideal circuit component used in a circuit model to represent the behavior of the actual electrical component to an acceptable degree of accuracy. The tools of **circuit analysis**, the focus of this book, are then applied to the circuit. Circuit analysis is based on mathematical techniques and is used to predict the behavior of the circuit model and its ideal circuit components. A comparison between the desired behavior, from the design specifications, and the predicted behavior, from circuit analysis, may lead to refinements in the circuit model and its ideal

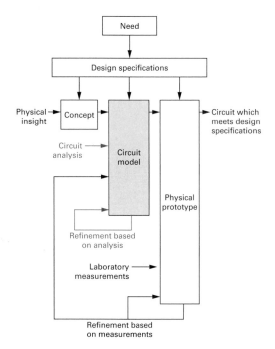

Figure 1.1 A conceptual model for electrical engineering design.

circuit elements. Once the desired and predicted behavior are in agreement, a physical prototype can be constructed.

The **physical prototype** is an actual electrical system, constructed from actual electrical components. Measurement techniques are used to determine the actual, quantitative behavior of the physical system. This actual behavior is compared with the desired behavior from the design specifications and the predicted behavior from circuit analysis. The comparisons may result in refinements to the physical prototype, the circuit model, or both. Eventually, this iterative process, in which models, components, and systems are continually refined, may produce a design that accurately matches the design specifications and thus meets the need.

From this description, it is clear that circuit analysis plays a very important role in the design process. Because circuit analysis is applied to circuit models, practicing engineers try to use mature circuit models so that the resulting designs will meet the design specifications in the first iteration. In this book, we use models that have been tested for between 20 and 100 years; you can assume that they are mature. The ability to model actual electrical systems with ideal circuit elements makes circuit theory extremely useful to engineers.

Saying that the interconnection of ideal circuit elements can be used to quantitatively predict the behavior of a system implies that we can describe the interconnection with mathematical equations. For the mathematical equations to be useful, we must write them in terms of measurable quantities. In the case of circuits, these quantities are voltage and current, which we discuss in Section 1.2. The study of circuit analysis involves understanding the behavior of each ideal circuit element in terms of its voltage and current and understanding the constraints imposed on the voltage and current as a result of interconnecting the ideal elements.

1.2 VOLTAGE, CURRENT, AND THE BASIC CIRCUIT ELEMENTS

Electric Charge

The concept of electric charge is the basis for describing all electrical phenomena. Let's review some important characteristics of electric charge.

- The charge is bipolar, meaning that electrical effects are described in terms of positive and negative charges.

- The electric charge exists in discrete quantities, which are integral multiples of the electronic charge, 1.6022×10^{-19} C.
- Electrical effects are attributed to both the separation of charge and charges in motion.

Voltage and Current

In circuit theory, the separation of charge creates an electric force (voltage), and the motion of charge creates an electric fluid (current).

The concepts of voltage and current are useful from an engineering point of view because they can be expressed quantitatively. Whenever positive and negative charges are separated, energy is expended. **Voltage** is the energy per unit charge created by the separation. We express this ratio in differential form as

$$v = \frac{dw}{dq},\qquad(1.1)$$

where

$$v = \text{the voltage in volts,}$$

$$w = \text{the energy in joules,}$$

$$q = \text{the charge in coulombs.}$$

The electrical effects caused by charges in motion depend on the rate of charge flow. The rate of charge flow is known as the **electric current**, which is expressed as

$$i = \frac{dq}{dt},\qquad(1.2)$$

where

$$i = \text{the current in amperes,}$$

$$q = \text{the charge in coulombs,}$$

$$t = \text{the time in seconds.}$$

Equations (1.1) and (1.2) are definitions for the magnitude of voltage and current, respectively. The bipolar nature of electric charge requires that we assign polarity references to these variables. We will do so in the next section.

Although current is made up of discrete, moving electrons, we do not need to consider them individually because of the enormous number of them. Rather, we can think of electrons and their corresponding charge as one smoothly flowing entity. Thus, i is treated as a continuous variable.

One advantage of using circuit models is that we can model a component strictly in terms of the voltage and current at its terminals. Thus two physically different components could have the same relationship between the terminal voltage and terminal current. If they do, for purposes of circuit analysis, they are identical. Once we know how a component behaves at its terminals, we can analyze its behavior in a circuit. However, when developing circuit models, we are interested in a component's internal behavior. We might want to know, for example, whether charge conduction is taking place because of free electrons moving through the crystal lattice structure of a metal or whether it is because of electrons moving within the covalent bonds of a semiconductor material. However, these concerns are beyond the realm of circuit theory. In this book we use circuit models that have already been developed; we do not discuss how component models are developed.

1.3   THE IDEAL BASIC CIRCUIT ELEMENT

An **ideal basic circuit element** has three attributes: (1) it has only two terminals, which are points of connection to other circuit components; (2) it is described mathematically in terms of current and/or voltage; and (3) it cannot be subdivided into other .elements. We use the word *ideal* to imply that a basic circuit element does not exist as a realizable physical component. We use the word *basic* to imply that the circuit element cannot be further reduced or subdivided into other elements. Thus the basic circuit elements form the building blocks for constructing circuit models, but they themselves cannot be modeled with any other type of element.

Figure 1.2 is a representation of an ideal basic circuit element. The box is blank because we are making no commitment at this time as to the type of circuit element it is. In Fig. 1.2, the voltage across the terminals of the box is denoted by v, and the current in the circuit element is denoted by i. The polarity reference for the voltage is indicated by the plus and minus signs, and the reference direction for the current is shown by the arrow placed alongside the current. The interpretation of these references given positive or negative numerical values of v and i is summarized in Table 1.1. Note that algebraically the notion of positive charge flowing in one direction is equivalent to the notion of negative charge flowing in the opposite direction.

The assignments of the reference polarity for voltage and the reference direction for current are entirely arbitrary. However, once you have assigned the references, you must write all subse-

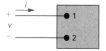

Figure 1.2 An ideal basic circuit element.

TABLE 1.1   **Interpretation of Reference Directions in Fig. 1.2.**

	POSITIVE VALUE	NEGATIVE VALUE
v	voltage drop from terminal 1 to terminal 2	voltage rise from terminal 1 to terminal 2
	or	*or*
	voltage rise from terminal 2 to terminal 1	voltage drop from terminal 2 to terminal 1
i	positive charge flowing from terminal 1 to terminal 2	positive charge flowing from terminal 2 to terminal 1
	or	*or*
	negative charge flowing from terminal 2 to terminal 1	negative charge flowing from terminal 1 to terminal 2

quent equations to agree with the chosen references. The most widely used sign convention applied to these references is called the **passive sign convention**, which we use throughout this book. The passive sign convention can be stated as follows:

> Whenever the reference direction for the current in an element is in the direction of the reference voltage drop across the element (as in Fig. 1.2), use a positive sign in any expression that relates the voltage to the current. Otherwise, use a negative sign.

We apply this sign convention in all the analyses that follow. Our purpose for introducing it even before we have introduced the different types of basic circuit elements is to impress on you the fact that the selection of polarity references along with the adoption of the passive sign convention is *not* a function of the basic elements or the type of interconnections made with the basic elements. We present the application and interpretation of the passive sign convention in power calculations in Section 1.4.

There are five ideal basic circuit elements: voltage sources, current sources, resistors, inductors, and capacitors. In this chapter we discuss the characteristics of voltage sources, current sources, and resistors. Although this may seem like a small number of elements with which to begin analyzing circuits, many practical systems can be modeled with just sources and resistors. They are also a useful starting point because of their relative simplicity; the mathematical relationships between voltage and current in sources and resistors are algebraic. Thus you will be able to begin learning the basic techniques of circuit analysis with only algebraic manipulations.

We will postpone introducing inductors and capacitors until Chapter 5, because their use requires that you solve integral and differential equations. However, the basic analytical techniques for solving circuits with inductors and capacitors are the same as those introduced in this chapter. So, by the time you need to begin manipulating more difficult equations, you should be very familiar with the methods of writing them.

DRILL EXERCISES

1.1   The current at the terminals of the element in Fig. 1.2 is

$$i = 0, \qquad\qquad t < 0;$$

$$i = 20e^{-5000t}\,\text{A}, \qquad t \geq 0.$$

Calculate the total charge (in microcoulombs) entering the element at its upper terminal.

ANSWER: 4000 μC.

1.2   The expression for the charge entering the upper terminal of Fig. 1.2 is

$$q = \frac{1}{\alpha^2} - \left(\frac{t}{\alpha} + \frac{1}{\alpha^2} \right) e^{-\alpha t}\ \text{C}.$$

Find the maximum value of the current entering the terminal if $\alpha = 0.03679\ \text{s}^{-1}$.

ANSWER: 10 A.

1.4 POWER AND ENERGY

Power and energy calculations also are important in circuit analysis. One reason is that although voltage and current are useful variables in the analysis and design of electrically based systems, the useful output of the system often is nonelectrical, and this output is conveniently expressed in terms of power or energy. Another reason is that all practical devices have limitations on the amount of power that they can handle. In the design process, therefore, voltage and current calculations by themselves are not sufficient.

We now relate power and energy to voltage and current and at the same time use the power calculation to illustrate the passive sign convention. Recall from basic physics that power is the time rate of expending or absorbing energy. (A water pump rated 75 kW can deliver more liters per second than one rated 7.5 kW.) Mathematically, energy per unit time is expressed in the form of a derivative, or

$$p = \frac{dw}{dt}, \qquad\qquad (1.3)$$

where

$$p = \text{the power in watts,}$$

$$w = \text{the energy in joules,}$$

$$t = \text{the time in seconds.}$$

Thus 1 W is equivalent to 1 J/s.

The power associated with the flow of charge follows directly from the definition of voltage and current in Eqs. (1.1) and (1.2), or

$$p = \frac{dw}{dt} = \left(\frac{dw}{dq}\right)\left(\frac{dq}{dt}\right) = vi, \tag{1.4}$$

where

$$p = \text{the power in watts,}$$

$$v = \text{the voltage in volts,}$$

$$i = \text{the current in amperes.}$$

Equation (1.4) shows that the **power** associated with a basic circuit element is simply the product of the current in the element and the voltage across the element. Therefore, power is a quantity associated with a pair of terminals, and we have to be able to tell from our calculation whether power is being delivered to the pair of terminals or extracted from it. This information comes from the correct application and interpretation of the passive sign convention.

If we use the passive sign convention, Eq. (1.4) is correct if the reference direction for the current is in the direction of the reference voltage drop across the terminals. Otherwise, Eq. (1.4) must be written with a minus sign. In other words, if the current reference is in the direction of a reference voltage rise across the terminals, the expression for the power is

$$p = -vi. \tag{1.5}$$

The algebraic sign of power is based on charge movement through voltage drops and rises. As positive charges move through a drop in voltage, they lose energy, and as they move through a rise in voltage, they gain energy. Figure 1.3 summarizes the relationship between the polarity references for voltage and current and the expression for power.

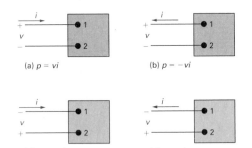

Figure 1.3 Polarity references and the expression for power.

We can now state the rule for interpreting the algebraic sign of power:

If the power is positive (that is, if $p > 0$), power is being delivered to the circuit inside the box. If the power is negative (that is, if $p < 0$), power is being extracted from the circuit inside the box.

For example, suppose that we have selected the polarity references shown in Fig. 1.3(b). Assume further that our calculations for the current and voltage yield the following numerical results:

$$i = 4 \text{ A} \quad \text{and} \quad v = -10 \text{ V}.$$

Then the power associated with the terminal pair 1,2 is

$$p = -(-10)(4) = 40 \text{ W}.$$

Thus the circuit inside the box is absorbing 40 W.

To take this analysis one step further, assume that a colleague is solving the same problem but has chosen the reference polarities shown in Fig. 1.3(c). The resulting numerical values are

$$i = -4 \text{ A}, \qquad v = 10 \text{ V}, \qquad \text{and} \qquad p = 40 \text{ W}.$$

Note that interpreting these results in terms of this reference system gives the same conclusions that we previously obtained— namely, that the circuit inside the box is absorbing 40 W. In fact, any of the reference systems in Fig. 1.3 yields this same result.

DRILL EXERCISES

1.3   Assume that a 20 V voltage drop occurs across an element from terminal 2 to terminal 1 and that a current of 4 A enters terminal 2.

   (a) Specify the values of v and i for the polarity references shown in Fig. 1.3(a)–(d).

   (b) State whether the circuit inside the box is absorbing or delivering power.

   (c) How much power is the circuit absorbing?

ANSWER: (a) Circuit 1.3(a): $v = -20$ V, $i = -4$A; circuit 1.3(b): $v = -20$ V, $i = 4$ A; circuit 1.3(c): $v = 20$ V, $i = -4$ A; circuit 1.3(d): $v = 20$ V, $i = 4$ A; (b) absorbing; (c) 80 W.

1.4 Assume that the voltage at the terminals of the element in Fig. 1.2 corresponding to the current in Drill Exercise 1.1 is

$$v = 0, \qquad\qquad t < 0;$$

$$v = 10e^{-5000t} \text{ kV}, \qquad t \geq 0.$$

Calculate the total energy (in joules) delivered to the circuit element.

1.5 A high-voltage direct-current (dc) transmission line between Celilo, Oregon and Sylmar, California is operating at 800 kV and carrying 1800 A, as shown. Calculate the power (in megawatts) at the Oregon end of the line and state the direction of power flow.

ANSWER: 1440 MW, Celilo to Sylmar.

ANSWER: 20 J.

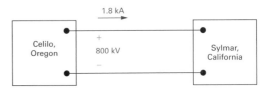

1.5 VOLTAGE AND CURRENT SOURCES

Before discussing ideal voltage and current sources, we need to consider the general nature of electrical sources. An **electrical source** is a device that is capable of converting nonelectric energy to electric energy and vice versa. A discharging battery converts chemical energy to electric energy, whereas a battery being charged converts electric energy to chemical energy. A dynamo is a machine that converts mechanical energy to electric energy and vice versa. If operating in the mechanical-to-electric mode, it is called a generator. If transforming from electric to mechanical energy, it is referred to as a motor. The important thing to remember about these sources is that they can either deliver or absorb electric power, generally maintaining either voltage or current. This behavior is of particular interest for circuit analysis and led to the creation of the ideal voltage source and the ideal current source as basic circuit elements. The challenge is to model practical sources in terms of the ideal basic circuit elements.

An **ideal voltage source** is a circuit element that maintains a prescribed voltage across its terminals regardless of the current flowing in those terminals. Similarly, an **ideal current source** is a circuit element that maintains a prescribed current through its terminals regardless of the voltage across those terminals. These circuit elements do not exist as practical devices—they are idealized models of actual voltage and current sources.

Using an ideal model for current and voltage sources places an important restriction on how we may describe them mathematically. Because an ideal voltage source provides a steady voltage, even if the current in the element changes, it is impossible to specify the current in an ideal voltage source as a function of its voltage. Likewise, if the only information you have about an ideal current source is the value of current supplied, it is impossible to determine the voltage across that current source. We have sacrificed our ability to relate voltage and current in a practical source for the simplicity of using ideal sources in circuit analysis.

Ideal voltage and current sources can be further described as either independent sources or dependent sources. An **independent source** establishes a voltage or current in a circuit without relying on voltages or currents elsewhere in the circuit. The value of the voltage or current supplied is specified by the value of the independent source alone. In contrast, a **dependent source** establishes a voltage or current whose value depends on the value of a voltage or current elsewhere in the circuit. You cannot specify the value of a dependent source unless you know the value of the voltage or current on which it depends.

The circuit symbols for the ideal independent sources are shown in Fig. 1.4. Note that a circle is used to represent an independent source. To completely specify an ideal independent voltage source in a circuit, you must include the value of the supplied voltage and the reference polarity, as shown in Fig. 1.4(a). Similarly, to completely specify an ideal independent current source, you must include the value of the supplied current and its reference direction, as shown in Fig. 1.4(b).

The circuit symbols for the ideal dependent sources are shown in Fig. 1.5. A diamond is used to represent a dependent source. Both the dependent current source and the dependent voltage source may be controlled by either a voltage or a current elsewhere in the circuit, so there are a total of four variations, as indicated by the symbols in Fig. 1.5. Dependent sources are sometimes called **controlled sources**.

To completely specify an ideal dependent voltage-controlled voltage source, you must identify the controlling voltage, the equation that permits you to compute the supplied voltage from the controlling voltage, and the reference polarity for the sup-

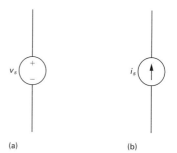

Figure 1.4 The circuit symbols for (a) an ideal independent voltage source and (b) an ideal independent current source.

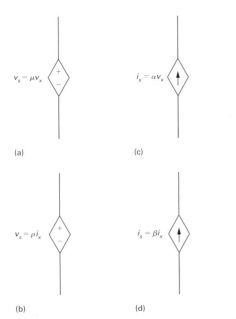

(a)

(c)

(b)

(d)

Figure 1.5 The circuit symbols for (a) an ideal dependent voltage-controlled voltage source, (b) an ideal dependent current-controlled voltage source, (c) an ideal dependent voltage-controlled current source, and (d) an ideal dependent current-controlled current source.

plied voltage. In Fig. 1.5(a), the controlling voltage is named v_x, the equation that determines the supplied voltage v_s is

$$v_s = \mu v_x,$$

and the reference polarity for v_s is as indicated. Note that μ is a multiplying constant that is dimensionless.

Similar requirements exist for completely specifying the other ideal dependent sources. In Fig. 1.5(b), the controlling current is i_x, the equation for the supplied voltage v_s is

$$v_s = \rho i_x,$$

the reference polarity is as shown, and the multiplying constant ρ has the dimension volts per ampere. In Fig. 1.5(c), the controlling voltage is v_x, the equation for the supplied current i_s is

$$i_s = \alpha v_x,$$

the reference direction is as shown, and the multiplying constant α has the dimension amperes per volt. In Fig. 1.5(d), the controlling current is i_x, the equation for the supplied current i_s is

$$i_s = \beta i_x,$$

the reference direction is as shown, and the multiplying constant β is dimensionless.

Finally, in our discussion of ideal sources, we note that they are examples of active circuit elements. An **active element** is one that models a device capable of generating electric energy. **Passive elements** model physical devices that cannot generate electric energy. Resistors, inductors, and capacitors are examples of passive circuit elements. Examples 1.1 and 1.2 illustrate how the characteristics of ideal independent and dependent sources limit the types of permissible interconnections of the sources.

EXAMPLE 1.1

Using the definitions of the ideal independent voltage and current sources, state which interconnections in Fig. 1.6 are permissible and which violate the constraints imposed by the ideal sources.

SOLUTION

Connection (a) is valid. Each source supplies voltage across the same pair of terminals, marked a,b. This requires that each source supply the same voltage with the same polarity, which they do.

Connection (b) is valid. Each source supplies current through the same pair of terminals, marked a,b. This requires that each source supply the same current in the same direction, which they do.

Connection (c) is not permissible. Each source supplies voltage across the same pair of terminals, marked a,b. This requires that each source supply the same voltage with the same polarity, which they do not.

Connection (d) is not permissible. Each source supplies current through the same pair of terminals, marked a,b. This requires that each source supply the same current in the same direction, which they do not.

Connection (e) is valid. The voltage source supplies voltage across the pair of terminals marked a,b. The current source supplies current through the same pair of terminals. Because an ideal voltage source supplies the same voltage regardless of the current, and an ideal current source supplies the same current regardless of the voltage, this is a permissible connection.

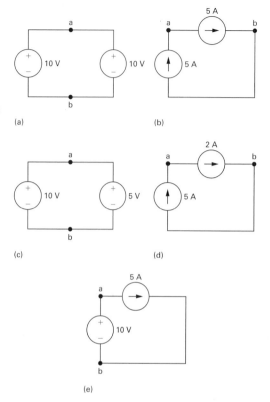

Figure 1.6 The circuits for Example 1.1.

EXAMPLE 1.2

Using the definitions of the ideal independent and dependent sources, state which interconnections in Fig. 1.7 are valid and which violate the constraints imposed by the ideal sources.

SOLUTION

Connection (a) is invalid. Both the independent source and the dependent source supply voltage across the same pair of terminals, labeled a,b. This requires that each source supply the same voltage with the same polarity. The independent source supplies 5 V, but the dependent source supplies 15 V.

Connection (b) is valid. The independent voltage source supplies voltage across the pair of terminals marked a,b. The dependent current source supplies current through the same pair of terminals. Because an ideal voltage source supplies the same

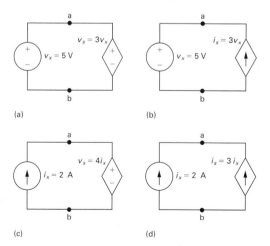

Figure 1.7 The circuits for Example 1.2.

voltage regardless of current, and an ideal current source supplies the same current regardless of voltage, this is an allowable connection.

Connection (c) is valid. The independent current source supplies current through the pair of terminals marked a,b. The dependent voltage source supplies voltage across the same pair of terminals. Because an ideal current source supplies the same current regardless of voltage, and an ideal voltage source supplies the same voltage regardless of current, this is an allowable connection.

Connection (d) is invalid. Both the independent source and the dependent source supply current through the same pair of terminals, labeled a,b. This requires that each source supply the same current in the same reference direction. The independent source supplies 2 A, but the dependent source supplies 6 A in the opposite direction.

1.6 ELECTRICAL RESISTANCE (OHM'S LAW)

Resistance is the capacity of materials to impede the flow of current or, more specifically, the flow of electric charge. The circuit element used to model this behavior is the **resistor**. Figure 1.8 shows the circuit symbol for the resistor, with R denoting the resistance value of the resistor.

Conceptually, we can understand resistance if we think about the moving electrons that make up electric current interacting with and being resisted by the atomic structure of the material through which they are moving. In the course of these interactions, some amount of electric energy is converted to thermal energy and dissipated in the form of heat. This effect may be undesirable. However, many useful electrical devices take advantage of resistance heating, including stoves, toasters, irons, and space heaters.

Most materials exhibit measurable resistance to current. The amount of resistance depends on the material. Metals such as copper and aluminum have small values of resistance, making them good choices for wiring used to conduct electric current. In fact, when represented in a circuit diagram, copper or aluminum wiring isn't usually modeled as a resistor; the resistance of the wire is so small compared to the resistance of other elements in the circuit that we can neglect the wiring resistance to simplify the diagram.

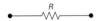

Figure 1.8 The circuit symbol for a resistor having a resistance R.

For purposes of circuit analysis, we must reference the current in the resistor to the terminal voltage. We can do so in two ways: either in the direction of the voltage drop across the resistor or in the direction of the voltage rise across the resistor, as shown in Fig. 1.9. If we choose the former, the relationship between the voltage and current is

$$v = iR, \tag{1.6}$$

where

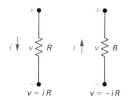

Figure 1.9 Two possible reference choices for the current and voltage at the terminals of a resistor, and the resulting equations.

$$v = \text{the voltage in volts,}$$

$$i = \text{the current in amperes,}$$

$$R = \text{the resistance in ohms.}$$

If we choose the second method, we must write

$$v = -iR, \tag{1.7}$$

where v, i, and R are, as before, measured in volts, amperes, and ohms, respectively. The algebraic signs used in Eqs. (1.6) and (1.7) are a direct consequence of the passive sign convention, which we introduced in Section 1.3.

Equations (1.6) and (1.7) are known as **Ohm's law** after Georg Simon Ohm, a German physicist who established its validity early in the nineteenth century. Ohm's law is the algebraic relationship between voltage and current for a resistor. In SI[1] units, resistance is measured in ohms. The Greek letter omega (Ω) is the standard symbol for an ohm. The circuit diagram symbol for an 8 Ω resistor is shown in Fig. 1.10.

Ohm's law expresses the voltage as a function of the current. However, expressing the current as a function of the voltage also is convenient. Thus, from Eq. (1.6),

Figure 1.10 The circuit symbol for an 8 Ω resistor.

$$i = \frac{v}{R}, \tag{1.8}$$

or, from Eq. (1.7),

$$i = -\frac{v}{R}. \tag{1.9}$$

The reciprocal of the resistance is referred to as **conductance**, is symbolized by the letter G, and is measured in siemens (S). Thus

$$G = \frac{1}{R}. \tag{1.10}$$

---

[1] SI units are discussed in Appendix A.

An 8 Ω resistor has a conductance value of 0.125 S. In much of the professional literature, the unit used for conductance is the mho (ohm spelled backward), which is symbolized by an inverted omega (℧). Therefore we may also describe an 8 Ω resistor as having a conductance of 0.125 mho, (℧).

We use ideal resistors in circuit analysis to model the behavior of physical devices. Using the qualifier *ideal* reminds us that the resistor model makes several simplifying assumptions about the behavior of actual resistive devices. The most important of these simplifying assumptions is that the resistance of the ideal resistor is constant and its value does not vary over time. Most actual resistive devices do not have constant resistance, and their resistance does vary over time. The ideal resistor model can be used to represent a physical device whose resistance doesn't vary much from some constant value over the time period of interest in the circuit analysis. In this book we assume that the simplifying assumptions about resistance devices are valid, and we thus use ideal resistors in circuit analysis.

We may calculate the power at the terminals of a resistor in several ways. The first approach is to use the defining equation and simply calculate the product of the terminal voltage and current. For the reference systems shown in Fig. 1.9, we write

$$p = vi \tag{1.11}$$

when $v = iR$ and

$$p = -vi \tag{1.12}$$

when $v = -iR$.

A second method of expressing the power at the terminals of a resistor expresses power in terms of the current and the resistance. Substituting Eq. (1.6) into Eq. (1.11), we obtain

$$p = vi = (iR)i = i^2R. \tag{1.13}$$

Likewise, substituting Eq. (1.7) into Eq. (1.12), we have

$$p = -vi = -(-iR)i = i^2R. \tag{1.14}$$

Equations (1.13) and (1.14) are identical and demonstrate clearly that, regardless of voltage polarity and current direction, the power at the terminals of a resistor is positive. Therefore, a resistor absorbs power from the circuit.

A third method of expressing the power at the terminals of a resistor is in terms of the voltage and resistance. The expression is independent of the polarity references, so

$$p = \frac{v^2}{R}. \tag{1.15}$$

Sometimes a resistor's value will be expressed as a conductance rather than as a resistance. Using the relationship between resistance and conductance given in Eq. (1.10), we may also write Eqs. (1.14) and (1.15) in terms of the conductance, or

$$p = \frac{i^2}{G},$$ (1.16)

$$p = v^2 G.$$ (1.17)

Equations (1.11)–(1.17) provide a variety of methods for calculating the power absorbed by a resistor. Each yields the same answer. In analyzing a circuit, look at the information provided and choose the power equation that uses that information directly.

Example 1.3 illustrates the application of Ohm's law in conjunction with an ideal source and a resistor. Power calculations at the terminals of a resistor also are illustrated.

EXAMPLE 1.3

In each circuit in Fig. 1.11, the value of either v or i is not known.

(a) Calculate the values of v and i.

(b) Determine the power dissipated in each resistor.

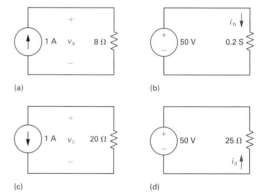

(a)   (b)

(c)   (d)

Figure 1.11 The circuits for Example 1.3.

SOLUTION

(a) The voltage v_a in Fig. 1.11(a) is a drop in the direction of the current in the resistor. Therefore,

$$v_a = (1)(8) = 8 \text{ V}.$$

The current i_b in the resistor with a conductance of 0.2 S in Fig. 1.11(b) is in the direction of the voltage drop across the resistor. Thus

$$i_b = (50)(0.2) = 10 \text{ A}.$$

The voltage v_c in Fig. 1.11(c) is a rise in the direction of the current in the resistor. Hence

$$v_c = -(1)(20) = -20 \text{ V}.$$

The current i_d in the 25 Ω resistor in Fig. 1.11(d) is in the direction of the voltage rise across the resistor. Therefore

$$i_d = \frac{-50}{25} = -2 \text{ A.}$$

(b) The power dissipated in each of the four resistors is

$$p_{8\Omega} = \frac{(8)^2}{8} = (1)^2(8) = 8 \text{ W,}$$

$$p_{0.2S} = (50)^2(0.2) = 500 \text{ W,}$$

$$p_{20\Omega} = \frac{(-20)^2}{20} = (1)^2(20) = 20 \text{ W,}$$

$$p_{25\Omega} = \frac{(50)^2}{25} = (-2)^2(25) = 100 \text{ W.}$$

1.7 KIRCHHOFF'S LAWS

A circuit is said to be solved when the voltage across and the current in every element have been determined. Ohm's law is an important equation for deriving such solutions.

In simple circuit structures, like those in Example 1.3, Ohm's law is sufficient for solving for the voltages across and the currents in every element. However, for more complex interconnections we need to use two more important algebraic relationships, known as Kirchhoff's laws, to solve for all the voltages and currents. To see how Kirchhoff's laws enter into circuit analysis consider the circuit shown in Fig. 1.12, where we have labeled the current and voltage variables associated with each resistor and the current associated with the voltage source. Labeling includes reference polarities, as always. For convenience, we attach the same subscript to the voltage and current labels as we do to the resistor labels. In Fig. 1.12, a **node** is a point where two or more circuit elements meet. It is necessary to identify nodes in order to use Kirchhoff's current law, as we will see in a moment. In Fig. 1.12, the nodes are labeled a, b, c, and d. Node d stretches all the way across the top of the diagram, though we label a single point for convenience.

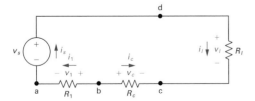

Figure 1.12 A circuit used to introduce Kirchhoff's laws. The voltage and current variables are assigned as shown.

For the circuit shown in Fig. 1.12, we can identify seven unknowns: i_s, i_1, i_c, i_l, v_1, v_c, and v_l. The variable v_s is a known voltage. The problem is to find the seven unknown variables. From algebra, you know that to find n unknown quantities you must solve n simultaneous independent equations. From our discussion of Ohm's law in Section 1.6, you know that three of the necessary equations are

$$v_1 = i_1 R_1, \tag{1.18}$$

$$v_c = i_c R_c, \tag{1.19}$$

$$v_l = i_l R_l. \tag{1.20}$$

What about the other four equations?

The interconnection of circuit elements imposes constraints on the relationship between the terminal voltages and currents. These constraints are referred to as Kirchhoff's laws, after Gustav Kirchhoff, who first stated them in a paper published in 1848. The two laws that state the constraints in mathematical form are known as Kirchhoff's current law and Kirchhoff's voltage law.

We can now state **Kirchhoff's current law**:

The algebraic sum of all the currents at any node in a circuit equals zero.

To use Kirchhoff's current law, an algebraic sign corresponding to a reference direction must be assigned to every current at the node. Assigning a positive sign to a current leaving a node requires assigning a negative sign to a current entering a node. Conversely, giving a negative sign to a current leaving a node requires giving a positive sign to a current entering a node.

Applying Kirchhoff's current law to the four nodes in the circuit shown in Fig. 1.12, using the convention that currents leaving a node are considered positive, yields four equations:

$$\text{node a} \qquad i_s - i_1 = 0, \tag{1.21}$$

$$\text{node b} \qquad i_1 + i_c = 0, \tag{1.22}$$

$$\text{node c} \qquad -i_c - i_l = 0, \tag{1.23}$$

$$\text{node d} \qquad i_l - i_s = 0. \tag{1.24}$$

Note that Eqs. (1.21)–(1.24) are not an independent set, because any one of the four can be derived from the other three. In any circuit with n nodes, $n - 1$ independent current equations can be derived from Kirchhoff's current law.[2] Let's disregard

---

[2] We say more about this observation in Chapter 3.

Eq. (1.24) so that we have six independent equations, namely, Eqs. (1.18)–(1.23). We need one more, which we can derive from Kirchhoff's voltage law.

Before we can state Kirchhoff's voltage law, we must define a **closed path** or **loop**. Starting at an arbitrarily selected node, we trace a closed path in a circuit through selected basic circuit elements and return to the original node without passing through any intermediate node more than once. The circuit shown in Fig. 1.12 has only one closed path or loop. For example, choosing node a as the starting point and tracing the circuit clockwise, we form the closed path by moving through nodes d, c, b, and back to node a. We can now state **Kirchhoff's voltage law**:

> The algebraic sum of all the voltages around any closed path in a circuit equals zero.

To use Kirchhoff's voltage law, we must assign an algebraic sign (reference direction) to each voltage in the loop. As we trace a closed path, a voltage will appear as either a rise or a drop in the tracing direction. Assigning a positive sign to a voltage rise requires assigning a negative sign to a voltage drop. Conversely, giving a negative sign to a voltage rise requires giving a positive sign to a voltage drop.

We now apply Kirchhoff's voltage law to the circuit shown in Fig. 1.12. We elect to trace the closed path clockwise, assigning a positive algebraic sign to voltage drops. Starting at node d leads to the expression

$$v_l - v_c + v_1 - v_s = 0, \tag{1.25}$$

which represents the seventh independent equation needed to find the seven unknown circuit variables mentioned earlier.

The thought of having to solve seven simultaneous equations to find the current delivered by a voltage source to a string of resistors is not very appealing. Thus in the coming chapters we introduce you to analytical techniques that will enable you to solve a simple one-loop circuit by writing a single equation. However, before moving on to a discussion of these circuit techniques, we need to make several observations about the detailed analysis of the circuit in Fig. 1.12. In general, these observations are true and therefore are important to the discussions in subsequent chapters.

First, note that if you know the current in a resistor, you also know the voltage across the resistor, because current and voltage are directly related through Ohm's law. Thus you can associate one unknown variable with each resistor, either the current or the voltage. Choose, say, the current as the unknown variable.

Then, once you solve for the unknown current in the resistor, you can find the voltage across the resistor. In general, if you know the current in a passive element, you can find the voltage across it, greatly reducing the number of simultaneous equations to be solved. For example, in the circuit in Fig. 1.12, we can eliminate the voltages v_c, v_l, and v_1 as unknowns. Thus at the outset we reduce the analytical task to solving four simultaneous equations rather than seven.

The second general observation relates to the consequences of connecting only two elements to form a node. According to Kirchhoff's current law, when only two elements connect to a node, if you know the current in one of the elements, you also know it in the second element. In other words, you need define only one unknown current for the two elements. When just two elements connect at a single node, the elements are said to be **in series**. The importance of this second observation is obvious when you note that each node in the circuit shown in Fig. 1.12 involves only two elements. Thus you need to define only one unknown current. The reason is that Eqs. (1.21)–(1.23) lead directly to

$$i_s = i_1 = -i_c = i_l, \qquad (1.26)$$

which states that if you know any one of the element currents, you know them all. For example, choosing to use i_s as the unknown eliminates i_1, i_c, and i_l. The problem is reduced to determining one unknown, namely, i_s.

Examples 1.4 and 1.5 illustrate how to write circuit equations based on Kirchhoff's laws. Example 1.6 illustrates how to use Kirchhoff's laws and Ohm's law to find an unknown current. Example 1.7 illustrates constructing a circuit model for a device whose terminal characteristics are known.

EXAMPLE 1.4

Sum the currents at each node in the circuit shown in Fig. 1.13. Note that there is no connection dot (●) in the center of the diagram, where the 4 Ω branch crosses the branch containing the ideal current source i_a.

SOLUTION

In writing the equations, we use a positive sign for a current leaving a node. The four equations are

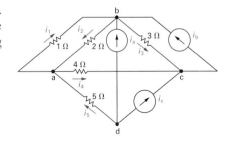

Figure 1.13 The circuit for Example 1.4.

$$\begin{array}{ll} \text{node a} & i_1 + i_4 - i_2 - i_5 = 0, \\ \text{node b} & i_2 + i_3 - i_1 - i_b - i_a = 0, \\ \text{node c} & i_b - i_3 - i_4 - i_c = 0, \\ \text{node d} & i_5 + i_a + i_c = 0. \end{array}$$

EXAMPLE 1.5

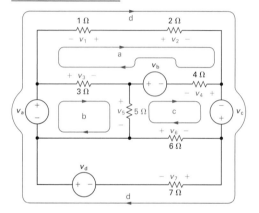

Figure 1.14 The circuit for Example 1.5.

Sum the voltages around each designated path in the circuit shown in Fig. 1.14.

SOLUTION

In writing the equations, we use a positive sign for a voltage drop. The four equations are

$$\begin{array}{ll} \text{path a} & -v_1 + v_2 + v_4 - v_b - v_3 = 0, \\ \text{path b} & -v_a + v_3 + v_5 = 0, \\ \text{path c} & v_b - v_4 - v_c - v_6 - v_5 = 0, \\ \text{path d} & -v_a - v_1 + v_2 - v_c + v_7 - v_d = 0. \end{array}$$

EXAMPLE 1.6

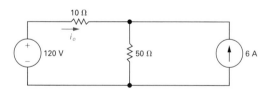

Figure 1.15 The circuit for Example 1.6.

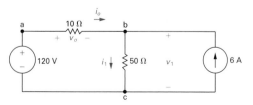

Figure 1.16 The circuit shown in Fig. 1.15, with the unknowns i_1, v_o, and v_1 defined.

(a) Use Kirchhoff's laws and Ohm's law to find i_o in the circuit shown in Fig. 1.15.

(b) Test the solution for i_o by verifying that the total power generated equals the total power dissipated.

SOLUTION

(a) We begin by redrawing the circuit and assigning an unknown current to the 50-Ω resistor and unknown voltages across the 10-Ω and 50-Ω resistors. Figure 1.16 shows the circuit. The nodes are labeled a, b, and c to aid the discussion.

    Because i_o also is the current in the 120 V source, we have two unknown currents and therefore must derive two simultaneous equations involving i_o and i_1. We obtain one of the equations by applying Kirchhoff's current law to either node b or c. Summing the currents at node b and assigning a positive sign to the currents leaving the node gives

$$i_1 - i_o - 6 = 0.$$

We obtain the second equation from Kirchhoff's voltage law in combination with Ohm's law. Noting from Ohm's law that v_o is $10i_o$ and v_1 is $50i_1$, we sum the voltages around the closed path cabc to obtain

$$-120 + 10i_o + 50i_1 = 0.$$

In writing this equation, we assigned a positive sign to voltage drops in the clockwise direction. Solving these two equations for i_o and i_1 yields

$$i_o = -3 \text{ A} \qquad \text{and} \qquad i_1 = 3 \text{ A}.$$

(b) The power dissipated in the 50 Ω resistor is

$$p_{50\Omega} = (3)^2 50 = 450 \text{ W}.$$

The power dissipated in the 10 Ω resistor is

$$p_{10\Omega} = (-3)^2(10) = 90 \text{ W}.$$

The power delivered to the 120 V source is

$$p_{120V} = -120i_o = -120(-3) = 360 \text{ W}.$$

The power delivered to the 6 A source is

$$p_{6A} = -v_1(6), \qquad \text{but} \qquad v_1 = 50i_1 = 150 \text{ V}.$$

Therefore

$$p_{6A} = -150(6) = -900 \text{ W}.$$

The 6 A source is delivering 900 W, and the 120 V source is absorbing 360 W. The total power absorbed is $360 + 450 + 90 = 900$ W. Therefore, the solution verifies that the power delivered equals the power absorbed.

EXAMPLE 1.7

The terminal voltage and terminal current were measured on the device shown in Fig. 1.17(a), and the values of v_t and i_t are tabulated in Fig. 1.17(b).

(a) Construct a circuit model of the device inside the box.

(b) Using this circuit model, predict the power this device will deliver to a 10 Ω resistor.

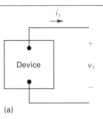

v_t (V)	i_t (A)
30	0
15	3
0	6

(a)   (b)

Figure 1.17 (a) Device and (b) data for Example 1.7.

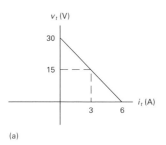

(a)

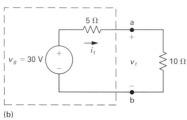

(b)

Figure 1.18 (a) The graph of v_t versus i_t for the device in Fig. 1.17(a). (b) The resulting circuit model for the device in Fig. 1.17(a), connected to a 10 Ω resistor.

SOLUTION

(a) Plotting the voltage as a function of the current yields the graph shown in Fig. 1.18(a). The equation of the line plotted is

$$v_t = 30 - 5i_t.$$

Now we need to identify the components of a circuit model that will produce the same relationship between voltage and current. Kirchhoff's voltage law tells us that the voltage drops across two components in series add. From the equation, one of those components produces a 30 V drop regardless of the current. This component can be modeled as an ideal independent voltage source.

The other component produces a positive voltage drop in the direction of the current i_t. Because the voltage drop is proportional to the current, Ohm's law tells us that this component can be modeled as an ideal resistor with a value of 5 Ω. The resulting circuit model is depicted in the dashed box in Fig. 1.18(b).

(b) Now we attach a 10 Ω resistor to the device in Fig. 1.18(b) to complete the circuit. Kirchhoff's current law tells us that the current in the 10 Ω resistor is the same as the current in the 5 Ω resistor. Using Kirchhoff's voltage law and Ohm's law, we can write the equation for the voltage drops around the circuit, starting at the voltage source and proceeding clockwise:

$$30 = 5i_t + 10i_t.$$

Solving for i_t, we get

$$i_t = 2 \text{ A}.$$

Because this is the value of current flowing in the 10 Ω resistor, we can use the power equation $p = i^2 R$ to compute the power delivered to this resistor:

$$p_{10\Omega} = (2)^2(10) = 40 \text{ W}.$$

DRILL EXERCISES

1.6 If the interconnection is valid find the total power developed in the circuit. If the interconnection is not valid, explain why.

ANSWER: Valid Interconnection, 3100 W

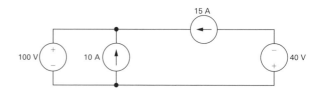

1.7   **(a)** Show that Eq. (1.25) can be written as

$$i_s R_l + i_s R_c + i_s R_1 - v_s = 0.$$

(b) Write the explicit expression for i_s in terms of v_s, R_1, R_c, and R_l.

ANSWER: (a) derivation;
(b) $i_s = v_s/(R_l + R_c + R_1)$.

1.8   For the circuit shown, calculate (a) i_5; (b) v_1;
(c) v_2; (d) v_5; and (e) the power delivered by the
24 V source.

ANSWER: (a) $i_5 = 2$ A; (b) $v_1 = -4$ V;
(c) $v_2 = 6$ V; (d) $v_5 = 14$ V; (e) 48 W.

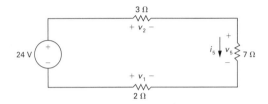

1.9   Use Ohm's law and Kirchhoff's law to find the
value of R in the circuit shown.

ANSWER: $R = 4\ \Omega$.

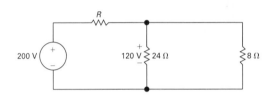

1.10   **(a)** The terminal voltage and terminal current were
measured on the device shown. The values of
v_t and i_t are provided in the table. Using these
values, create the straight line plot of v_t versus
i_t. Compute the equation of the line and use
the equation to construct a circuit model for
the device using an ideal voltage source and a
resistor.

(b) Use the model constructed in (a) to predict the
power that the device will deliver to a 25 Ω
resistor.

ANSWER: (a) A 25 V source in series with a
100 Ω resistor; (b) 1 W.

v_t (V)	i_t (A)
25	0
15	0.1
5	0.2
0	0.25

(a)   (b)

1.11   Repeat Drill Exercise 1.10 but use the equation of
the graphed line to construct a circuit model containing an ideal current source and a resistor.

ANSWER: (a) A 0.25 A current source connected
between the terminals of a 100 Ω resistor; (b) 1 W.

1.8 ANALYSIS OF A CIRCUIT CONTAINING DEPENDENT SOURCES

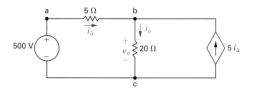

Figure 1.19 A circuit with a dependent source.

We conclude this introduction to elementary circuit analysis with a discussion of a circuit that contains a dependent source, as depicted in Fig. 1.19.

We want to use Kirchhoff's laws and Ohm's law to find v_o in this circuit. Before writing equations, it is good practice to examine the circuit diagram closely. This will help us identify the information that is known and the information we must calculate. It may also help us devise a strategy for solving the circuit using only a few calculations.

A look at the circuit in Fig. 1.19 reveals that

- Once we know i_o, we can calculate v_o using Ohm's law.

- Once we know i_Δ, we also know the current supplied by the dependent source $5i_\Delta$.

- The current in the 500 V source is i_Δ.

There are thus two unknown currents, i_Δ and i_o. We need to construct and solve two independent equations involving these two currents to produce a value for v_o.

From the circuit, notice the closed path containing the voltage source, the 5 Ω resistor, and the 20 Ω resistor. We can apply Kirchhoff's voltage law around this closed path. The resulting equation contains the two unknown currents:

$$500 = 5i_\Delta + 20i_o. \tag{1.27}$$

Now we need to generate a second equation containing these two currents. Consider the closed path formed by the 20 Ω resistor and the dependent current source. If we attempt to apply Kirchhoff's voltage law to this loop, we fail to develop a useful equation, because we don't know the value of the voltage across the dependent current source. In fact, the voltage across the dependent source is v_o, which is the voltage we are trying to compute. Writing an equation for this loop does not advance us toward a solution. For this same reason, we do not use the closed path containing the voltage source, the 5 Ω resistor, and the dependent source.

There are three nodes in the circuit, so we turn to Kirchhoff's current law to generate the second equation. Node a connects the voltage source and the 5 Ω resistor; as we have already observed, the current in these two elements is the same. Either node b or node c can be used to construct the second equation from Kirchhoff's current law. We select node b and produce the following equation:

$$i_o = i_\Delta + 5i_\Delta = 6i_\Delta. \qquad (1.28)$$

Solving Eqs. (1.27) and (1.28) for the currents, we get

$$i_\Delta = 4 \text{ A},$$

$$i_o = 24 \text{ A}. \qquad (1.29)$$

Using Eq. (1.29) and Ohm's law for the 20 Ω resistor, we can solve for the voltage v_o:

$$v_o = 20i_o = 480 \text{ V}.$$

Think about a circuit analysis strategy before beginning to write equations. As we have demonstrated, not every closed path provides an opportunity to write a useful equation based on Kirchhoff's voltage law. Not every node provides for a useful application of Kirchhoff's current law. Some preliminary thinking about the problem can help in selecting the most fruitful approach and the most useful analysis tools for a particular problem. Choosing a good approach and the appropriate tools will usually reduce the number and complexity of equations to be solved.

Example 1.8 illustrates another application of Ohm's law and Kirchhoff's laws to a circuit with a dependent source.

EXAMPLE 1.8

(a) Use Kirchhoff's laws and Ohm's law to find the voltage v_o as shown in Fig. 1.20.

(b) Show that your solution is consistent with the constraint that the total power developed in the circuit equals the total power dissipated.

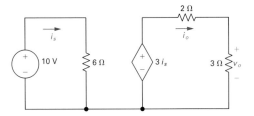

Figure 1.20 The circuit for Example 1.8.

SOLUTION

(a) A close look at the circuit in Fig. 1.20 reveals that:

- There are two closed paths, the one on the left with the current i_s and the one on the right with the current i_o.
- Once i_o is known, we can compute v_o.

We need two equations for the two currents. Because there are two closed paths and both have voltage sources, we can apply Kirchhoff's voltage law to each to give the following equations:

$$10 = 6i_s,$$

$$3i_s = 2i_o + 3i_o.$$

Solving for the currents yields

$$i_s = 1.67 \text{ A},$$

$$i_o = 1 \text{ A}.$$

Applying Ohm's law to the 3 Ω resistor gives the desired voltage:

$$v_o = 3i_o = 3 \text{ V}.$$

(b) To compute the power delivered to the voltage sources, we use the power equation in the form $p = vi$. The power delivered to the independent voltage source is

$$p = (10)(-1.67) = -16.7 \text{ W}.$$

The power delivered to the dependent voltage source is

$$p = (3i_s)(-i_o) = (5)(-1) = -5 \text{ W}.$$

Both sources are developing power, and the total developed power is 21.7 W.

To compute the power delivered to the resistors, we use the power equation in the form $p = i^2 R$. The power delivered to the 6 Ω resistor is

$$p = (1.67)^2(6) = 16.7 \text{ W}.$$

The power delivered to the 2 Ω resistor is

$$p = (1)^2(2) = 2 \text{ W}.$$

The power delivered to the 3 Ω resistor is

$$p = (1)^2(3) = 3 \text{ W}.$$

The resistors all dissipate power, and the total power dissipated is 21.7 W, equal to the total power developed in the sources.

SUMMARY

- Circuit analysis is based on the variables of voltage and current.

- **Voltage** is the energy per unit charge created by charge separation and has the SI unit of volt ($v = dw/dq$).

- **Current** is the rate of charge flow and has the SI unit of ampere ($i = dq/dt$).

- The **ideal basic circuit element** is a two-terminal component that cannot be subdivided; it can be described mathematically in terms of its terminal voltage and current.

- The **passive sign convention** uses a positive sign in the expression that relates the voltage and current at the terminals of an element when the reference direction for the current through the element is in the direction of the reference voltage drop across the element.

- **Power** is energy per unit of time and is equal to the product of the terminal voltage and current; it has the SI unit of watt ($p = dw/dt = vi$).

- The algebraic sign of power is interpreted as follows:

    - If $p > 0$, power is being delivered to the circuit or circuit component.

    - If $p < 0$, power is being extracted from the circuit or circuit component.

- The circuit elements introduced in this chapter are voltage sources, current sources, and resistors:

    - An **ideal voltage source** maintains a prescribed voltage regardless of the current in the device. An **ideal current source** maintains a prescribed current regardless of the voltage across the device. Voltage and current sources are either **independent**, that is, not influenced by any other current or voltage in the circuit; or **dependent**, that is, determined by some other current or voltage in the circuit.

    - A **resistor** constrains its voltage and current to be proportional to each other. The value of the proportional constant relating voltage and current in a resistor is called its **resistance** and is measured in ohms.

- **Ohm's law** establishes the proportionality of voltage and current in a resistor. Specifically,

$$v = iR$$

if the current flow in the resistor is in the direction of the voltage drop across it, or

$$v = -iR$$

if the current flow in the resistor is in the direction of the voltage rise across it.

- By combining the equation for power, $p = vi$, with Ohm's law, we can determine the power absorbed by a resistor:

$$p = i^2 R = v^2/R.$$

- Circuits are described by nodes and closed paths. A **node** is a point where two or more circuit elements join. When just two elements connect to form a node, they are said to be **in series**. A **closed path** is a loop traced through connecting elements, starting and ending at the same node and encountering intermediate nodes only once each.

- The voltages and currents of interconnected circuit elements obey Kirchhoff's laws:

  - **Kirchhoff's current law** states that the algebraic sum of all the currents at any node in a circuit equals zero.

  - **Kirchhoff's voltage law** states that the algebraic sum of all the voltages around any closed path in a circuit equals zero.

- A circuit is solved when the voltage across and the current in every element have been determined. By combining an understanding of independent and dependent sources, Ohm's law, and Kirchhoff's laws, we can solve many simple circuits.

PROBLEMS

1.1. The current entering the upper terminal of Fig. 1.2 is

$$i = 24\cos 4000t \ \text{A}.$$

Assume the charge at the upper terminal is zero at the instant the current is passing through its maximum value. Find the expression for $q(t)$.

1.2. Four series connected 1.5 V batteries supply 100 mA to a portable CD player. How much energy do the batteries supply in 3 h?

1.3. Two electric circuits, represented by boxes A and B, are connected as shown in Fig. P1.3. The reference direction for the current i in the interconnection and the reference polarity for the voltage v across the interconnection are as shown in the figure. For each of the following sets of numerical values, calculate the power in the interconnection and state whether the power is flowing from A to B or vice versa.

 (a) $i = 5$ A, $v = 120$ V

 (b) $i = -8$ A, $v = 250$ V

 (c) $i = 16$ A, $v = -150$ V

 (d) $i = -10$ A, $v = -480$ V

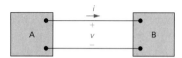

Figure P1.3

1.4. The references for the voltage and current at the terminal of a circuit element are as shown in Fig. 1.3(d). The numerical values for v and i are 40 V and -10 A.

 (a) Calculate the power at the terminals and state whether the power is being absorbed or delivered by the element in the box.

 (b) Given that the current is due to electron flow, state whether the electrons are entering or leaving terminal 2.

 (c) Do the electrons gain or lose energy as they pass through the element in the box?

1.5. Repeat Problem 1.4 with a voltage of -60 V.

1.6. When a car has a dead battery, it can often be started by connecting the battery from another car across its terminals. The positive terminals are connected together as are the negative terminals. The connection is illustrated in Fig. P1.6. Assume the current i in Fig. P1.6 is measured and found to be 30 A.

 (a) Which car has the dead battery?

 (b) If this connection is maintained for 1 min, how much energy is transferred to the dead battery?

1.7. The manufacturer of a 6 V dry-cell flashlight battery says that the battery will deliver 15 mA for 60 continuous hours. During that time the voltage will drop from 6 V to 4 V.

　　Assume the drop in voltage is linear with time. How much energy does the battery deliver in this 60 h interval?

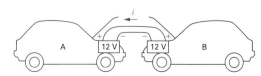

Figure P1.6

1.8. The voltage and current at the terminals of the circuit element in Fig. 1.2 are zero for $t < 0$. For $t \geq 0$ they are

$$v = e^{-500t} - e^{-1500t} \ \text{V},$$

$$i = 30 - 40e^{-500t} + 10e^{-1500t} \ \text{mA}.$$

(a) Find the power at $t = 1$ ms.

(b) How much energy is delivered to the circuit element between 0 and 1 ms?

(c) Find the total energy delivered to the element.

 1.9. The voltage and current at the terminals of the circuit element in Fig. 1.2 are zero for $t < 0$. For $t \geq 0$ they are

$$v = 100e^{-50t} \sin 150t \ \text{V},$$

$$i = 20e^{-50t} \sin 150t \ \text{A}.$$

(a) Find the power absorbed by the element at $t = 20$ ms.

(b) Find the total energy (in millijoules) absorbed by the element.

 1.10. The voltage and current at the terminals of the circuit element in Fig. 1.2 are shown in Fig. P1.10(a) and (b), respectively.

(a) Sketch the power versus t plot for $0 \leq t \leq 100$ ms.

(b) Calculate the energy delivered to the circuit element at $t = 25, 60, 90,$ and 100 ms.

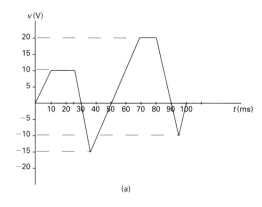

(a)

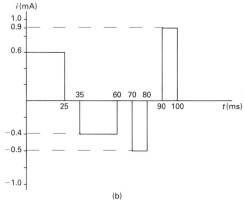

(b)

Figure P1.10

1.11. The voltage and current at the terminals of the circuit element in Fig. 1.2 are zero for $t < 0$. For $t \geq 0$ they are

$$v = 100e^{-500t} \text{ V,}$$

$$i = 20 - 20e^{-500t} \text{ mA.}$$

(a) Find the maximum value of the power delivered to the circuit.

(b) Find the total energy delivered to the element.

1.12. The voltage and current at the terminals of the element in Fig. 1.2 are

$$v = 36 \sin 200\pi t \text{ V,}$$

$$i = 25 \cos 200\pi t \text{ A.}$$

(a) Find the maximum value of the power being delivered to the element.

(b) Find the maximum value of the power being extracted from the element.

(c) Find the average value of p in the interval $0 \leq t \leq 5$ ms.

(d) Find the average value of p in the interval $0 \leq t \leq 6.25$ ms.

1.13. The voltage and current at the terminals of an automobile battery during a charge cycle are shown in Fig. P1.13.

(a) Calculate the total charge transferred to the battery.

(b) Calculate the total energy transferred to the battery.

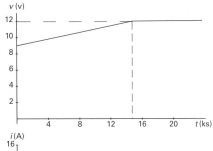

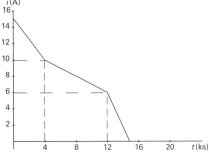

Figure P1.13

1.14. The voltage and current at the terminals of the circuit element in Fig. 1.2 are zero for $t < 0$. For $t \geq 0$ they are

$$v = (10,000t + 5)e^{-400t} \text{ V.}$$

$$i = (40t + 0.05)e^{-400t} \text{ A.}$$

(a) At what instant of time is maximum power delivered to the element?

(b) Find the maximum power in watts.

(c) Find the total energy delivered to the element in millijoules.

1.15. The voltage and current at the terminals of the circuit element in Fig. 1.2 are zero for $t < 0$ and $t > 3$ s. In the interval between 0 and 3 s the expressions are

$$v = t(3 - t) \text{ V,} \qquad 0 < t < 3 \text{ s;}$$

$$i = 6 - 4t \text{ mA,} \qquad 0 < t < 3 \text{ s.}$$

(a) At what instant of time is the power being delivered to the circuit element maximum?

(b) What is the power at the time found in part (a)?

(c) At what instant of time is the power being extracted from the circuit element maximum?

(d) What is the power at the time found in part (c)?

(e) Calculate the net energy delivered to the circuit at 0, 1, 2, and 3 s.

1.16. The voltage and current at the terminals of the circuit element in Fig. 1.2 are zero for $t < 0$. For $t \geq 0$ they are

$$v = 80,000te^{-500t} \text{ V,} \quad t \geq 0;$$

$$i = 15te^{-500t} \text{ A,} \qquad t \geq 0.$$

(a) Find the time (in milliseconds) when the power delivered to the circuit element is maximum.

(b) Find the maximum value of p in milliwatts.

(c) Find the total energy delivered to the circuit element in microjoules.

1.17. Assume you are an engineer in charge of a project and one of your subordinate engineers reports that the interconnection in Fig. P1.17 does not pass the power check. The data for the interconnection are given in Table P1.17.

(a) Is the subordinate correct? Explain your answer.

(b) If the subordinate is correct, can you find the error in the data?

TABLE P1.17

ELEMENT	VOLTAGE (V)	CURRENT (A)
a	46.16	6.00
b	14.16	4.72
c	−32.00	−6.40
d	22.00	1.28
e	33.60	1.68
f	66.00	−0.40
g	2.56	1.28
h	−0.40	0.40

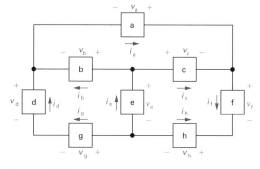

Figure P1.17

1.18. One method of checking calculations involving interconnected circuit elements is to see that the total power delivered equals the total power absorbed (conservation-of-energy principle). With this thought in mind, check the interconnection in Fig. P1.18 and state whether it satisfies this power check. The current and voltage values for each element are given in Table P1.18.

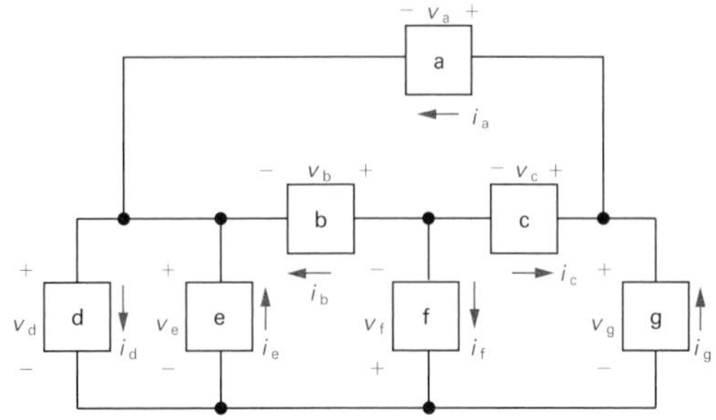

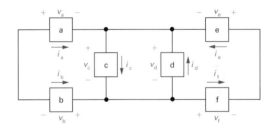

Figure P1.18

TABLE P1.18

ELEMENT	VOLTAGE (V)	CURRENT (A)
a	48	12
b	18	−4
c	30	−10
d	36	16
e	36	8
f	−54	14
g	84	22

1.19. **(a)** In the circuit shown in Fig. P1.19, identify which elements have the voltage and current reference polarities defined using the passive sign convention.

(b) The numerical values of the currents and voltages for each element are given in Table P1.19. How much total power is absorbed and how much is delivered in this circuit?

Figure P1.19

TABLE P1.19

ELEMENT	VOLTAGE (V)	CURRENT (A)
a	−8	7
b	−2	−7
c	10	15
d	10	5
e	−6	3
f	−4	3

1.20. A simplified circuit model for an industrial wiring system is shown in Fig. P1.20.

  (a) How many basic circuit elements are there in this model?

  (b) How many nodes are there in the circuit?

  (c) How many of the nodes connect three or more basic elements?

  (d) Identify the circuit elements that form a series pair.

  (e) What is the minimum number of unknown currents?

  (f) Describe seven closed paths in the circuit.

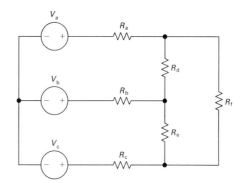

Figure P1.20

1.21. If the interconnection in Fig. P1.21 is valid, find the total power developed in the circuit. If the interconnection is not valid, explain why.

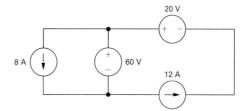

Figure P1.21

1.22. **(a)** Is the interconnection in Fig. P1.22 valid? Explain.

  (b) Can you find the total energy developed in the circuit? Explain.

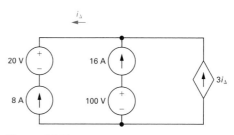

Figure P1.22

1.23. **(a)** Is the interconnection of ideal sources in the circuit in Fig. P1.23 valid? Explain.

  (b) Identify which sources are developing power and which sources are absorbing power.

  (c) Verify that the total power developed in the circuit equals the total power absorbed.

  (d) Repeat (a)–(c), reversing the polarity of the 30 V source.

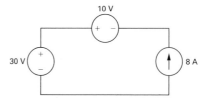

Figure P1.23

1.24. If the interconnection in Fig. P1.24 is valid, find the total power developed in the circuit. If the interconnection is not valid, explain why.

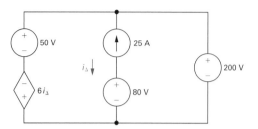

Figure P1.24

1.25. If the interconnection in Fig. P1.25 is valid, find the total power developed in the circuit. If the interconnection is not valid, explain why.

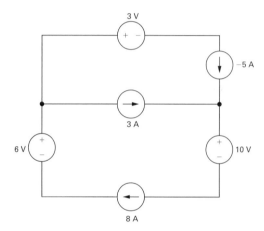

Figure P1.25

1.26. Find the total power developed in the circuit in Fig. P1.26 if $v_o = 100$ V.

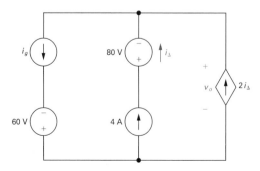

Figure P1.26

1.27. If the interconnection in Fig. P1.27 is valid, find the total power developed by the voltage sources. If the interconnection is not valid, explain why.

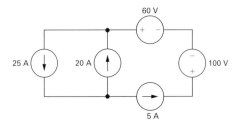

Figure P1.27

1.28. The interconnection of ideal sources can lead to an indeterminate solution. With this thought in mind, explain why the solutions for v_1 and v_2 in the circuit in Fig. P1.28 are not unique.

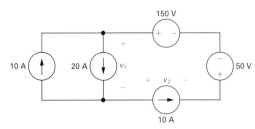

Figure P1.28

1.29. **(a)** Find the currents i_g and i_a in the circuit in Fig. P1.29.

(b) Find the voltage v_g.

(c) Verify that the total power developed equals the total power dissipated.

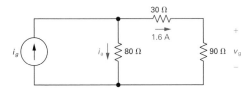

Figure P1.29

1.30. The current i_a in the circuit shown in Fig. P1.30 is 20 A. Find (a) i_o; (b) i_g; and (c) the power delivered by the independent current source.

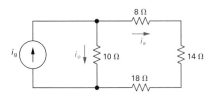

Figure P1.30

1.31. Given the circuit shown in Fig. P1.31, find

    (a) the value of i_a,

    (b) the value of i_b,

    (c) the value of v_o,

    (d) the power dissipated in each resistor,

    (e) the power delivered by the 50 V source.

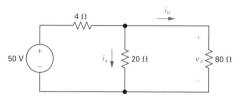

Figure P1.31

1.32. The currents i_1 and i_2 in the circuit in Fig. P1.32 are 20 A and 15 A, respectively.

    (a) Find the power supplied by each voltage source.

    (b) Show that the total power supplied equals the total power dissipated in the resistors.

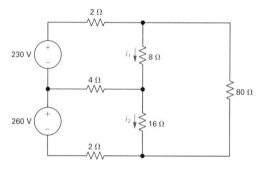

Figure P1.32

1.33. The currents i_a and i_b in the circuit in Fig. P1.33 are 4 A and 2 A, respectively.

    (a) Find i_g.

    (b) Find the power dissipated in each resistor.

    (c) Find v_g.

    (d) Show that the power delivered by the current source is equal to the power absorbed by all the other elements.

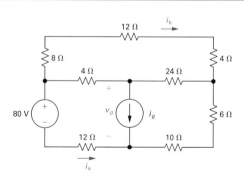

Figure P1.33

1.34. The current i_o in the circuit in Fig. P1.34 is 4 A.

   (a) Find i_1.

   (b) Find the power dissipated in each resistor.

   (c) Verify that the total power dissipated in the circuit equals the power developed by the 180 V source.

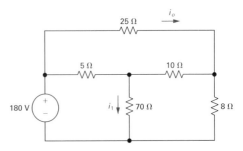

Figure P1.34

1.35. The table in Fig. P1.35(a) gives the relationship between the terminal current and voltage of the practical constant current source shown in Fig. P1.35(b).

   (a) Plot i_s versus v_s.

   (b) Construct a circuit model of this current source that is valid for $0 \leq v_s \leq 30$ V, based on the equation of the line plotted in (a).

   (c) Use your circuit model to predict the current delivered to a 3 kΩ resistor.

   (d) Use your circuit model to predict the open-circuit voltage of the current source.

   (e) What is the actual open-circuit voltage?

   (f) Explain why the answers to (d) and (e) are not the same.

i_s (mA)	v_s (V)
40	0
35	10
30	20
25	30
18	40
8	50
0	55

(a)

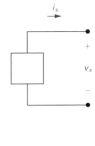

(b)

Figure P1.35

1.36. The voltage and current were measured at the terminals of the device shown in Fig. P1.36(a). The results are tabulated in Fig. 1.36(b).

   (a) Construct a circuit model for this device using an ideal current source and a resistor.

   (b) Use the model to predict the value of i_t when a 20 Ω resistor is connected across the terminals of the device.

(a)

Figure P1.36

v_t (V)	i_t (A)
50	0
65	3
80	6
95	9
110	12
125	15

(b)

1.37. The table in Fig. P1.37(a) gives the relationship between the terminal voltage and current of the practical constant voltage source shown in Fig. P1.37(b).

  (a) Plot v_s versus i_s.

  (b) Construct a circuit model of the practical source that is valid for $0 \le i_s \le 24$ A, based on the equation of the line plotted in (a). (Use an ideal voltage source in series with an ideal resistor.)

  (c) Use your circuit model to predict the current delivered to a 1 Ω resistor connected to the terminals of the practical source.

  (d) Use your circuit model to predict the current delivered to a short circuit connected to the terminals of the practical source.

  (e) What is the actual short-circuit current?

  (f) Explain why the answers to (d) and (e) are not the same.

v_s (V)	i_s (A)
24	0
22	8
20	16
18	24
15	32
10	40
0	48

(a)

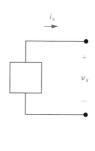

(b)

Figure P1.37

1.38. The voltage and current were measured at the terminals of the device shown in Fig. P1.38(a). The results are tabulated in Fig. P1.38(b).

  (a) Construct a circuit model for this device using an ideal voltage source and a resistor.

  (b) Use the model to predict the amount of power the device will deliver to a 20 Ω resistor.

Device v_t

(a)

v_t (V)	i_t (A)
100	0
180	4
260	8
340	12
420	16

(b)

Figure P1.38

 1.39. For the circuit shown in Fig. P1.39, find (a) R and (b) the power supplied by the 250 V source.

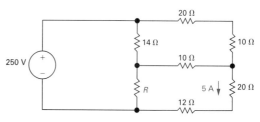

Figure P1.39

1.40. The variable resistor R in the circuit in Fig. P1.40 is adjusted until i_a equals 1 A. Find the value of R.

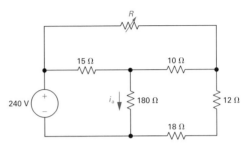

Figure P1.40

1.41. **(a)** Find the voltage vy in the circuit in Fig. P1.41.

(b) Show that the total power generated in the circuit equals the total power absorbed.

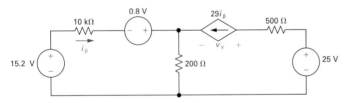

Figure P1.41

1.42. Find (a) i_2, (b) i_1, and (c) i_o in the circuit in Fig. P1.42.

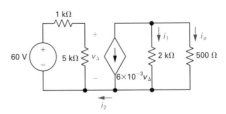

Figure P1.42

❖ **1.43.** It is often desirable in designing an electric wiring system to be able to control a single appliance from two or more locations, for example, to control a lighting fixture from both the top and bottom of a stairwell. In home wiring systems, this type of control is implemented with three-way and four-way switches. A three-way switch is a three-terminal, two-position switch, and a four-way switch is a four-terminal, two-position switch. The switches are shown schematically in Fig. P1.43(a), which illustrates a three-way switch, and P1.43(b), which illustrates a four-way switch.

(a) Show how two three-way switches can be connected between a and b in the circuit in Fig. P1.43(c) so that the lamp *l* can be turned ON or OFF from two locations.

(b) If the lamp (appliance) is to be controlled from more than two locations, four-way switches are used in conjunction with two three-way switches. One four-way switch is required for each location in excess of two. Show how one four-way switch plus two three-way switches can be connected between a and b in Fig. P1.43(c) to control the lamp from three locations. (*Hint:* The four-way switch is placed between the three-way switches.)

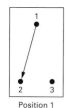 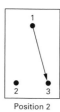  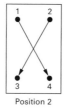 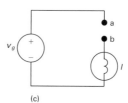

(a) (b) (c)

Figure P1.43

Digital-to-Analog Resistive Ladder

For digital computers and systems to be useful we need to be able to convert digital signals to analog signals and vice versa. The resistive ladder network shown in the accompanying figure will convert a digital signal to an analog signal, the analog signal in this instance being the output voltage v_o while the digital signal controls the position of the n switches. We shall discuss the operation of this digital-to-analog circuit after we have discussed the circuit simplification techniques introduced in this chapter.

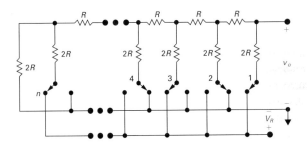

Digital-to-analog resistive ladder.

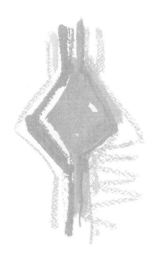

2 Some Circuit Simplification Techniques

Chapter Contents

In Chapter 3 we will introduce you to two analytical techniques (the node-voltage method and the mesh-current method) that have been developed to aid in the analysis of circuit interconnections more complex than the relatively simple circuits we have analyzed thus far. Before embarking on these more sophisticated techniques, however, we want to start introducing some circuit simplification techniques. We have two reasons for doing this. First, the simplification techniques allow us to reduce more complex structures that can be analyzed using only Ohm's law in conjunction with Kirchhoff's laws. Second, studying these circuit simplification techniques will give us a chance to acquaint ourselves thoroughly with the laws underlying the more sophisticated methods.

2.1 COMBINING RESISTORS IN SERIES AND PARALLEL

Resistors in Series

In Chapter 1, we said that when just two elements connect at a single node, they are said to be in series. **Series-connected circuit elements** carry the same current. The resistors in the circuit shown in Fig. 2.1 are connected in series. We can show that these resistors carry the same current by applying Kirchhoff's current law to each node in the circuit. The series interconnection in Fig. 2.1 requires that

$$i_s = i_1 = -i_2 = i_3 = i_4 = -i_5 = -i_6 = i_7, \qquad (2.1)$$

which states that if we know any one of the seven currents, we know them all. Thus we can redraw Fig. 2.1 as shown in Fig. 2.2, retaining the identity of the single current i_s.

To find i_s, we apply Kirchhoff's voltage law around the single closed loop. Defining the voltage across each resistor as a drop in the direction of i_s gives

$$-v_s + i_s R_1 + i_s R_2 + i_s R_3 + i_s R_4 + i_s R_5 + i_s R_6 + i_s R_7 = 0, \quad (2.2)$$

or

$$v_s = i_s (R_1 + R_2 + R_3 + R_4 + R_5 + R_6 + R_7). \qquad (2.3)$$

The significance of Eq. (2.3) for calculating i_s is that the seven resistors can be replaced by a single resistor whose numerical value is the sum of the individual resistors, that is,

$$R_{eq} = R_1 + R_2 + R_3 + R_4 + R_5 + R_6 + R_7 \qquad (2.4)$$

and

$$v_s = i_s R_{eq}. \qquad (2.5)$$

Thus we can redraw Fig. 2.2 as shown in Fig. 2.3.

In general, if k resistors are connected in series, the equivalent single resistor has a resistance equal to the sum of the k resistances, or

$$R_{eq} = \sum_{i=1}^{k} R_i = R_1 + R_2 + \cdots + R_k. \qquad (2.6)$$

Note that the resistance of the equivalent resistor is always larger than that of the largest resistor in the series connection.

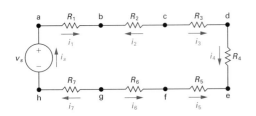

Figure 2.1 Resistors connected in series.

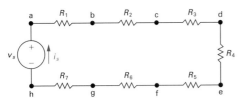

Figure 2.2 Series resistors with a single unknown current i_s.

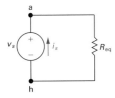

Figure 2.3 A simplified version of the circuit shown in Fig. 2.2.

Another way to think about this concept of an equivalent resistance is to visualize the string of resistors as being inside a black box. (An electrical engineer uses the term **black box** to imply an opaque container; that is, the contents are hidden from view. The engineer is then challenged to model the contents of the box by studying the relationship between the voltage and current at its terminals.) Determining whether the box contains k resistors or a single equivalent resistor is impossible. Figure 2.4 illustrates this method of studying the circuit shown in Fig. 2.2.

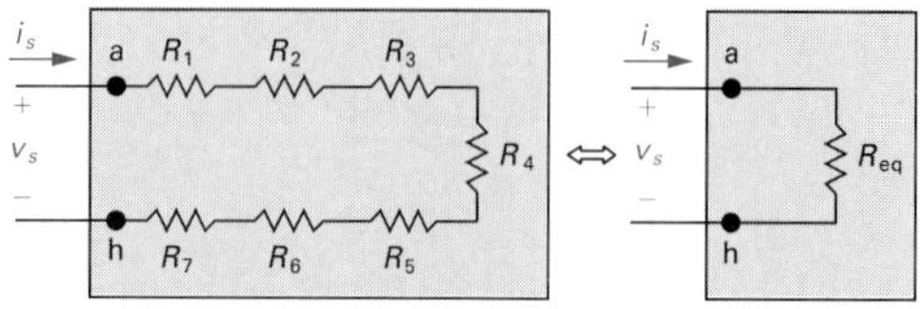

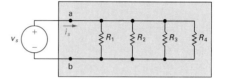

Figure 2.4 The black box equivalent of the circuit shown in Fig. 2.2.

Resistors in Parallel

When two elements connect at a single node pair, they are said to be in parallel. **Parallel-connected circuit elements** have the same voltage across their terminals. The circuit shown in Fig. 2.5 illustrates resistors connected in parallel. Don't make the mistake of assuming that two elements are parallel connected merely because they are lined up in parallel in a circuit diagram. The defining characteristic of parallel-connected elements is that they have the same voltage across their terminals. In Fig. 2.6, you can see that R_1 and R_3 are not parallel connected because, between their respective terminals, another resistor dissipates some of the voltage.

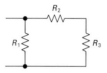

Figure 2.5 Resistors in parallel.

Resistors in parallel can be reduced to a single equivalent resistor using Kirchhoff's current law and Ohm's law, as we now demonstrate. In the circuit shown in Fig. 2.5, we let the currents i_1, i_2, i_3, and i_4 be the currents in the resistors R_1 through R_4, respectively. We also let the positive reference direction for each resistor current be down through the resistor, that is, from node a to node b. From Kirchhoff's current law,

$$i_s = i_1 + i_2 + i_3 + i_4. \qquad (2.7)$$

Figure 2.6 Nonparallel resistors.

The parallel connection of the resistors means that the voltage across each resistor must be the same. Hence, from Ohm's law,

$$i_1 R_1 = i_2 R_2 = i_3 R_3 = i_4 R_4 = v_s. \qquad (2.8)$$

Therefore,

$$i_1 = \frac{v_s}{R_1}, \quad i_2 = \frac{v_s}{R_2}, \quad i_3 = \frac{v_s}{R_3}, \quad \text{and} \quad i_4 = \frac{v_s}{R_4}. \qquad (2.9)$$

Substituting Eq. (2.9) into Eq. (2.7) yields

$$i_s = v_s \left(\frac{1}{R_1} + \frac{1}{R_2} + \frac{1}{R_3} + \frac{1}{R_4} \right), \qquad (2.10)$$

from which

$$\frac{i_s}{v_s} = \frac{1}{R_{eq}} = \frac{1}{R_1} + \frac{1}{R_2} + \frac{1}{R_3} + \frac{1}{R_4}. \tag{2.11}$$

Equation (2.11) is what we set out to show: that the four resistors in the circuit shown in Fig. 2.5 can be replaced by a single equivalent resistor. The circuit shown in Fig. 2.7 illustrates the substitution. For k resistors connected in parallel, Eq. (2.11) becomes

$$\frac{1}{R_{eq}} = \sum_{i=1}^{k} \frac{1}{R_i} = \frac{1}{R_1} + \frac{1}{R_2} + \cdots + \frac{1}{R_k}. \tag{2.12}$$

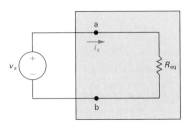

Figure 2.7 Replacing the four parallel resistors shown in Fig. 2.5 with a single equivalent resistor.

Note that the resistance of the equivalent resistor is always smaller than the resistance of the smallest resistor in the parallel connection. Sometimes, using conductance when dealing with resistors connected in parallel is more convenient. In that case, Eq. (2.12) becomes

$$G_{eq} = \sum_{i=1}^{k} G_i = G_1 + G_2 + \cdots + G_k. \tag{2.13}$$

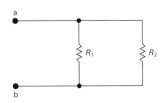

Figure 2.8 Two resistors connected in parallel.

Many times only two resistors are connected in parallel. Figure 2.8 illustrates this special case. We calculate the equivalent resistance from Eq. (2.12):

$$\frac{1}{R_{eq}} = \frac{1}{R_1} + \frac{1}{R_2} = \frac{R_2 + R_1}{R_1 R_2}, \tag{2.14}$$

or

$$R_{eq} = \frac{R_1 R_2}{R_1 + R_2}. \tag{2.15}$$

Thus for just two resistors in parallel the equivalent resistance equals the product of the resistances divided by the sum of the resistances. Remember that you can only use this result in the special case of just two resistors in parallel. Example 2.1 illustrates the usefulness of these results.

EXAMPLE 2.1

Find i_s, i_1, and i_2 in the circuit shown in Fig. 2.9.

SOLUTION

We begin by noting that the 3 Ω resistor is in series with the 6 Ω resistor. We therefore replace this series combination with a 9 Ω resistor, reducing the circuit to the one shown in Fig. 2.10(a). We now can replace the parallel combination of the 9 Ω and 18 Ω resistors with a single resistance of $(18 \times 9)/(18 + 9)$, or 6 Ω. Figure 2.10(b) shows this further reduction of the circuit. The nodes x and y marked on all diagrams facilitate tracing through the reduction of the circuit.

From Fig. 2.10(b) you can verify that i_s equals 120/10, or 12 A. Figure 2.11 shows the result at this point in the analysis. We added the voltage v_1 to help clarify the subsequent discussion. Using Ohm's law we compute the value of v_1:

$$v_1 = (12)(6) = 72 \text{ V}. \tag{2.16}$$

But v_1 is the voltage drop from node x to node y, so we can return to the circuit shown in Fig. 2.10(a) and again use Ohm's law to calculate i_1 and i_2. Thus,

$$i_1 = \frac{v_1}{18} = \frac{72}{18} = 4 \text{ A}, \tag{2.17}$$

$$i_2 = \frac{v_1}{9} = \frac{72}{9} = 8 \text{ A}. \tag{2.18}$$

We have found the three specified currents by using series-parallel reductions in combination with Ohm's law.

Before leaving Example 2.1, we suggest that you take the time to show that the solution satisfies Kirchhoff's current law at every node and Kirchhoff's voltage law around every closed path. (Note that there are three closed paths that can be tested.) Showing that the power delivered by the voltage source equals the total power dissipated in the resistors also is informative. (See Problems 2.1 and 2.2.)

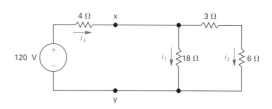

Figure 2.9 The circuit for Example 2.1.

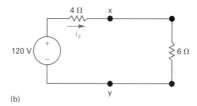

(a)

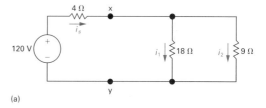

(b)

Figure 2.10 A simplification of the circuit shown in Fig. 2.9.

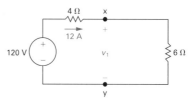

Figure 2.11 The circuit of Fig. 2.10(b) showing the numerical value of i_s.

DRILL EXERCISE

2.1 For the circuit shown, find (a) the voltage v, (b) the power delivered to the circuit by the current source, and (c) the power dissipated in the 10 Ω resistor.

   ANSWER: (a) 60 V; (b) 300 W; (c) 57.6 W.

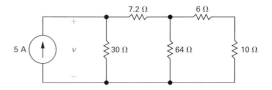

2.2 VOLTAGE AND CURRENT DIVIDERS

The Voltage-Divider Circuit

At times—especially in electronic circuits—developing more than one voltage level from a single voltage supply is necessary. One way of doing this is by using a **voltage-divider circuit**, such as the one in Fig. 2.12.

We analyze this circuit by directly applying Ohm's law and Kirchhoff's laws. To aid the analysis, we introduce the current i as shown in Fig. 2.12(b). From Kirchhoff's current law, R_1 and R_2 carry the same current. Applying Kirchhoff's voltage law around the closed loop yields

$$v_s = i R_1 + i R_2, \tag{2.19}$$

or

$$i = \frac{v_s}{R_1 + R_2}. \tag{2.20}$$

Now we can use Ohm's law to calculate v_1 and v_2:

$$v_1 = i R_1 = v_s \frac{R_1}{R_1 + R_2}, \tag{2.21}$$

$$v_2 = i R_2 = v_s \frac{R_2}{R_1 + R_2}. \tag{2.22}$$

Equations (2.21) and (2.22) show that v_1 and v_2 are fractions of v_s. Each fraction is the ratio of the resistance across which the divided voltage is defined to the sum of the two resistances.

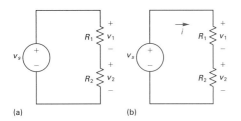

Figure 2.12 (a) A voltage-divider circuit and (b) the voltage-divider circuit with current i indicated.

Because this ratio is always less than 1.0, the divided voltages v_1 and v_2 are always less than the source voltage v_s.

If you desire a particular value of v_2, and v_s is specified, an infinite number of combinations of R_1 and R_2 yield the proper ratio. For example, suppose that v_s equals 15 V and v_2 is to be 5 V. Then $v_2/v_s = \frac{1}{3}$ and, from Eq. (2.22), we find that this ratio is satisfied whenever $R_2 = \frac{1}{2}R_1$. Other factors that may enter into the selection of R_1, and hence R_2, include the power losses that occur in dividing the source voltage and the effects of connecting the voltage-divider circuit to other circuit components.

Consider connecting a resistor R_L in parallel with R_2, as shown in Fig. 2.13. The resistor R_L acts as a load on the voltage-divider circuit. A **load** on any circuit consists of one or more circuit elements that draw power from the circuit. With the load R_L connected, the expression for the output voltage becomes

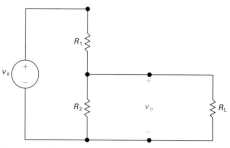

Figure 2.13 A voltage divider connected to a load R_L.

$$v_o = \frac{R_{eq}}{R_1 + R_{eq}} v_s, \tag{2.23}$$

where

$$R_{eq} = \frac{R_2 R_L}{R_2 + R_L}. \tag{2.24}$$

Substituting Eq. (2.24) into Eq. (2.23) yields

$$v_o = \frac{R_2}{R_1[1 + (R_2/R_L)] + R_2} v_s. \tag{2.25}$$

Note that Eq. (2.25) reduces to Eq. (2.22) as $R_L \to \infty$, as it should. Equation (2.25) shows that, as long as $R_L \gg R_2$, the voltage ratio v_o/v_s is essentially undisturbed by the addition of the load on the divider.

Another characteristic of the voltage-divider circuit of interest is the sensitivity of the divider to the tolerances of the resistors. By *tolerance* we mean a range of possible values. The resistances of commercially available resistors always vary within some percentage of their stated value. Example 2.2 illustrates the effect of resistor tolerances in a voltage-divider circuit.

EXAMPLE 2.2

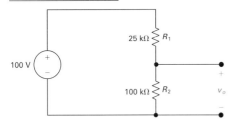

Figure 2.14 The circuit for Example 2.2.

The resistors used in the voltage-divider circuit shown in Fig. 2.14 have a tolerance of ±10%. Find the maximum and minimum value of v_o.

SOLUTION

From Eq. (2.22), the maximum value of v_o occurs when R_2 is 10% high and R_1 is 10% low, and the minimum value of v_o occurs when R_2 is 10% low and R_1 is 10% high. Therefore

$$v_o(\text{max}) = \frac{(100)(110)}{110 + 22.5} = 83.02 \text{ V},$$

$$v_o(\text{min}) = \frac{(100)(90)}{90 + 27.5} = 76.60 \text{ V}.$$

Thus, in making the decision to use 10% resistors in this voltage divider, we recognize that the no-load output voltage will lie between 76.60 and 83.02 V.

DRILL EXERCISE

2.2 (a) Find the no-load value of v_o in the circuit shown.

   (b) Find v_o when R_L is 150 kΩ.

   (c) How much power is dissipated in the 25 kΩ resistor if the load terminals are accidentally short-circuited?

   (d) What is the maximum power dissipated in the 75 kΩ resistor?

   ANSWER: (a) 150 V; (b) 133.33 V; (c) 1.6 W; (d) 0.3 W.

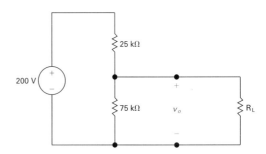

The Current-Divider Circuit

The **current-divider circuit** shown in Fig. 2.15 consists of two resistors connected in parallel across a current source. The current divider is designed to divide the current i_s between R_1 and R_2. We find the relationship between the current i_s and the current in each resistor (that is, i_1 and i_2) by directly applying Ohm's law and Kirchhoff's current law. The voltage across the parallel resistors is

$$v = i_1 R_1 = i_2 R_2 = \frac{R_1 R_2}{R_1 + R_2} i_s. \qquad (2.26)$$

From Eq. (2.26),

$$i_1 = \frac{R_2}{R_1 + R_2} i_s, \qquad (2.27)$$

$$i_2 = \frac{R_1}{R_1 + R_2} i_s. \qquad (2.28)$$

Equations (2.27) and (2.28) show that the current divides between two resistors in parallel such that the current in one resistor equals the current entering the parallel pair multiplied by the other resistance and divided by the sum of the resistances. Example 2.3 illustrates the use of the current-divider equation.

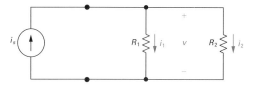

Figure 2.15 The current-divider circuit.

EXAMPLE 2.3

Find the power dissipated in the 6 Ω resistor shown in Fig. 2.16.

SOLUTION

First, we must find the current in the resistor by simplifying the circuit with series-parallel reductions. Thus, the circuit shown in Fig. 2.16 reduces to the one shown in Fig. 2.17. We find the current i_o by using the formula for current division:

$$i_o = \frac{(10)(16)}{16 + 4} = 8 \text{ A}.$$

Note that i_o is the current in the 1.6 Ω resistor in Fig. 2.16. We now can further divide i_o between the 6 Ω and 4 Ω resistors. The current in the 6 Ω resistor is

$$i_6 = \frac{(8)(4)}{10} = 3.2 \text{ A},$$

and the power dissipated in the 6 Ω resistor is

$$p = (3.2)^2 (6) = 61.44 \text{ W}.$$

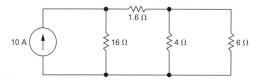

Figure 2.16 The circuit for Example 2.3.

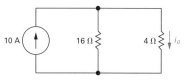

Figure 2.17 A simplification of the circuit shown in Fig. 2.16.

2.3 SOURCE TRANSFORMATIONS

A **source transformation**, shown in Fig. 2.18, allows a voltage source in series with a resistor to be replaced by a current source in parallel with the same resistor or vice versa. The double-headed arrow emphasizes that a source transformation is bilateral; that is, we can start with either configuration and derive the other.

We need to find the relationship between v_s and i_s that guarantees the two configurations in Fig. 2.18 are equivalent with respect to nodes a,b. Equivalence is achieved if any resistor R_L experiences the same current flow, and thus the same voltage drop, whether connected between nodes a,b in Fig. 2.18(a) or Fig. 2.18(b).

Suppose R_L is connected between nodes a,b in Fig. 2.18(a). Using Ohm's law, the current in R_L is

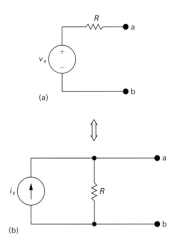

(a)

(b)

Figure 2.18 Source transformations.

$$i_L = \frac{v_s}{R + R_L}. \tag{2.29}$$

Now suppose the same resistor R_L is connected between nodes a,b in Fig. 2.18(b). Combining the parallel resistors R and R_L and then applying Ohm's law, we find the current in R_L is

$$i_L = \frac{R}{R + R_L} i_s. \tag{2.30}$$

If the two circuits in Fig. 2.18 are equivalent, these resistor currents must be the same. Equating the right-hand sides of Eqs. (2.29) and (2.30) and simplifying,

$$i_s = \frac{v_s}{R}. \tag{2.31}$$

When Eq. (2.31) is satisfied for the circuits in Fig. 2.18, the current in R_L is the same for both circuits in the figure for all values of R_L. If the current through R_L is the same in both circuits, then the voltage drop across R_L is the same in both circuits, and the circuits are equivalent at nodes a,b. If the polarity of v_s is reversed, the orientation of i_s must be reversed to maintain equivalence.

Example 2.4 illustrates the usefulness of making source transformations to simplify a circuit-analysis problem.

___

EXAMPLE 2.4

(a) For the circuit shown in Fig. 2.19, find the power associated
with the 6 V source.

(b) State whether the 6 V source is absorbing or delivering the
power calculated in (a).

SOLUTION

(a) To find the power associated with the 6 V source we need to
find the current in the source. If we study the circuit shown
in Fig. 2.19, we see we have six unknown currents, i.e., the
current in each resistor. Note the current in the 6 V source is
the same as the current in the 4 Ω resistor. If we approach the
problem by applying Ohm's and Kirchhoff's laws, we must
write and solve six simultaneous equations. Using source
transformations, we can first simplify the circuit so that we
can determine the 6 V source current by solving a single
equation.

We must reduce the circuit in a way that preserves the
identity of the branch containing the 6 V source. We have no
reason to preserve the identity of the branch containing the
40 V source. Beginning with this branch, we can transform
the 40 V source in series with the 5 Ω resistor into an 8 A
current source in parallel with a 5 Ω resistor, as shown in
Fig. 2.20(a). Next, we can replace the parallel combination
of the 20 Ω and 5 Ω resistors with a 4 Ω resistor. This 4 Ω
resistor is in parallel with the 8 A source and therefore can
be replaced with a 32 V source in series with a 4 Ω resistor, as
shown in Fig. 2.20(b). The 32 V source is in series with 20 Ω
of resistance and, hence, can be replaced by a current source
of 1.6 A in parallel with 20 Ω, as shown in Fig. 2.20(c). The
20 Ω and 30 Ω parallel resistors can be reduced to a single
12 Ω resistor. The parallel combination of the 1.6 A current
source and the 12 Ω resistor transforms into a voltage source
of 19.2 V in series with 12 Ω. Figure 2.20(d) shows the result

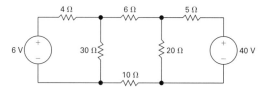

Figure 2.19 The circuit for Example 2.4.

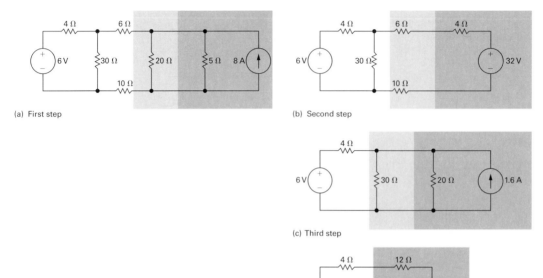

(a) First step

(b) Second step

(c) Third step

(d) Fourth step

Figure 2.20 Step-by-step simplification of the circuit shown in Fig. 2.19

of this last transformation. The current in the direction of the voltage drop across the 6 V source is $(19.2 - 6)/16$, or 0.825 A. Therefore the power associated with the 6 V source is

$$p_{6V} = (0.825)(6) = 4.95 \text{ W}.$$

(b) The voltage source is absorbing power.

DRILL EXERCISE

2.3 Use the fact that once you know the current in the 4 Ω resistor in Example 2.4 you can find the current in each of the remaining five resistors by a systematic application of Ohm's and Kirchhoff's laws. Assume the remaining five currents are oriented as follows: 30 Ω down; 6 Ω right-to-left; 10 Ω left to right; 20 Ω down; and 5 Ω right-to-left. Find these currents.

ANSWER: $i_{30\Omega} = 0.31$ A; $i_{6\Omega} = 1.135$ A; $i_{10\Omega} = 1.135$ A; $i_{20\Omega} = 1.373$ A; and $i_{5\Omega} = 2.508$ A;

A question that arises from use of the source transformation depicted in Fig. 2.20 is, "What happens if there is a resistance R_p in parallel with the voltage source or a resistance R_s in series with the current source?" In both cases, the resistance has no effect on the equivalent circuit that predicts behavior with respect to terminals a,b. Figure 2.21 summarizes this observation.

The two circuits depicted in Fig. 2.21(a) are equivalent with respect to terminals a,b because they produce the same voltage and current in any resistor R_L inserted between nodes a,b. The same can be said for the circuits in Fig. 2.21(b). Example 2.5 illustrates an application of the equivalent circuits depicted in Fig. 2.21.

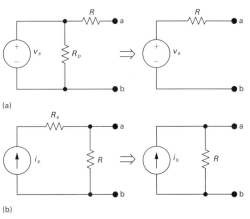

Figure 2.21 Equivalent circuits containing a resistance in parallel with a voltage source or in series with a current source.

EXAMPLE 2.5

(a) Use source transformations to find the voltage v_o in the circuit shown in Fig. 2.22.

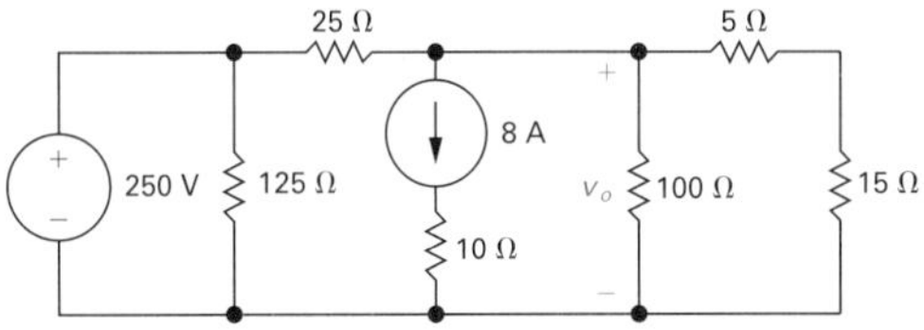

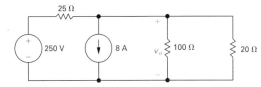

Figure 2.22 The circuit for Example 2.5.

(b) Find the power developed by the 250 V voltage source.

(c) Find the power developed by the 8 A current source.

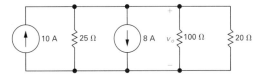

Figure 2.23 A simplified version of the circuit shown in Fig. 2.22.

SOLUTION

(a) We begin by removing the 125 Ω and 10 Ω resistors, because the 125 Ω resistor is connected across the 250 V voltage source and the 10 Ω resistor is connected in series with the 8 A current source. We also combine the series-connected resistors into a single resistance of 20 Ω. Figure 2.23 shows the simplified circuit.

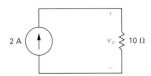

Figure 2.24 The circuit shown in Fig. 2.23 after a source transformation.

We now use a source transformation to replace the 250 V source and 25 Ω resistor with a 10 A source in parallel with the 25 Ω resistor, as shown in Fig. 2.24. We can now simplify the circuit shown in Fig. 2.24 by using Kirchhoff's current law to combine the parallel current sources into a single source. The parallel resistors combine into a single resistor. Figure 2.25 shows the result. Hence $v_o = 20$ V.

Figure 2.25 The circuit shown in Fig. 2.24. after combining sources and resistors.

(b) The current supplied by the 250 V source equals the current in the 125 Ω resistor plus the current in the 25 Ω resistor. Thus

$$i_s = \frac{250}{125} + \frac{250 - 20}{25} = 11.2 \text{ A}.$$

Therefore the power developed by the voltage source is

$$p_{250V}(\text{developed}) = (250)(11.2) = 2800 \text{ W}.$$

(c) To find the power developed by the 8 A current source, we first find the voltage across the source. If we let v_s represent the voltage across the source, positive at the upper terminal of the source, we obtain

$$v_s + 8(10) = v_o = 20,$$

or

$$v_s = -60 \text{ V},$$

and the power developed by the 8 A source is 480 W. Note that the 125 Ω and 10 Ω resistors do not affect the value of v_o but do affect the power calculations.

DRILL EXERCISE

2.4 **(a)** Use a series of source transformations to find the voltage v in the circuit shown.

(b) How much power does the 120 V source deliver to the circuit?

ANSWER: (a) 48 V; (b) 374.4 W.

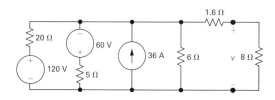

2.4 SUPERPOSITION

A linear system obeys the principle of **superposition**, which states that whenever a linear system is excited, or driven, by more than one independent source of energy, the total response is the sum of the individual responses. An individual response is the result of an independent source acting alone. Because we are dealing with circuits made up of interconnected linear-circuit elements, we can apply the principle of superposition directly to the analysis of such circuits when they are driven by more than one independent energy source. At present, we restrict the discussion to simple resistive networks; however, the principle is applicable to any linear system.

We demonstrate the superposition principle by using it to find the branch currents in the circuit shown in Fig. 2.26. We begin by finding the branch currents resulting from the 120 V voltage source. We denote those currents with a prime. Replacing the ideal current source with an open circuit deactivates it; Fig. 2.27 shows this. The branch currents in this circuit are the result of only the voltage source.

With the current source deactivated we have 6 Ω in parallel with 3 Ω, and this parallel combination is in series with 6 Ω. Thus the current i_1' is

$$i_1' = \frac{120}{6+2} = 15 \text{ A.} \tag{2.32}$$

The voltage drop across the 3 Ω resistor is

$$v_{3\Omega} = 120 - 15(6) = 30 \text{ V.} \tag{2.33}$$

It follows that

$$i_2' = \frac{30}{3} = 10 \text{ A} \tag{2.34}$$

and

$$i_3' = i_4' = \frac{30}{6} = 5 \text{ A.} \tag{2.35}$$

To find the component of the branch currents resulting from the current source, we deactivate the ideal voltage source and solve the circuit shown in Fig. 2.28. Note that we deactivate an ideal voltage source by replacing it with a short circuit. The double-prime notation for the currents indicates they are the components of the total current resulting from the ideal current source.

With the voltage source deactivated we leave it to the reader to show that the equivalent resistance across the 12 A source is 2 Ω. From Ohm's law we know the voltage across the current

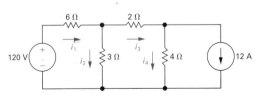

Figure 2.26 A circuit used to illustrate superposition.

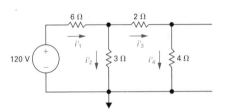

Figure 2.27 The circuit shown in Fig. 2.26 with the current source deactivated.

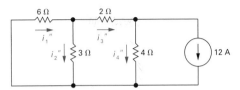

Figure 2.28 The circuit shown in Fig. 2.26 with the voltage source deactivated.

source will be 24 V, positive at the lower terminal. It follows that

$$i_4'' = \frac{-24}{4} = -6 \text{ A}. \tag{2.36}$$

Using Kirchhoff's current law, we can write

$$i_3'' = 12 + i_4'' = 12 - 6 = 6 \text{ A}. \tag{2.37}$$

It follows that the voltage across the 3 Ω resistor, positive at the lower terminal, is $24 - 6(2)$ or 12 V. Hence

$$i_2'' = \frac{-12}{3} = -4 \text{ A} \tag{2.38}$$

and

$$i_1'' = \frac{12}{6} = 2 \text{ A}. \tag{2.39}$$

To find the branch currents in the original circuit, that is, the currents i_1, i_2, i_3, and i_4 in Fig. 2.26, we simply add the currents given by Eqs. (2.36)–(2.39) to the currents given by Eqs. (2.32), (2.34), and (2.35):

$$i_1 = i_1' + i_1'' = 15 + 2 = 17 \text{ A}, \tag{2.40}$$

$$i_2 = i_2' + i_2'' = 10 - 4 = 6 \text{ A}, \tag{2.41}$$

$$i_3 = i_3' + i_3'' = 5 + 6 = 11 \text{ A}, \tag{2.42}$$

$$i_4 = i_4' + i_4'' = 5 - 6 = -1 \text{ A}. \tag{2.43}$$

When applying superposition to linear circuits containing both independent and dependent sources, you must recognize that the dependent sources are never deactivated. Example 2.6 illustrates the application of superposition when a circuit contains both dependent and independent sources.

EXAMPLE 2.6

Use the principle of superposition to find v_o in the circuit shown in Fig. 2.29.

SOLUTION

We begin by finding the component of v_o resulting from the 10 V source. Figure 2.30 shows the circuit. With the 5 A source deactivated, v'_Δ must equal $(-0.4v'_\Delta)(10)$. Hence, v'_Δ must be zero, the branch containing the two dependent sources is open, and

$$v'_o = \frac{10}{25}(20) = 8 \text{ V.}$$

When the 10 V source is deactivated, the circuit reduces to the one shown in Fig. 2.31. Using Kirchhoff's current law, we find the current i''_{10} in Fig. 2.31 is

$$i''_{10} = 5 - 0.4v''_\Delta$$

From Ohm's law we have

$$v''_\Delta = 10(5 - 0.4v''_\Delta) = 50 - 4v''_\Delta.$$

Solving for v''_Δ yields

$$v''_\Delta = 10 \text{ V.}$$

With the independent voltage source deactivated v''_o is the voltage across an equivalent resistance of 4 Ω. Thus

$$v''_o = (0.4v''_\Delta)4 = 1.6v''_\Delta = 16 \text{ V.}$$

The value of v_o is the sum of v'_o and v''_o or 24 V.

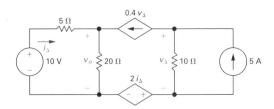

Figure 2.29 The circuit for Example 2.6.

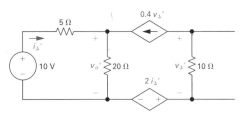

Figure 2.30 The circuit shown in Fig. 2.29 with the 5 A source deactivated.

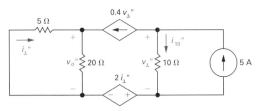

Figure 2.31 The circuit shown in Fig. 2.29 with the 10 V source deactivated.

DRILL EXERCISES

2.5   **(a)** Use the principle of superposition to find the voltage v in the circuit shown.

   (b) Find the power dissipated in the 10 Ω resistor.

   ANSWER: (a) 50 V; (b) 250 W.

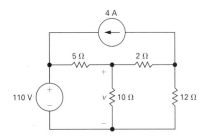

2.6 Use the principle of superposition to find the voltage v in the circuit shown.

ANSWER: 30 V.

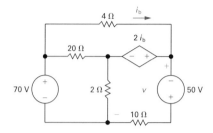

Practical Perspective

Digital-to-Analog Resistive Ladder

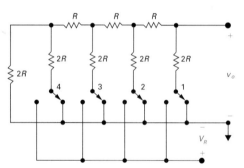

Figure 2.32 Digital-to-analog resistive ladder with four switches.

Before analyzing the digital-to-analog resistive ladder, let us review the conversion of binary-to-decimal numbers. Each position of the binary number system represents a particular value of 2^n. Thus the decimal equivalent of the binary number 0111 is $0 + 2^2 + 2^1 + 2^0$ or $0 + 4 + 2 + 1$, which is 7. The decimal equivalent of 1111 is $2^3 + 2^2 + 2^1 + 2^0$ or 15. Table 2.1 gives the binary-to-decimal conversion for 0 through 15.

We have limited Table 2.1 to four binary positions in order to limit the number of switches in the binary-to-analog ladder circuit to four. After analyzing the ladder circuit with four switches the extension to n switches, corresponding to n binary positions will become apparent. The digital-to-analog resistive ladder with four switches is shown in Fig. 2.32.

TABLE 2.1 Binary-to-Decimal Conversion

BINARY	DECIMAL	BINARY	DECIMAL
0000	0	1000	8
0001	1	1001	9
0010	2	1010	10
0011	3	1011	11
0100	4	1100	12
0101	5	1101	13
0110	6	1110	14
0111	7	1111	15

Our goal is to show that, depending on the positions of the n switches, the output voltage v_o corresponds to the decimal value of the binary input. Observe that each switch either connects a resistor labeled $2R$ to ground or to a reference voltage V_R. The position of a switch is controlled by the digital input to the ladder circuit. For example, if the logic level to a switch is high representing a binary 1, the switch is connected to V_R. On the other hand, if the logic level to a switch is low representing a binary 0, the switch is connected to ground.

If all the switches are connected to ground, as shown in Fig. 2.32, the output voltage is zero. The reader can verify this by starting at the left end of the ladder circuit and by parallel-series simplifications reduce the circuit to that shown in Fig. 2.33.

If switch 1 connects to V_R and the other switches remain connected to ground, the output voltage is

$$v_o = \frac{1}{2}V_R \qquad (2.44)$$

Equation (2.44) can be derived once the ladder is simplified. With switch 1 connected to V_R and the remaining switches connected to ground, the ladder circuit becomes as shown in Fig. 2.34.

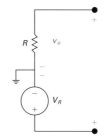

Figure 2.33 Simplification of the circuit shown in Fig. 2.32.

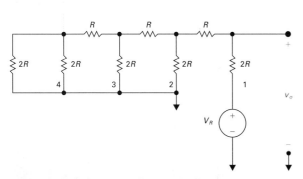

Figure 2.34 The ladder circuit when switch 1 connected to V_R and the remaining switches connected to ground.

Now by parallel-series reductions the circuit in Fig. 2.34 simplifies to that shown in Fig. 2.35. Observe from Fig. 2.35 that v_o will equal one-half V_R, thus confirming Eq. (2.44).

If switch 2 is connected to V_R and the remaining switches are connected to ground, the circuit reduces to that shown in Fig. 2.36. Now the output voltage is

$$v_o = \frac{1}{4}V_R \qquad (2.45)$$

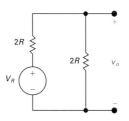

Figure 2.35 Simplified version of Fig. 2.28.

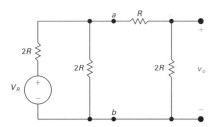

Figure 2.36 The ladder circuit when switch 2 connected to V_R and the remaining switches connected to ground.

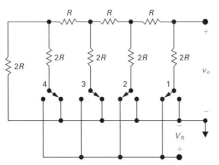

Figure 2.37 The ladder circuit when switches 1 and 2 connect to V_R and switches 3 and 4 connect to ground.

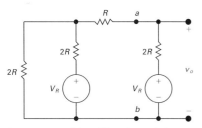

Figure 2.38 Simplification of the circuit in Fig. 2.37 using parallel-series combinations of resistors.

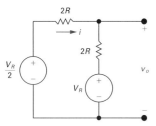

Figure 2.39 Further simplification of the circuit in Fig. 2.38.

Equation (2.45) is easily verified by first making a couple of source transformations to the left of the terminals a, b in Fig. 2.36.

We will leave to the reader via Problems 2.29 and 2.30 that if switch 3 is connected to V_R and the remaining switches are connected to ground that

$$v_o = \frac{1}{8} V_R \qquad (2.46)$$

and if switch 4 is connected to V_R and the remaining switches are connected to ground that

$$v_o = \frac{1}{16} V_R \qquad (2.47)$$

We are now ready to investigate the relationship between v_o and V_R when more than one switch is connected to V_R. We begin by assuming switches 1 and 2 are connected to V_R and switches 3 and 4 are connected to ground. This arrangement of switches is depicted in Fig. 2.37.

The circuit depicted in Fig. 2.37 can be reduced to that shown in Fig. 2.38.

The circuit in Fig. 2.38 can be further simplified by making a couple of source transformations to the left of the terminals a, b. The result of this simplification is shown in Fig. 2.39.

It follows from the circuit in Fig. 2.39 that

$$i = -\frac{V_R}{8R} \qquad (2.48)$$

and

$$v_o = V_R + 2Ri = \frac{3}{4} V_R \qquad (2.49)$$

Now we make the important observation that Eq. 2.49 could have been derived using the principle of superposition. That is, we could combine the derivations of Eqs. (2.44) and (2.45). Thus

$$v_o = \frac{1}{2} V_R + \frac{1}{4} V_R = \frac{3}{4} V_R \qquad (2.50)$$

We are now in a position to write down the relationship between v_o and V_R for any positions of the switches by superimposing the results given by Eqs. (2.44), (2.45), (2.46), and (2.47). For example, if switches 1 and 3 are connected to V_R and switches 2 and 4 are connected to ground, the output voltage is obtained by combining Eqs. (2.44) and (2.46). Thus

$$v_o = \frac{1}{2} V_R + \frac{1}{8} V_R = \frac{5}{8} V_R \qquad (2.51)$$

TABLE 2.2 **The Relationship between Switch Positions and Output Voltage for the Digital-Analog Ladder Circuit Assuming $V_R = 16$ V.**

SWITCH POSITIONS				v_o (V)	SWITCH POSITIONS				v_o (V)
1	2	3	4		1	2	3	4	
0	0	0	0	0	V_R	0	0	0	8
0	0	0	V_R	1	V_R	0	0	V_R	9
0	0	V_R	0	2	V_R	0	V_R	0	10
0	0	V_R	V_R	3	V_R	0	V_R	V_R	11
0	V_R	0	0	4	V_R	V_R	0	0	12
0	V_R	0	V_R	5	V_R	V_R	0	V_R	13
0	V_R	V_R	0	6	V_R	V_R	V_R	0	14
0	V_R	V_R	V_R	7	V_R	V_R	V_R	V_R	15

Finally we can now demonstrate that v_o is the decimal equivalent of the binary input to the switches. For convenience let us assume V_R is 16 V. Table 2.2 gives the relationship between v_o and the switch positions. In Table 2.2 a switch position labeled 0 denotes the switch is connected to ground and V_R denotes the switch is connected to the reference voltage. We leave to the reader the task of confirming the entries in Table 2.2 via Problem 2.31.

SUMMARY

- **Series resistors** can be combined to obtain a single equivalent resistance according to the equation

$$R_{eq} = \sum_{i=1}^{k} R_i = R_1 + R_2 + \cdots + R_k.$$

- **Parallel resistors** can be combined to obtain a single equivalent resistance according to the equation

$$\frac{1}{R_{eq}} = \sum_{i=1}^{k} \frac{1}{R_i} = \frac{1}{R_1} + \frac{1}{R_2} + \cdots + \frac{1}{R_k}.$$

When just two resistors are in parallel, the equation for equivalent resistance can be simplified to give

$$R_{eq} = \frac{R_1 R_2}{R_1 + R_2}.$$

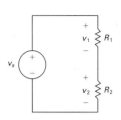

● When voltage is divided between series resistors, as shown in the figure, the voltage across each resistor can be found according to the equations

$$v_1 = \frac{R_1}{R_1 + R_2} v_s,$$

$$v_2 = \frac{R_2}{R_1 + R_2} v_s.$$

● When current is divided between parallel resistors, as shown in the figure, the current through each resistor can be found according to the equations

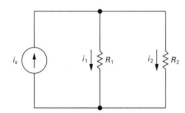

$$i_1 = \frac{R_2}{R_1 + R_2} i_s,$$

$$i_2 = \frac{R_1}{R_1 + R_2} i_s.$$

● **Source transformations** allow us to exchange a voltage source (v_s) and a series resistor (R) for a current source (i_s) and a parallel resistor (R) and vice versa. The combinations must be equivalent in terms of their terminal voltage and current. Terminal equivalence holds provided that

$$i_s = \frac{v_s}{R}.$$

● In a circuit with multiple independent sources, **superposition** allows us to activate one source at a time and sum the resulting voltages and currents to determine the voltages and currents that exist when all independent sources are active. Dependent sources are never deactivated when applying superposition.

PROBLEMS

2.1. **(a)** Find the power dissipated in each resistor in the circuit shown in Fig. 2.9.

(b) Find the power delivered by the 120 V source.

(c) Show that the power delivered equals the power dissipated.

2.2. **(a)** Show that the solution of the circuit in Fig. 2.9 (see Example 2.1) satisfies Kirchhoff's current law at junctions x and y.

(b) Show that the solution of the circuit in Fig. 2.9 satisfies Kirchhoff's voltage law around every closed loop.

2.3. Find the power dissipated in the 5 Ω resistor in the circuit in Fig. P2.3.

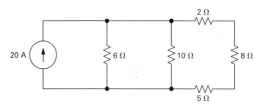

Figure P2.3

2.4. For the circuit in Fig. P2.4 calculate

(a) v_o and i_o

(b) the power dissipated in the 15 Ω resistor

(c) the power developed by the voltage source.

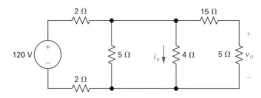

Figure P2.4

2.5. (a) Find an expression for the equivalent resistance of two resistors of value R in parallel.

(b) Find an expression for the equivalent resistance of n resistors of value R in parallel.

(c) Using the results of (b), design a resistive network with an equivalent resistance of 700 Ω using 1 kΩ resistors.

(d) Using the results of (b), design a resistive network with an equivalent resistance of 5.5 kΩ using 2 kΩ resistors.

2.6. Find the equivalent resistance R_{ab} for each of the circuits in Fig. P2.6.

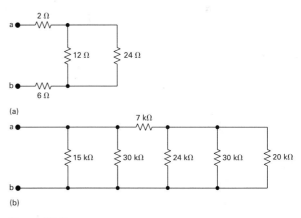

Figure P2.6

2.7. (a) In the circuits in Fig. P2.7(a)–(c), find the equivalent resistance R_{ab}.

(b) For each circuit find the power delivered by the source.

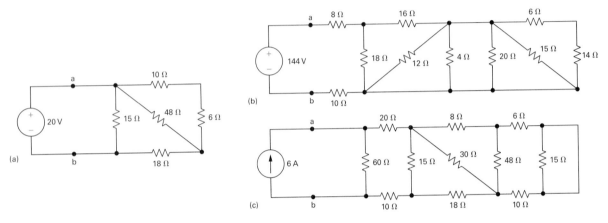

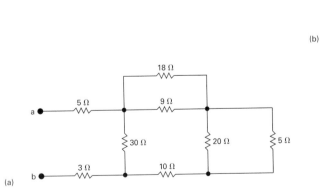

Figure P2.7

2.8. Find the equivalent resistance R_{ab} for each of the circuits in Fig. P2.8.

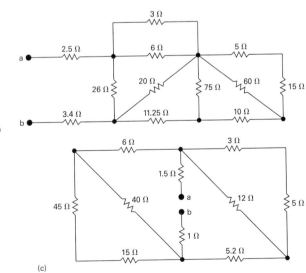

Figure P2.8

2.9. The high-voltage direct-current transmission line introduced in Drill Exercise 1.5 is 845 mi long. Each side of the circuit consists of two conductors in parallel. The resistance of each conductor is 0.0397 Ω/mi. The arrangement is illustrated in Fig. P2.9.

   (a) The voltage at the Oregon terminal of the line is 800 kV. Each conductor is carrying 1000 A, as shown in the figure. Calculate the power received at the California end of the line and the efficiency of the power transmission from Oregon to California.

   (b) Repeat (a) with the voltage at Celilo raised to 1000 kV and the current remaining at 1000 A conductor.

   (c) Repeat (b) with a third conductor added to each side of the circuit and the current remaining at 1000 A/conductor.

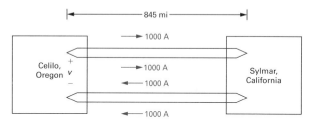

Figure P2.9

2.10. Find v_o in the circuit in Fig. P2.10.

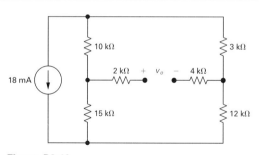

Figure P2.10

2.11. **(a)** Find the voltage v_x in the circuit in Fig. P2.11.

   (b) Replace the 30 V source with a general voltage source equal to V_s. Assume V_s is positive at the upper terminal. Find v_x as a function of V_s.

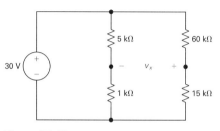

Figure P2.11

2.12. Find v_o and v_g in the circuit in Fig. P2.12.

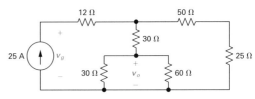

Figure P2.12

2.13. Find i_o and i_g in the circuit in Fig. P2.13.

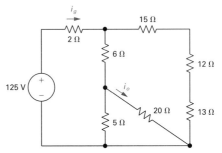

Figure P2.13

2.14. For the circuit in Fig. P2.14, calculate (a) i_o and (b) the power dissipated in the 10 Ω resistor.

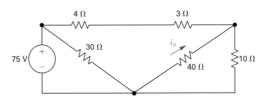

Figure P2.14

2.15. The current in the 9 Ω resistor in the circuit in Fig. P2.15 is 1 A, as shown.

   (a) Find v_g.

   (b) Find the power dissipated in the 20 Ω resistor.

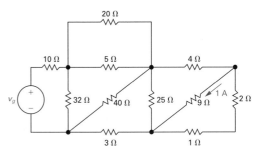

Figure P2.15

2.16. (a) The circuit shown in Fig. P2.16 is a voltage divider circuit. Show that the voltage across the kth resistor in the series string of n resistors is

$$v_k = \frac{R_k}{R_1 + R_2 + \cdots R_k + \ldots R_n} v_s$$

(b) Assume $v_s = 200$ V; $R_1 = 5\ \Omega$; $R_2 = 15\ \Omega$; $R_3 = 30\ \Omega$; $R_4 = 10\ \Omega$; and $R_5 = 40\ \Omega$. Find v_1; v_2; v_3; v_4; and v_5.

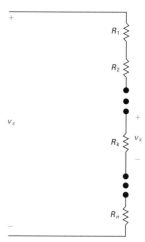

Figure P2.16

2.17. When the switch in Fig. P2.17 is open, the voltage divider is operating at no load. When the switch is closed, the divider is said to be loaded.

(a) Calculate the no load value of v_o.

(b) Calculate the load value of v_o.

(c) What is the ratio $v_o/25$ V at no load?

(d) What is the ratio $v_o/25$ V under load?

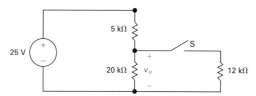

Figure P2.17

❖ **2.18.** In the voltage-divider circuit in Fig. P2.17 replace the 25 V source with a source denoted as V_s; the 5 kΩ resistor with a resistor denoted as R_1; the 20 kΩ resistor with a resistor denoted R_2; and the 12 kΩ resistor denoted as R_L. The divider is to be designed so that $v_o = \sigma V_s$ at no load (the switch is open) and $v_o = \beta V_s$ under load (the switch is closed). Note by definition that $\beta < \sigma < 1$.

(a) Show that

$$R_1 = \frac{(\sigma - \beta)}{\sigma \beta} R_L \text{ and } R_2 = \frac{(\sigma - \beta)}{\beta(1 - \sigma)} R_L.$$

(b) Specify the numerical values of R_1 and R_2 if $\sigma = 0.9$, $\beta = 0.70$, and $R_L = 126$ kΩ.

2.19. **(a)** Show that the current in the kth branch of the circuit in Fig. P2.19(a) is equal to the source current i_g times the conductance of the kth branch divided by the sum of the conductances, that is,

$$i_k = \frac{i_g G_k}{[G_1 + G_2 + G_3 + \cdots + G_k + \cdots + G_N]}.$$

(b) Use the result derived in (a) to calculate the current in the 6.25 Ω resistor in the circuit in Fig. P2.19(b).

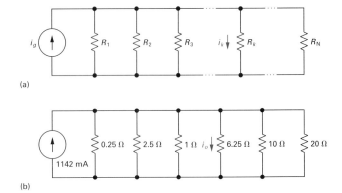

(a)

(b)

Figure P2.19

❖ 2.20. Specify the resistors in the circuit in Fig. P2.20 to meet the following design criteria:

$i_g = 8$ mA; $v_g = 4$ V; $i_1 = 2i_2$; $i_2 = 10i_3$; and $i_3 = i_4$.

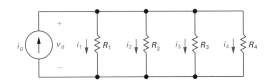

Figure P2.20

2.21. **(a)** Find the current in the 5 kΩ resistor in the circuit in Fig. P2.21 by making a succession of appropriate source transformations.

(b) Using the result obtained in (a), work back through the circuit to find the power developed by the 75 V source.

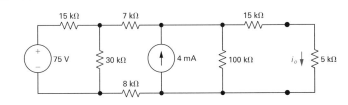

Figure P2.21

2.22. Use a series of source transformations to find the current i_o in the circuit in Fig. P2.22.

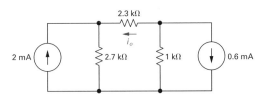

Figure P2.22

2.23. **(a)** Use source transformations to find v_o in the circuit in Fig. P2.23.

(b) Find the power developed by the 300 V source.

(c) Find the power developed by the 10 A current source.

(d) Verify that the total power developed equals the total power dissipated.

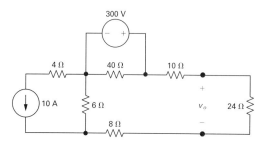

Figure P2.23

2.24. Use a series of source transformations to find i_o in the circuit in Fig. P2.24.

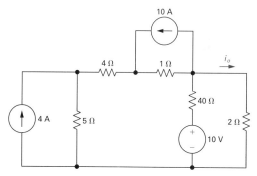

Figure P2.24

2.25. Determine i_o and v_o in the circuit shown in Fig. P2.25 when R_o is 0, 2, 6, 10, 15, 20, 30, 40, 50, and 70 Ω.

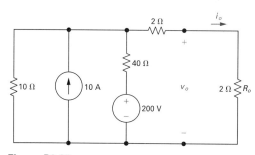

Figure P2.25

2.26. Use the principle of superposition to find the voltage v_o in the circuit in Fig. P2.26.

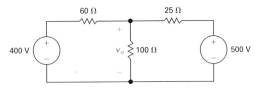

Figure P2.26

2.27. Use the principle of superposition to find the current i_o in the circuit in Fig. P2.27.

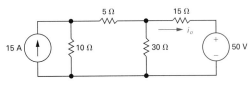

Figure P2.27

2.28. Use the principle of superposition to find v_o in the circuit in Fig. P2.28.

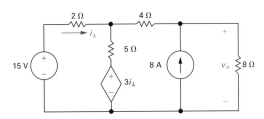

Figure P2.28

◆ **2.29.** **(a)** Show that if switch 3 in the circuit in Fig. 2.32 is connected to V_R and the remaining switches are connected to ground, the circuit reduces to that shown in Fig. P2.29.

(b) Using source transformations show that

$$v_o = \frac{1}{8} V_R.$$

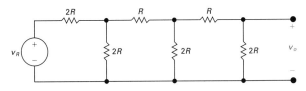

Figure P2.29

◆ **2.30.** **(a)** Show that if switch 4 in the circuit in Fig. 2.32 is connected to V_R and the remaining switches are connected to ground, the circuit reduces to that shown in Fig. P2.30.

(b) Using source transformations show that

$$v_o = \frac{1}{16} V_R.$$

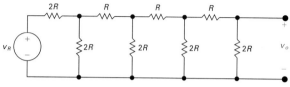

Figure P2.30

◆ **2.31.** Using Equations (2.44)–(2.47), along with the principle of superposition, confirm the entries in Table 2.2.

Circuits with Realistic Resistors

In the last chapter we began to explore the effect of imprecise resistor values on the performance of a circuit—specifically, on the performance of a voltage divider. Resistors are manufactured for only a small number of discrete values, and any given resistor from a batch of resistors will vary from its stated value within some tolerance. Resistors with tighter tolerance, say 1%, are more expensive than those with greater tolerance, say 10%. Therefore, in a circuit that uses many resistors, it would be important to understand which resistor's value has the greatest impact on the expected performance of the circuit. In other words, we would like to predict the effect of varying each resistor's value. If we know that a particular resistor must be very close to its stated value for the circuit to function correctly, we can then decide to spend the extra money necessary to achieve a tighter tolerance on that resistor's value.

Exploring the effect of a circuit component's value on the circuit's output is known as **sensitivity analysis**. Once we have presented additional circuit analysis techniques, we will examine the topic of sensitivity analysis.

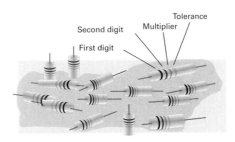

3 Techniques of Circuit Analysis

Chapter Contents

So far, we have analyzed relatively simple resistive circuits by applying Kirchhoff's laws in combination with Ohm's law. We can use this approach for all circuits, but as they become structurally more complicated and involve more and more elements, this direct method soon becomes cumbersome. In this chapter we introduce two powerful techniques of circuit analysis that aid in the analysis of complex circuit structures: the node-voltage method and the mesh-current method. These techniques give us two systematic methods of describing circuits with the minimum number of simultaneous equations.

In addition to these two general analytical methods, in this chapter we also discuss another technique for simplifying circuits. We have already demonstrated how to use series-parallel reductions and source transformations to simplify a circuit's structure. We now add Thévenin and Norton equivalent circuits to those techniques.

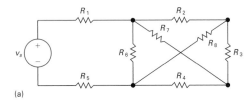

(a)

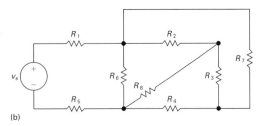

(b)

Figure 3.1 (a) A planar circuit. (b) The same circuit redrawn to verify that it is planar.

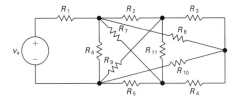

Figure 3.2 A nonplanar circuit.

The final topic in this chapter considers the conditions necessary to ensure that the power delivered to a resistive load by a source is maximized. Thévenin equivalent circuits are used in establishing the maximum power transfer conditions.

3.1 TERMINOLOGY

To discuss the more-involved methods of circuit analysis, we must define a few basic terms. So far, all the circuits presented have been **planar circuits**—that is, those circuits that can be drawn on a plane with no crossing branches. A circuit that is drawn with crossing branches still is considered planar if it can be redrawn with no crossover branches. For example, the circuit shown in Fig. 3.1(a) can be redrawn as Fig. 3.1(b); the circuits are equivalent because all the node connections have been maintained. Therefore, Fig. 3.1(a) is a planar circuit because it can be redrawn as one. Figure 3.2 shows a nonplanar circuit—it cannot be redrawn in such a way that all the node connections are maintained and no branches overlap. The node-voltage method is applicable to both planar and nonplanar circuits, whereas the mesh-current method is limited to planar circuits.

Describing a Circuit—The Vocabulary

In Section 1.3 we defined an ideal basic circuit element. When basic circuit elements are interconnected to form a circuit, the resulting interconnection is described in terms of nodes, paths, branches, loops, and meshes. We defined both a node and a closed path, or loop, in Section 1.7. Here we restate those definitions and then define the terms *path, branch,* and *mesh.* For your convenience, all of these definitions are presented in Table 3.1. Table 3.1 also includes examples of each definition taken from the circuit in Fig. 3.3, which are developed in Example 3.1.

EXAMPLE 3.1

For the circuit in Fig. 3.3, identify
(a) all nodes
(b) all essential nodes
(c) all branches
(d) all essential branches
(e) all meshes
(f) two paths that are not loops or essential branches
(g) two loops that are not meshes

TABLE 3.1   **Terms for Describing Circuits**

NAME	DEFINITION	EXAMPLE FROM FIG. 3.3
node	A point where two or more circuit elements join	a
essential node	A node where three or more circuit elements join	b
path	A trace of adjoining basic elements with no elements included more than once	$v_1 - R_1 - R_5 - R_6$
branch	A path that connects two nodes	R_1
essential branch	A path which connects two essential nodes without passing through an essential node	$v_1 - R_1$
loop	A path whose last node is the same as the starting node	$v_1 - R_1 - R_5 - R_6 - R_4 - v_2$
mesh	A loop that does not enclose any other loops	$v_1 - R_1 - R_5 - R_3 - R_2$
planar circuit	A circuit that can be drawn on a plane with no crossing branches	Fig. 3.3 is a planar circuit.
		Fig. 3.2 is a nonplanar circuit.

SOLUTION

(a) The nodes are a, b, c, d, e, f, and g.

(b) The essential nodes are b, c, e, and g.

(c) The branches are v_1, v_2, R_1, R_2, R_3, R_4, R_5, R_6, R_7, and I.

(d) The essential branches are v_1-R_1, R_2-R_3, v_2-R_4, R_5, R_6, R_7, and I.

(e) The meshes are $v_1-R_1-R_5-R_3-R_2$, $v_2-R_2-R_3-R_6-R_4$, $R_5-R_7-R_6$, and R_7-I.

(f) $R_1-R_5-R_6$ is a path, but it is not a loop (because it does not have the same starting and ending nodes), nor is it an essential branch (because it does not connect two essential nodes). v_2-R_2 is also a path but is neither a loop nor an essential branch, for the same reasons.

(g) $v_1-R_1-R_5-R_6-R_4-v_2$ is a loop but is not a mesh, because there are two loops within it. $I-R_5-R_6$ is also a loop but not a mesh.

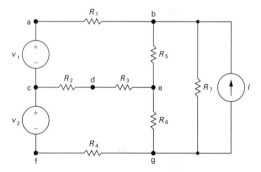

Figure 3.3 A circuit illustrating nodes, branches, meshes, paths, and loops.

Simultaneous Equations—How Many?

The number of unknown currents in a circuit equals the number of branches, b, where the current is not known. For example, the circuit shown in Fig. 3.3 has nine branches in which the current is unknown. Recall that we must have b independent equations to solve a circuit with b unknown currents. If we let n represent the number of nodes in the circuit, we can derive $n - 1$ independent equations by applying Kirchhoff's current law to any set of $n - 1$ nodes. (Application of the current law to the nth node does not generate an independent equation, because this equation can be derived from the previous $n - 1$ equations. See Drill Exercise 3.2.) Because we need b equations to describe a given circuit and because we can obtain $n - 1$ of these equations from Kirchhoff's current law, we must apply Kirchhoff's voltage law to loops or meshes to obtain the remaining $b - (n - 1)$ equations.

Thus by counting nodes, meshes, and branches where the current is unknown, we have established a systematic method for writing the necessary number of equations to solve a circuit. Specifically, we apply Kirchhoff's current law to $n - 1$ nodes and Kirchhoff's voltage law to $b - (n - 1)$ loops (or meshes). These observations also are valid in terms of essential nodes and essential branches. Thus if we let n_e represent the number of essential nodes and b_e the number of essential branches where the current is unknown, we can apply Kirchhoff's current law at $n_e - 1$ nodes and Kirchhoff's voltage law around $b_e - (n_e - 1)$ loops or meshes. In circuits, the number of essential nodes is less than or equal to the number of nodes, and the number of essential branches is less than or equal to the number of branches. Thus it is often convenient to use essential nodes and essential branches when analyzing a circuit, because they produce fewer independent equations to solve.

A circuit may consist of disconnected parts. An example of such a circuit is examined in Drill Exercise 3.4. The statements pertaining to the number of equations that can be derived from Kirchhoff's current law, $n - 1$, and voltage law, $b - (n - 1)$, apply to connected circuits. If a circuit has n nodes and b branches and is made up of s parts, the current law can be applied $n - s$ times, and the voltage law $b - n + s$ times. Any two separate parts can be connected by a single conductor. This connection always causes two nodes to form one node. Moreover, no current exists in the single conductor, so any circuit made up of s disconnected parts can always be reduced to a connected circuit.

The Systematic Approach—An Illustration

We now illustrate this systematic approach by using the circuit shown in Fig. 3.4. We write the equations on the basis of essential nodes and branches. The circuit has four essential nodes and six essential branches, denoted i_1–i_6, for which the current is unknown.

We derive three of the six simultaneous equations needed by applying Kirchhoff's current law to any three of the four essential

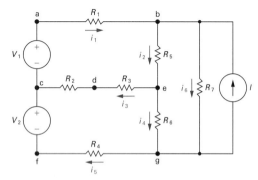

Figure 3.4 The circuit shown in Fig. 3.3 with six unknown branch currents defined.

nodes. We use the nodes b, c, and e to get

$$-i_1 + i_2 + i_6 - I = 0,$$

$$i_1 - i_3 - i_5 = 0,$$

$$i_3 + i_4 - i_2 = 0. \tag{3.1}$$

We derive the remaining three equations by applying Kirchhoff's voltage law around three meshes. Because the circuit has four meshes, we need to dismiss one mesh. We choose R_7–I, because we don't know the voltage across I.[1]
    Using the other three meshes gives

$$R_1 i_1 + R_5 i_2 + i_3(R_2 + R_3) - v_1 = 0,$$

$$-i_3(R_2 + R_3) + i_4 R_6 + i_5 R_4 - v_2 = 0,$$

$$-i_2 R_5 + i_6 R_7 - i_4 R_6 = 0. \tag{3.2}$$

Rearranging Eqs. (3.1) and (3.2) to facilitate their solution yields the set

$$-i_1 + i_2 + 0i_3 + 0i_4 + 0i_5 + i_6 = I,$$

$$i_1 + 0i_2 - i_3 + 0i_4 - i_5 + 0i_6 = 0,$$

$$0i_1 - i_2 + i_3 + i_4 + 0i_5 + 0i_6 = 0,$$

$$R_1 i_1 + R_5 i_2 + (R_2 + R_3)i_3 + 0i_4 + 0i_5 + 0i_6 = v_1,$$

$$0i_1 + 0i_2 - (R_2 + R_3)i_3 + R_6 i_4 + R_4 i_5 + 0i_6 = v_2,$$

$$0i_1 - R_5 i_2 + 0i_3 - R_6 i_4 + 0i_5 + R_7 i_6 = 0. \tag{3.3}$$

Note that summing the current at the nth node (g in this example) gives

$$i_5 - i_4 - i_6 + I = 0. \tag{3.4}$$

Equation (3.4) is not independent, because we can derive it by summing Eqs. (3.1) and then multiplying the sum by -1. Thus Eq. (3.4) is a linear combination of Eqs. (3.1) and therefore is not independent of them. We now carry the procedure one step further. By introducing new variables, we can describe a circuit with just $n - 1$ equations or just $b - (n - 1)$ equations. Therefore these new variables allow us to obtain a solution by manipulating

---

[1] We say more about this decision in Section 3.7.

fewer equations, a desirable goal even if a computer is to be used to obtain a numerical solution.

The new variables are known as node voltages and mesh currents. The node-voltage method enables us to describe a circuit in terms of $n_e - 1$ equations; the mesh-current method enables us to describe a circuit in terms of $b_e - (n_e - 1)$ equations. We begin in Section 3.2 with the node-voltage method.

DRILL EXERCISES

3.1   For the circuit shown, state the numerical value of the number of (a) branches, (b) branches where the current is unknown, (c) essential branches, (d) essential branches where the current is unknown, (e) nodes, (f) essential nodes, and (g) meshes.

ANSWER: (a) 11; (b) 9; (c) 9; (d) 7; (e) 6; (f) 4; (g) 6.

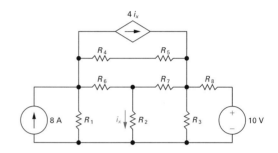

3.2   A current leaving a node is defined as positive.
   (a) Sum the currents at each node in the circuit shown.
   (b) Show that any one of the equations in (a) can be derived from the remaining two equations.

ANSWER: (a) 1: $i_1 - i_g + i_2 = 0$; 2: $i_3 + i_4 - i_2 = 0$; 3: $i_g - i_1 - i_3 - i_4 = 0$. (b) To derive any one equation from the other two, simply add the two equations and then multiply the resulting sum by -1.

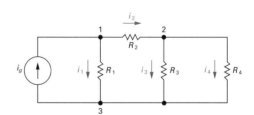

3.3   **(a)** If only the essential nodes and branches are identified in the circuit of Drill Exercise 3.1, how many simultaneous equations are needed to describe the circuit?
   (b) How many of these equations can be derived using Kirchhoff's current law?

   (c) How many must be derived using Kirchhoff's voltage law?
   (d) What two meshes should be avoided in applying the voltage law?

ANSWER: (a) 7; (b) 3; (c) 4; (d) $R_4 - R_5 - 4i_x$ and 8 A$-R_1$.

3.4   **(a)** How many separate parts does the circuit shown have?

(b) How many nodes?

(c) How many independent current equations can be written?

(d) How many branches are there?

(e) How many branches are there where the current is unknown?

(f) How many equations must be written using the voltage law?

(g) Assume that the lower node in each part of the circuit is joined by a single conductor. Repeat the calculations in (a)–(f).

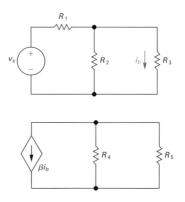

ANSWER: (a) 2; (b) 5; (c) 2; (d) 7; (e) 6; (f) 3; (g) 1, 4, 2, 7, 6, 3.

3.2 INTRODUCTION TO THE NODE-VOLTAGE METHOD

We introduce the node-voltage method by using the essential nodes of the circuit. The first step is to make a neat layout of the circuit so that no branches cross over and to mark clearly the essential nodes on the circuit diagram, as in Fig. 3.5. This circuit has three essential nodes ($n_e = 3$); therefore, we need two ($n_e - 1$) node-voltage equations to describe the circuit. The next step is to select one of the three essential nodes as a reference node. Although theoretically the choice is arbitrary, practically the choice for the reference node often is obvious. For example, the node with the most branches is usually a good choice. The optimum choice of the reference node (if one exists) will become apparent after you have gained some experience using this method. In the circuit shown in Fig. 3.5, the lower node connects the most branches, so we use it as the reference node. We flag the chosen reference node with the symbol ▼, as in Fig. 3.6.

After selecting the reference node, we define the node voltages on the circuit diagram. A **node voltage** is defined as the voltage rise from the reference node to a nonreference node. For this circuit, we must define two node voltages, which are denoted v_1 and v_2 in Fig. 3.6.

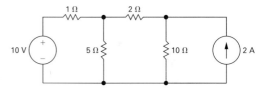

Figure 3.5 A circuit used to illustrate the node-voltage method of circuit analysis.

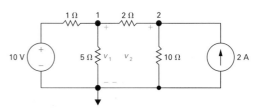

Figure 3.6 The circuit shown in Fig. 3.5 with a reference node and the node voltages.

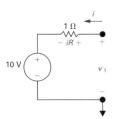

Figure 3.7 Computation of the branch current i.

We are now ready to generate the node-voltage equations. We do so by first writing the current leaving each branch connected to a nonreference node as a function of the node voltages and then summing these currents to zero in accordance with Kirchhoff's current law. For the circuit in Fig. 3.6, the current away from node 1 through the 1 Ω resistor is the voltage drop across the resistor divided by the resistance (Ohm's law). The voltage drop across the resistor, in the direction of the current away from the node, is $v_1 - 10$. Therefore the current in the 1 Ω resistor is $(v_1 - 10)/1$. Figure 3.7 depicts these observations. It shows the 10 V–1 Ω branch, with the appropriate voltages and current.

This same reasoning yields the current in every branch where the current is unknown. Thus the current away from node 1 through the 5 Ω resistor is $v_1/5$, and the current away from node 1 through the 2 Ω resistor is $(v_1 - v_2)/2$. The sum of the three currents leaving node 1 must equal zero; therefore the node-voltage equation derived at node 1 is

$$\frac{v_1 - 10}{1} + \frac{v_1}{5} + \frac{v_1 - v_2}{2} = 0. \tag{3.5}$$

The node-voltage equation derived at node 2 is

$$\frac{v_2 - v_1}{2} + \frac{v_2}{10} - 2 = 0. \tag{3.6}$$

Note that the first term in Eq. (3.6) is the current away from node 2 through the 2 Ω resistor, the second term is the current away from node 2 through the 10 Ω resistor, and the third term is the current away from node 2 through the current source.

Equations (3.5) and (3.6) are the two simultaneous equations that describe the circuit shown in Fig. 3.6 in terms of the node voltages v_1 and v_2. Solving for v_1 and v_2 yields

$$v_1 = \frac{100}{11} = 9.09 \text{ V},$$

$$v_2 = \frac{120}{11} = 10.91 \text{ V}.$$

Once the node voltages are known, all the branch currents can be calculated. Once these are known, the branch voltages and powers can be calculated. Example 3.2 illustrates the use of the node-voltage method.

EXAMPLE 3.2

(a) Use the node-voltage method of circuit analysis to find the branch currents i_a, i_b, and i_c in the circuit shown in Fig. 3.8.

(b) Find the power associated with each source, and state whether the source is delivering or absorbing power.

SOLUTION

(a) We begin by noting that the circuit has two essential nodes; thus we need to write a single node-voltage expression. We select the lower node as the reference node and define the unknown node voltage as v_1. Figure 3.9 illustrates these decisions. Summing the currents away from node 1 generates the node-voltage equation

$$\frac{v_1 - 50}{5} + \frac{v_1}{10} + \frac{v_1}{40} - 3 = 0.$$

Solving for v_1 gives

$$v_1 = 40 \text{ V}.$$

Hence

$$i_a = \frac{50 - 40}{5} = 2 \text{ A},$$

$$i_b = \frac{40}{10} = 4 \text{ A},$$

$$i_c = \frac{40}{40} = 1 \text{ A}.$$

(b) The power associated with the 50 V source is

$$p_{50V} = -50i_a = -100 \text{ W (delivering)}.$$

The power associated with the 3 A source is

$$p_{3A} = -3v_1 = -3(40) = -120 \text{ W (delivering)}.$$

We check these calculations by noting that the total delivered power is 220 W. The total power absorbed by the three resistors is $4(5) + 16(10) + 1(40)$, or 220 W, as we calculated and as it must be.

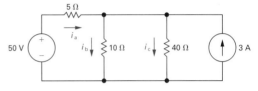

Figure 3.8 The circuit for Example 3.2.

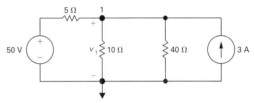

Figure 3.9 The circuit shown in Fig. 3.8 with a reference node and the unknown node voltage v_1.

DRILL EXERCISES

3.5 **(a)** For the circuit shown, use the node-voltage method to find v_1, v_2, and i_1.

(b) How much power is delivered to the circuit by the 15 A source?

(c) Repeat (b) for the 5 A source.

ANSWER: (a) 60 V, 10 V, 10 A; (b) 900 W; (c) −50 W.

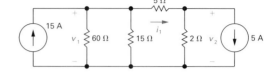

3.6 Use the node-voltage method to find v in the circuit shown.

ANSWER: 15 V.

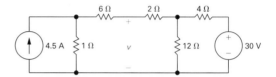

3.3 THE NODE-VOLTAGE METHOD AND DEPENDENT SOURCES

If the circuit contains dependent sources, the node-voltage equations must be supplemented with the constraint equations imposed by the presence of the dependent sources. Example 3.3 illustrates the application of the node-voltage method to a circuit containing a dependent source.

EXAMPLE 3.3

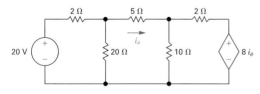

Figure 3.10 The circuit for Example 3.3.

Use the node-voltage method to find the power dissipated in the 5 Ω resistor in the circuit shown in Fig. 3.10.

SOLUTION

We begin by noting that the circuit has three essential nodes. Hence we need two node-voltage equations to describe the circuit. Four branches terminate on the lower node, so we select it as the reference node. The two unknown node voltages are defined on the circuit shown in Fig. 3.11. Summing the currents away from node 1 generates the equation

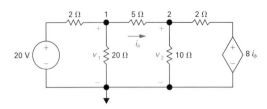

$$\frac{v_1 - 20}{2} + \frac{v_1}{20} + \frac{v_1 - v_2}{5} = 0.$$

Figure 3.11 The circuit shown in Fig. 3.10, with a reference node and the node voltages.

Summing the currents away from node 2 yields

$$\frac{v_2 - v_1}{5} + \frac{v_2}{10} + \frac{v_2 - 8i_\phi}{2} = 0.$$

As written, these two node-voltage equations contain three unknowns, namely, v_1, v_2, and i_ϕ. To eliminate i_ϕ we must express this controlling current in terms of the node voltages, or

$$i_\phi = \frac{v_1 - v_2}{5}.$$

Substituting this relationship into the node 2 equation simplifies the two node-voltage equations to

$$0.75v_1 - 0.2v_2 = 10,$$

$$-v_1 + 1.6v_2 = 0.$$

Solving for v_1 and v_2 gives

$$v_1 = 16 \text{ V} \quad \text{and} \quad v_2 = 10 \text{ V}.$$

Then,

$$i_\phi = \frac{16 - 10}{5} = 1.2 \text{ A}$$

$$p_{5\Omega} = (1.44)(5) = 7.2 \text{ W}.$$

A good exercise to build your problem-solving intuition is to reconsider this example, using node 2 as the reference node. Does it make the analysis easier or harder?

DRILL EXERCISE

3.7 **(a)** Use the node-voltage method to find the power delivered by each source in the circuit shown.

(b) State whether the source is delivering power to the circuit or extracting power from the circuit.

ANSWER: (a) $p_{50V} = 150$ W, $p_{3i_1} = 144$ W, $p_{5A} = 80$ W; (b) all sources are delivering power to the circuit.

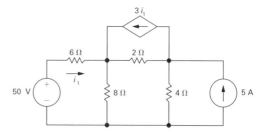

3.4 THE NODE-VOLTAGE METHOD: SOME SPECIAL CASES

When a voltage source is the only element between two essential nodes, the node-voltage method is simplified. As an example, look at the circuit in Fig. 3.12. There are three essential nodes in this circuit, which means that two simultaneous equations are needed. From these three essential nodes, a reference node has been chosen and two other nodes have been labeled. But the 100 V source constrains the voltage between node 1 and the reference node to 100 V. This means that there is only one unknown node voltage (v_2). Solution of this circuit thus involves only a single node-voltage equation at node 2:

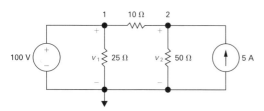

Figure 3.12 A circuit with a known node voltage.

$$\frac{v_2 - v_1}{10} + \frac{v_2}{50} - 5 = 0. \tag{3.7}$$

But $v_1 = 100$ V, so Eq. (3.7) can be solved for v_2:

$$v_2 = 125 \text{ V.} \tag{3.8}$$

Knowing v_2, we can calculate the current in every branch. You should verify that the current into node 1 in the branch containing the independent voltage source is 1.5 A.

In general, when you use the node-voltage method to solve circuits that have voltage sources connected directly between essential nodes, the number of unknown node voltages is reduced. The reason is that, whenever a voltage source connects two essential nodes, it constrains the difference between the node voltages at these nodes to equal the voltage of the source. Taking the time to see if you can reduce the number of unknowns in this way will simplify circuit analysis.

Suppose that the circuit shown in Fig. 3.13 is to be analyzed using the node-voltage method. The circuit contains four essential nodes, so we anticipate writing three node-voltage equations. However, two essential nodes are connected by an independent voltage source, and two other essential nodes are connected by a current-controlled dependent voltage source. Hence, there actually is only one unknown node voltage.

Choosing which node to use as the reference node involves several possibilities. Either node on each side of the dependent voltage source looks attractive because, if chosen, one of the node voltages would be known to be either $+10i_\phi$ (left node is the reference) or $-10i_\phi$ (right node is the reference). The lower node looks even better because one node voltage is immediately known (50 V) and five branches terminate there. We therefore opt for the lower node as the reference.

Figure 3.14 shows the redrawn circuit, with the reference node flagged and the node voltages defined. Also, we introduce the current i because we cannot express the current in the dependent voltage source branch as a function of the node voltages v_2 and v_3. Thus at node 2

$$\frac{v_2 - v_1}{5} + \frac{v_2}{50} + i = 0, \tag{3.9}$$

and at node 3

$$\frac{v_3}{100} - i - 4 = 0. \tag{3.10}$$

We eliminate i simply by adding Eqs. (3.9) and (3.10) to get

$$\frac{v_2 - v_1}{5} + \frac{v_2}{50} + \frac{v_3}{100} - 4 = 0. \tag{3.11}$$

The Concept of a Supernode

Equation (3.11) may be written directly, without resorting to the intermediate step represented by Eqs. (3.9) and (3.10). To do so, we consider nodes 2 and 3 to be a single node and simply sum the currents away from the node in terms of the node voltages v_2 and v_3. Figure 3.15 illustrates this approach.

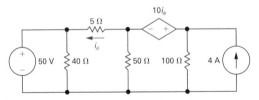

Figure 3.13 A circuit with a dependent voltage source connected between nodes.

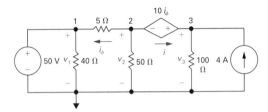

Figure 3.14 The circuit shown in Fig. 3.13, with the selected node voltages defined.

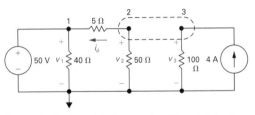

Figure 3.15 Considering nodes 2 and 3 to be a supernode.

When a voltage source is between two essential nodes, we can combine those nodes to form a **supernode**. Obviously, Kirchhoff's current law must hold for the supernode. In Fig. 3.15, starting with the 5 Ω branch and moving counterclockwise around the supernode, we generate the equation

$$\frac{v_2 - v_1}{5} + \frac{v_2}{50} + \frac{v_3}{100} - 4 = 0, \tag{3.12}$$

which is identical to Eq. (3.11). Creating a supernode at nodes 2 and 3 has made the task of analyzing this circuit easier. It is therefore always worth taking the time to look for this type of shortcut before writing any equations.

After Eq. (3.12) has been derived, the next step is to reduce the expression to a single unknown node voltage. First we eliminate v_1 from the equation because we know that $v_1 = 50$ V. Next we express v_3 as a function of v_2:

$$v_3 = v_2 + 10i_\phi. \tag{3.13}$$

We now express the current controlling the dependent voltage source as a function of the node voltages:

$$i_\phi = \frac{v_2 - 50}{5}. \tag{3.14}$$

Using Eqs. (3.13) and (3.14) and $v_1 = 50$ V reduces Eq. (3.12) to

$$v_2 \left(\frac{1}{50} + \frac{1}{5} + \frac{1}{100} + \frac{10}{500} \right) = 10 + 4 + 1$$

$$v_2(0.25) = 15$$

$$v_2 = 60 \text{ V.}$$

From Eqs. (3.13) and (3.14):

$$i_\phi = \frac{60 - 50}{5} = 2 \text{ A}$$

$$v_3 = 60 + 20 = 80 \text{ V.}$$

DRILL EXERCISES

3.8 Use the node-voltage method to find v in the circuit shown.

ANSWER: 8 V.

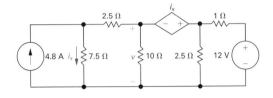

3.9 Use the node-voltage method to find v_1 in the circuit shown.

ANSWER: 48 V.

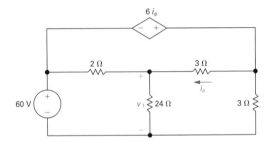

3.10 Use the node-voltage method to find v_o in the circuit shown.

ANSWER: 24 V.

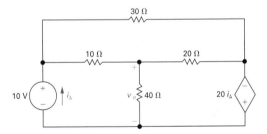

3.5 INTRODUCTION TO THE MESH-CURRENT METHOD

As stated in Section 3.1, the mesh-current method of circuit analysis enables us to describe a circuit in terms of $b_e - (n_e - 1)$ equations. Recall that a mesh is a loop with no other loops inside it. The circuit in Fig. 3.1(b) is shown again in Fig. 3.16, with current arrows inside each loop to distinguish it. Recall also that the mesh-current method is applicable only to planar circuits. The

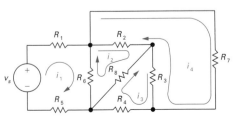

Figure 3.16 The circuit shown in Fig. 3.1(b), with the mesh currents defined.

circuit in Fig. 3.16 contains seven essential branches where the current is unknown and four essential nodes. Therefore, to solve it via the mesh-current method, we must write four $[7 - (4 - 1)]$ mesh-current equations.

A **mesh current** is the current that exists only in the perimeter of a mesh. On a circuit diagram it appears as either a closed solid line or an almost-closed solid line that follows the perimeter of the appropriate mesh. An arrowhead on the solid line indicates the reference direction for the mesh current. Figure 3.16 shows the four mesh currents that describe the circuit in Fig. 3.1(b). Note that by definition, mesh currents automatically satisfy Kirchhoff's current law. That is, at any node in the circuit, a given mesh current both enters and leaves the node.

Figure 3.16 also shows that identifying a mesh current in terms of a branch current is not always possible. For example, the mesh current i_2 is not equal to any branch current, whereas mesh currents i_1, i_3, and i_4 can be identified with branch currents. Thus measuring a mesh current is not always possible; note that there is no place where an ammeter can be inserted to measure the mesh current i_2. The fact that a mesh current can be a fictitious quantity doesn't mean that it is a useless concept. On the contrary, the mesh-current method of circuit analysis evolves quite naturally from the branch-current equations.

We can use the circuit in Fig. 3.17 to show the evolution of the mesh-current technique. We begin by using the branch currents $(i_1, i_2,$ and $i_3)$ to formulate the set of independent equations. For this circuit, $b_e = 3$ and $n_e = 2$. We can write only one independent current equation, so we need two independent voltage equations. Applying Kirchhoff's current law to the upper node and Kirchhoff's voltage law around the two meshes generates the following set of equations:

$$i_1 = i_2 + i_3, \tag{3.15}$$

$$v_1 = i_1 R_1 + i_3 R_3, \tag{3.16}$$

$$-v_2 = i_2 R_2 - i_3 R_3. \tag{3.17}$$

We reduce this set of three equations to a set of two equations by solving Eq. (3.15) for i_3 and then substituting this expression into Eqs. (3.16) and (3.17):

$$v_1 = i_1(R_1 + R_3) - i_2 R_3, \tag{3.18}$$

$$-v_2 = -i_1 R_3 + i_2(R_2 + R_3). \tag{3.19}$$

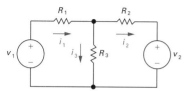

Figure 3.17 A circuit used to illustrate development of the mesh-current method of circuit analysis.

We can solve Eqs. (3.18) and (3.19) for i_1 and i_2 to replace the solution of three simultaneous equations with the solution of two simultaneous equations. We derived Eqs. (3.18) and (3.19) by substituting the $n_e - 1$ current equations into the $b_e - (n_e - 1)$ voltage equations. The value of the mesh-current method is that, by defining mesh currents, we automatically eliminate the $n_e - 1$ current equations. Thus the mesh-current method is equivalent to a systematic substitution of the $n_e - 1$ current equations into the $b_e - (n_e - 1)$ voltage equations. The mesh currents in Fig. 3.17 that are equivalent to eliminating the branch current i_3 from Eqs. (3.16) and (3.17) are shown in Fig. 3.18. We now apply Kirchhoff's voltage law around the two meshes, expressing all voltages across resistors in terms of the mesh currents, to get the equations

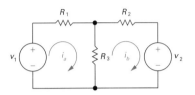

Figure 3.18 Mesh currents i_a and i_b.

$$v_1 = i_a R_1 + (i_a - i_b)R_3, \qquad (3.20)$$

$$-v_2 = (i_b - i_a)R_3 + i_b R_2. \qquad (3.21)$$

Collecting the coefficients of i_a and i_b in Eqs. (3.20) and (3.21) gives

$$v_1 = i_a(R_1 + R_3) - i_b R_3, \qquad (3.22)$$

$$-v_2 = -i_a R_3 + i_b(R_2 + R_3). \qquad (3.23)$$

Note that Eqs. (3.22) and (3.23) and Eqs. (3.18) and (3.19) are identical in form, with the mesh currents i_a and i_b replacing the branch currents i_1 and i_2. Note also that the branch currents shown in Fig. 3.17 can be expressed in terms of the mesh currents shown in Fig. 3.18, or

$$i_1 = i_a, \qquad (3.24)$$

$$i_2 = i_b, \qquad (3.25)$$

$$i_3 = i_a - i_b. \qquad (3.26)$$

The ability to write Eqs. (3.24)–(3.26) by inspection is crucial to the mesh-current method of circuit analysis. Once you know the mesh currents, you also know the branch currents. And once you know the branch currents, you can compute any voltages or powers of interest.

Example 3.4 illustrates how the mesh-current method is used to find source powers and a branch voltage.

EXAMPLE 3.4

(a) Use the mesh-current method to determine the power associated with each voltage source in the circuit shown in Fig. 3.19.

(b) Calculate the voltage v_o across the 8 Ω resistor.

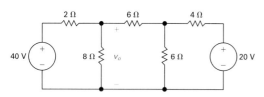

Figure 3.19 The circuit for Example 3.4.

SOLUTION

(a) To calculate the power associated with each source, we need to know the current in each source. The circuit indicates that these source currents will be identical to mesh currents. Also, note that the circuit has seven branches where the current is unknown and five nodes. Therefore we need three $[b - (n - 1) = 7 - (5 - 1)]$ mesh-current equations to describe the circuit. Figure 3.20 shows the three mesh currents used to describe the circuit in Fig. 3.19. If we assume that the voltage drops are positive, the three mesh equations are

$$-40 + 2i_a + 8(i_a - i_b) = 0,$$

$$8(i_b - i_a) + 6i_b + 6(i_b - i_c) = 0,$$

$$6(i_c - i_b) + 4i_c + 20 = 0. \tag{3.27}$$

Your calculator can probably solve these equations, or you can use a computer tool. Reorganizing Eqs. (3.27) in anticipation of using your calculator or a computer program gives

$$10i_a - 8i_b + 0i_c = 40;$$

$$-8i_a + 20i_b - 6i_c = 0;$$

$$0i_a - 6i_b + 10i_c = -20. \tag{3.28}$$

The three mesh currents are

$$i_a = 5.6 \text{ A},$$

$$i_b = 2.0 \text{ A},$$

$$i_c = -0.80 \text{ A}.$$

The mesh current i_a is identical with the branch current in the 40 V source, so the power associated with this source is

$$p_{40V} = -40i_a = -224 \text{ W.}$$

The minus sign means that this source is delivering power to the network. The current in the 20 V source is identical to the mesh current i_c; therefore

$$p_{20V} = 20i_c = -16 \text{ W.}$$

The 20 V source also is delivering power to the network.

(b) The branch current in the 8 Ω resistor in the direction of the voltage drop v_o is $i_a - i_b$. Therefore

$$v_o = 8(i_a - i_b) = 8(3.6) = 28.8 \text{ V.}$$

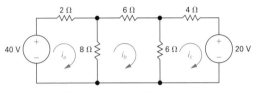

Figure 3.20 The three mesh currents used to analyze the circuit shown in Fig. 3.19.

DRILL EXERCISE

3.11 Use the mesh-current method to find (a) the power delivered by the 80 V source to the circuit shown and (b) the power dissipated in the 8 Ω resistor.

ANSWER: (a) 400 W; (b) 50 W.

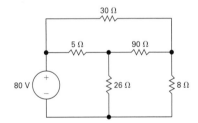

3.6 THE MESH-CURRENT METHOD AND DEPENDENT SOURCES

If the circuit contains dependent sources, the mesh-current equations must be supplemented by the appropriate constraint equations. Example 3.5 illustrates the application of the mesh-current method when the circuit includes a dependent source.

EXAMPLE 3.5

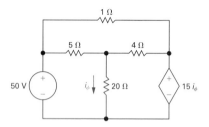

Figure 3.21 The circuit for Example 3.5.

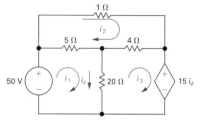

Figure 3.22 The circuit shown in Fig. 3.21 with the three mesh currents.

Use the mesh-current method of circuit analysis to determine the power dissipated in the 4 Ω resistor in the circuit shown in Fig. 3.21.

SOLUTION

This circuit has six branches where the current is unknown and four nodes. Therefore we need three mesh currents to describe the circuit. They are defined on the circuit shown in Fig. 3.22. The three mesh-current equations are

$$50 = 5(i_1 - i_2) + 20(i_1 - i_3),$$

$$0 = 5(i_2 - i_1) + 1i_2 + 4(i_2 - i_3),$$

$$0 = 20(i_3 - i_1) + 4(i_3 - i_2) + 15i_\phi. \tag{3.29}$$

We now express the branch current controlling the dependent voltage source in terms of the mesh currents as

$$i_\phi = i_1 - i_3, \tag{3.30}$$

which is the supplemental equation imposed by the presence of the dependent source. Substituting Eq. (3.30) into Eqs. (3.29) and collecting the coefficients of i_1, i_2, and i_3 in each equation generates

$$50 = 25i_1 - 5i_2 - 20i_3,$$

$$0 = -5i_1 + 10i_2 - 4i_3,$$

$$0 = -5i_1 - 4i_2 + 9i_3.$$

Because we are calculating the power dissipated in the 4 Ω resistor, we compute the mesh currents i_2 and i_3:

$$i_2 = 26 \text{ A},$$

$$i_3 = 28 \text{ A}.$$

The current in the 4 Ω resistor oriented from left to right is $i_3 - i_2$, or 2 A. Therefore the power dissipated is

$$p_{4\Omega} = (i_3 - i_2)^2(4) = (2)^2(4) = 16 \text{ W}.$$

What if you had not been told to use the mesh-current method? Would you have chosen the node-voltage method? It reduces the problem to finding one unknown node voltage because of the presence of two voltage sources between essential nodes. We present more about making such choices later.

DRILL EXERCISES

3.12 **(a)** Determine the number of mesh-current equations needed to solve the circuit shown.

(b) Use the mesh-current method to find how much power is being delivered to the dependent voltage source.

ANSWER: (a) 3; (b) −36 W.

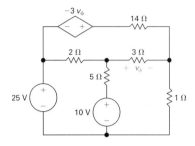

3.13 Use the mesh-current method to find v_o in the circuit shown.

ANSWER: 16 V.

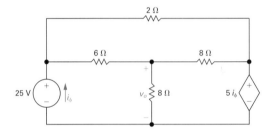

3.7 THE MESH-CURRENT METHOD: SOME SPECIAL CASES

When a branch includes a current source, the mesh-current method requires some additional manipulations. The circuit shown in Fig. 3.23 depicts the nature of the problem.

We have defined the mesh currents i_a, i_b, and i_c, as well as the voltage across the 5 A current source, to aid the discussion. Note that the circuit contains five essential branches where the current is unknown and four essential nodes. Hence we need to write two $[5 - (4 - 1)]$ mesh-current equations to solve the circuit. The presence of the current source reduces the three unknown mesh currents to two, because it constrains the difference between i_a and i_c to equal 5 A. Hence, if we know i_a, we know i_c, and vice versa.

However, when we attempt to sum the voltages around either mesh a or mesh c, we must introduce into the equations the unknown voltage across the 5 A current source. Thus, for

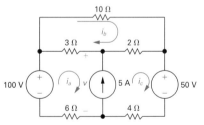

Figure 3.23 A circuit illustrating mesh analysis when a branch contains an independent current source.

mesh a:

$$100 = 3(i_a - i_b) + v + 6i_a, \tag{3.31}$$

and for mesh c:

$$-50 = 4i_c - v + 2(i_c - i_b). \tag{3.32}$$

We now add Eqs. (3.31) and (3.32) to eliminate v and obtain

$$50 = 9i_a - 5i_b + 6i_c. \tag{3.33}$$

Summing voltages around mesh b gives

$$0 = 3(i_b - i_a) + 10i_b + 2(i_b - i_c). \tag{3.34}$$

We reduce Eqs. (3.33) and (3.34) to two equations and two unknowns by using the constraint that

$$i_c - i_a = 5. \tag{3.35}$$

We leave to you the verification that, when Eq. (3.35) is combined with Eqs. (3.33) and (3.34), the solutions for the three mesh currents are

$$i_a = 1.75 \text{ A}, \quad i_b = 1.25 \text{ A}, \quad \text{and} \quad i_c = 6.75 \text{ A}.$$

The Concept of a Supermesh

We can derive Eq. (3.33) without introducing the unknown voltage v by using the concept of a supermesh. To create a supermesh, we mentally remove the current source from the circuit by simply avoiding this branch when writing the mesh-current equations. We express the voltages around the supermesh in terms of the original mesh currents. Figure 3.24 illustrates the supermesh concept. When we sum the voltages around the supermesh (denoted by the dashed line), we obtain the equation

$$-100 + 3(i_a - i_b) + 2(i_c - i_b) + 50 + 4i_c + 6i_a = 0, \tag{3.36}$$

which reduces to

$$50 = 9i_a - 5i_b + 6i_c. \tag{3.37}$$

Note that Eqs. (3.37) and (3.33) are identical. Thus the supermesh has eliminated the need for introducing the unknown voltage across the current source. Once again, taking time to look carefully at a circuit to identify a shortcut such as this provides a big payoff in simplifying the analysis.

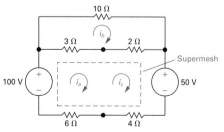

Figure 3.24 The circuit shown in Fig. 3.23, illustrating the concept of a supermesh.

DRILL EXERCISES

3.14 Use the mesh-current method to find the power dissipated in the 2 Ω resistor in the circuit shown.

ANSWER: 72 W.

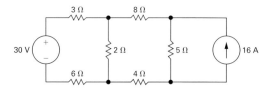

3.15 Use the mesh-current method to find the mesh current i_a in the circuit shown.

ANSWER: 15 A.

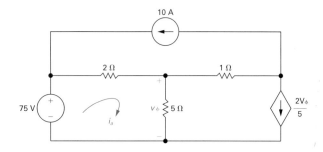

3.16 Use the mesh-current method to find the power dissipated in the 1 Ω resistor in the circuit shown.

ANSWER: 36 W.

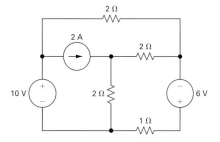

3.8 THE NODE-VOLTAGE METHOD VERSUS THE MESH-CURRENT METHOD

The greatest advantage of both the node-voltage and mesh-current methods is that they reduce the number of simultaneous equations that must be manipulated. They also require the analyst to be quite systematic in terms of organizing and writing these equations. It is natural to ask, then, "When is the node-voltage method preferred to the mesh-current method and vice versa?" As you might suspect, there is no clear-cut answer. Asking a number of questions, however, may help you identify the more efficient method before plunging into the solution process:

- Does one of the methods result in fewer simultaneous equations to solve?
- Does the circuit contain supernodes? If so, using the node-voltage method will permit you to reduce the number of equations to be solved.
- Does the circuit contain supermeshes? If so, using the mesh-current method will permit you to reduce the number of equations to be solved.
- Will solving some portion of the circuit give the requested solution? If so, which method is most efficient for solving just the pertinent portion of the circuit?

Perhaps the most important observation is that, for any situation, some time spent thinking about the problem in relation to the various analytical approaches available is time well spent. Examples 3.6 and 3.7 illustrate the process of deciding between the node-voltage and mesh-current methods.

EXAMPLE 3.6

Find the power dissipated in the 300 Ω resistor in the circuit shown in Fig. 3.25.

SOLUTION

To find the power dissipated in the 300 Ω resistor, we need to find either the current in the resistor or the voltage across it. The mesh-current method yields the current in the resistor; this approach requires solving five simultaneous mesh equations, as depicted in Fig. 3.26. In writing the five equations, we must include the constraint $i_\Delta = -i_b$.

Before going further, let's also look at the circuit in terms of the node-voltage method. Note that, once we know the node voltages, we can calculate either the current in the 300 Ω resistor or the voltage across it. The circuit has four essential nodes, and therefore only three node-voltage equations are required to describe the circuit. Because of the dependent voltage source between two essential nodes, we have to sum the currents at only two nodes. Hence the problem is reduced to writing two node-voltage equations and a constraint equation. Because the node-voltage method requires only three simultaneous equations, it is the more attractive approach.

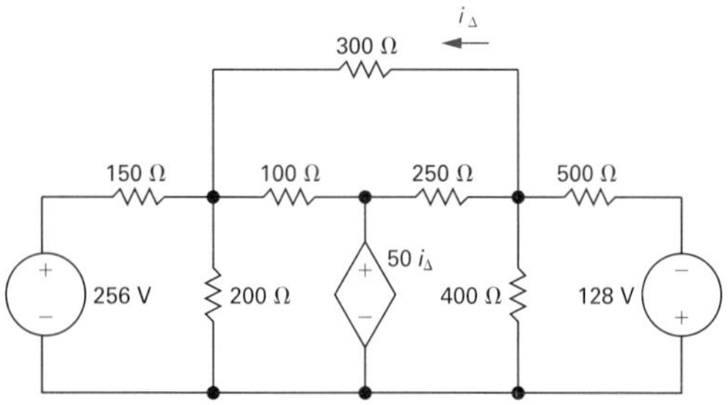

Figure 3.25 The circuit for Example 3.6.

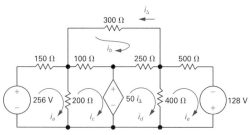

Figure 3.26 The circuit shown in Fig. 3.25, with the five mesh currents.

Once the decision to use the node-voltage method has been made, the next step is to select a reference node. Two essential nodes in the circuit in Fig. 3.25 merit consideration. The first is the reference node in Fig. 3.27. If this node is selected, one of the unknown node voltages is the voltage across the 300 Ω resistor, namely, v_2 in Fig. 3.27. Once we know this voltage, we calculate the power in the 300 Ω resistor by using the expression

$$p_{300\Omega} = v_2^2/300.$$

Note that, in addition to selecting the reference node, we defined the three node voltages v_1, v_2, and v_3 and indicated that nodes 1 and 3 form a supernode, because they are connected by a dependent voltage source. It is understood that a node voltage is a rise from the reference node; therefore, in Fig. 3.27, we have not placed the node voltage polarity references on the circuit diagram.

The second node that merits consideration as the reference node is the lower node in the circuit, as shown in Fig. 3.28. It is attractive because it has the most branches connected to it, and the node-voltage equations are thus easier to write. However, to find either the current in the 300 Ω resistor or the voltage across it requires an additional calculation, once we know the node voltages v_a and v_c. For example, the current in the 300 Ω resistor is $(v_c - v_a)/300$, whereas the voltage across the resistor is $v_c - v_a$.

We compare these two possible reference nodes by means of the following sets of equations. The first set pertains to the circuit shown in Fig. 3.27, and the second set is based on the circuit shown in Fig. 3.28.

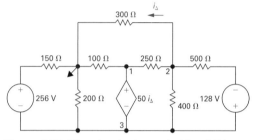

Figure 3.27 The circuit shown in Fig. 3.25, with a reference node.

- Set 1 (Fig. 3.27)
   At the supernode,

$$\frac{v_1}{100} + \frac{v_1 - v_2}{250} + \frac{v_3}{200} + \frac{v_3 - v_2}{400} + \frac{v_3 - (v_2 + 128)}{500}$$
$$+ \frac{v_3 + 256}{150} = 0.$$

At v_2,

$$\frac{v_2}{300} + \frac{v_2 - v_1}{250} + \frac{v_2 - v_3}{400} + \frac{v_2 + 128 - v_3}{500} = 0.$$

From the supernode, the constraint equation is

$$v_3 = v_1 - 50i_\Delta = v_1 - \frac{v_2}{6}.$$

- Set 2 (Fig. 3.28)
  At v_a,

$$\frac{v_a}{200} + \frac{v_a - 256}{150} + \frac{v_a - v_b}{100} + \frac{v_a - v_c}{300} = 0.$$

At v_c,

$$\frac{v_c}{400} + \frac{v_c + 128}{500} + \frac{v_c - v_b}{250} + \frac{v_c - v_a}{300} = 0.$$

From the supernode, the constraint equation is

$$v_b = 50i_\Delta = \frac{50(v_c - v_a)}{300} = \frac{v_c - v_a}{6}.$$

You should verify that the solution of either set leads to a power calculation of 16.57 W dissipated in the 300 Ω resistor.

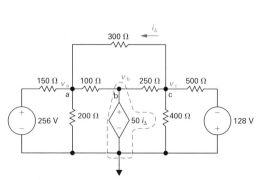

Figure 3.28 The circuit shown in Fig. 3.25 with an alternative reference node.

EXAMPLE 3.7

Find the voltage v_o in the circuit shown in Fig. 3.29.

SOLUTION

At first glance, the node-voltage method looks appealing, because we may define the unknown voltage as a node voltage by choosing the lower terminal of the dependent current source as the reference node. The circuit has four essential nodes and two voltage-controlled dependent sources, so the node-voltage method requires manipulation of three node-voltage equations and two constraint equations.

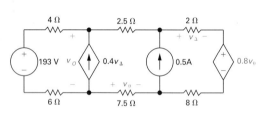

Figure 3.29 The circuit for Example 3.7.

Let's now turn to the mesh-current method for finding v_o. The circuit contains three meshes, and we can use the leftmost one to calculate v_o. If we let i_a denote the leftmost mesh current, then $v_o = 193 - 10i_a$. The presence of the two current sources reduces the problem to manipulating a single supermesh equation and two constraint equations. Hence the mesh-current method is the more attractive technique here.

To help you compare the two approaches, we summarize both methods. The mesh-current equations are based on the circuit shown in Fig. 3.30, and the node-voltage equations are based on the circuit shown in Fig. 3.31. The supermesh equation is

$$193 = 10i_a + 10i_b + 10i_c + 0.8v_\theta,$$

and the constraint equations are

$$i_b - i_a = 0.4v_\Delta = 0.8i_c; \quad v_\theta = -7.5i_b; \quad \text{and} \quad i_c - i_b = 0.5.$$

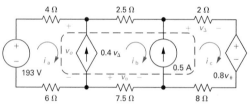

Figure 3.30 The circuit shown in Fig. 3.29 with the three mesh currents.

We use the constraint equations to write the supermesh equation in terms of i_a:

$$160 = 80i_a, \quad \text{or} \quad i_a = 2 \text{ A},$$

$$v_o = 193 - 20 = 173 \text{ V}.$$

The node-voltage equations are

$$\frac{v_o - 193}{10} - 0.4v_\Delta + \frac{v_o - v_a}{2.5} = 0,$$

$$\frac{v_a - v_o}{2.5} - 0.5 + \frac{v_a - (v_b + 0.8v_\theta)}{10} = 0,$$

$$\frac{v_b}{7.5} + 0.5 + \frac{v_b + 0.8v_\theta - v_a}{10} = 0.$$

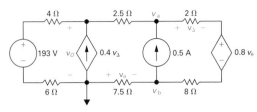

Figure 3.31 The circuit shown in Fig. 3.29 with node voltages.

The constraint equations are

$$v_\theta = -v_b,$$

$$v_\Delta = \left[\frac{v_a - (v_b + 0.8v_\theta)}{10}\right]2.$$

We use the constraint equations to reduce the node-voltage equations to three simultaneous equations involving v_o, v_a, and v_b. You should verify that the node-voltage approach also gives $v_o = 173$ V.

DRILL EXERCISES

3.17 Find the power delivered by the 2 A current source in the circuit shown.

   ANSWER: 70 W.

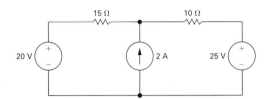

3.18 Find the power delivered by the 4 A current source in the circuit shown.

   ANSWER: 40 W.

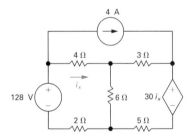

3.9   THÉVENIN AND NORTON EQUIVALENTS

At times in circuit analysis we want to concentrate on what happens at a specific pair of terminals. For example, when we plug a toaster into an outlet, we are interested primarily in the voltage and current at the terminals of the toaster. We have little or no interest in the effect that connecting the toaster has on voltages or currents elsewhere in the circuit supplying the outlet. We can expand this interest in terminal behavior to a set of appliances, each requiring a different amount of power. We then are interested in how the voltage and current delivered at the outlet change as we change appliances. In other words, we want to focus on the behavior of the circuit supplying the outlet, but only at the outlet terminals.

Thévenin and Norton equivalents are circuit simplification techniques that focus on terminal behavior and thus are extremely valuable aids in analysis. Although here we discuss them as they pertain to resistive circuits, Thévenin and Norton equivalent circuits may be used to represent any circuit made up of linear elements.

We can best describe a Thévenin equivalent circuit by reference to Fig. 3.32, which represents any circuit made up of sources (both independent and dependent) and resistors. The letters a and b denote the pair of terminals of interest. Figure 3.32(b) shows the Thévenin equivalent. Thus, a **Thévenin equivalent circuit** is an independent voltage source V_{Th} in series with a resistor R_{Th}, which replaces an interconnection of sources and resistors. This series combination of V_{Th} and R_{Th} is equivalent to the original circuit in the sense that, if we connect the same load across the terminals a,b of each circuit, we get the same voltage and current at the terminals of the load. This equivalence holds for all possible values of load resistance.

To represent the original circuit by its Thévenin equivalent, we must be able to determine the Thévenin voltage V_{Th} and the Thévenin resistance R_{Th}. First, we note that if the load resistance is infinitely large, we have an open-circuit condition. The open-circuit voltage at the terminals a,b in the circuit shown in Fig. 3.32(b) is V_{Th}. By hypothesis, this must be the same as the open-circuit voltage at the terminals a,b in the original circuit. Therefore, to calculate the Thévenin voltage V_{Th}, we simply calculate the open-circuit voltage in the original circuit.

Reducing the load resistance to zero gives us a short-circuit condition. If we place a short circuit across the terminals a,b of the Thévenin equivalent circuit, the short-circuit current directed from a to b is

$$i_{sc} = \frac{V_{Th}}{R_{Th}}. \tag{3.38}$$

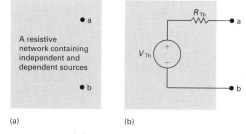

Figure 3.32 (a) A general circuit. (b) The Thévenin equivalent circuit.

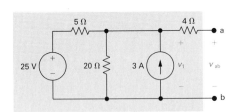

Figure 3.33 A circuit used to illustrate a Thévenin equivalent.

By hypothesis, this short-circuit current must be identical to that which exists in a short circuit placed across the terminals a,b of the original network. From Eq. (3.38),

$$R_{Th} = \frac{V_{Th}}{i_{sc}}. \qquad (3.39)$$

Thus the Thévenin resistance is the ratio of the open-circuit voltage to the short-circuit current.

Finding a Thévenin Equivalent

To find the Thévenin equivalent of the circuit shown in Fig. 3.33, we first calculate the open-circuit voltage of v_{ab}. Note that when the terminals a,b are open, there is no current in the 4 Ω resistor. Therefore the open-circuit voltage v_{ab} is identical to the voltage across the 3 A current source, labeled v_1. We find the voltage v_1 by solving a single node-voltage equation. Choosing the lower node as the reference node, we get

$$\frac{v_1 - 25}{5} + \frac{v_1}{20} - 3 = 0. \qquad (3.40)$$

Solving for v_1 yields

$$v_1 = 32 \text{ V}. \qquad (3.41)$$

Hence the Thévenin voltage for the circuit is 32 V.

The next step is to place a short circuit across the terminals and calculate the resulting short-circuit current. Figure 3.34 shows the circuit with the short in place. Note that the short-circuit current is in the direction of the open-circuit voltage drop across the terminals a,b. If the short-circuit current is in the direction of the open-circuit voltage rise across the terminals, a minus sign must be inserted in Eq. (3.39).

The short-circuit current (i_{sc}) is easily found once v_2 is known. Therefore the problem reduces to finding v_2 with the short in place. Again, if we use the lower node as the reference node, the equation for v_2 becomes

$$\frac{v_2 - 25}{5} + \frac{v_2}{20} - 3 + \frac{v_2}{4} = 0. \qquad (3.42)$$

Solving Eq. (3.42) for v_2 gives

$$v_2 = 16 \text{ V}. \qquad (3.43)$$

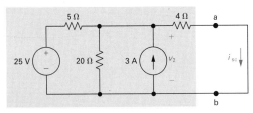

Figure 3.34 The circuit shown in Fig. 3.33 with terminals a and b short-circuited.

Hence, the short-circuit current is

$$i_{\text{sc}} = \frac{16}{4} = 4 \text{ A}. \tag{3.44}$$

We now find the Thévenin resistance by substituting the numerical results from Eqs. (3.41) and (3.44) into Eq. (3.39):

$$R_{\text{Th}} = \frac{V_{\text{Th}}}{i_{sc}} = \frac{32}{4} = 8 \ \Omega. \tag{3.45}$$

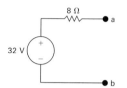

Figure 3.35 The Thévenin equivalent of the circuit shown in Fig. 3.33.

Figure 3.35 shows the Thévenin equivalent for the circuit shown in Fig. 3.33.

You should verify that, if a 24 Ω resistor is connected across the terminals a,b in Fig. 3.33, the voltage across the resistor will be 24 V and the current in the resistor will be 1 A, as would be the case with the Thévenin circuit in Fig. 3.35. This same equivalence between the circuit in Figs. 3.33 and 3.35 holds for any resistor value connected between nodes a,b.

The Norton Equivalent

A **Norton equivalent circuit** consists of an independent current source in parallel with the Norton equivalent resistance. We can derive it from a Thévenin equivalent circuit simply by making a source transformation. Thus the Norton current equals the short-circuit current at the terminals of interest, and the Norton resistance is identical to the Thévenin resistance.

Using Source Transformations

Sometimes we can make effective use of source transformations to derive a Thévenin or Norton equivalent circuit. For example, we can derive the Thévenin and Norton equivalents of the circuit shown in Fig. 3.33 by making the series of source transformations shown in Fig. 3.36. This technique is most useful when the network contains only independent sources. The presence of dependent sources requires retaining the identity of the controlling voltages and/or currents, and this constraint usually prohibits continued reduction of the circuit by source transformations. We discuss the problem of finding the Thévenin equivalent when a circuit contains dependent sources in Example 3.8.

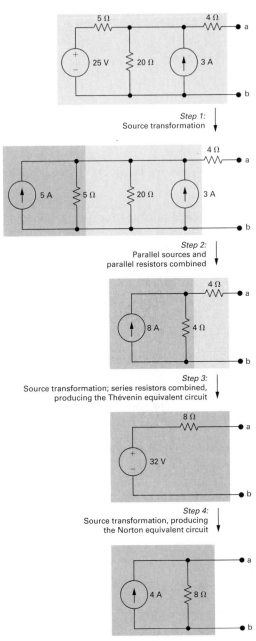

Figure 3.36 Step-by-step derivation of the Thévenin and Norton equivalents of the circuit shown in Fig. 3.33.

EXAMPLE 3.8

Find the Thévenin equivalent for the circuit containing dependent sources shown in Fig. 3.37.

SOLUTION

The first step in analyzing the circuit in Fig. 3.37 is to recognize that the current labeled i_x must be zero. (Note the absence of a return path for i_x to enter the left-hand portion of the circuit.) The open-circuit, or Thévenin, voltage will be the voltage across the 25 Ω resistor. With $i_x = 0$,

$$V_{\text{Th}} = v_{\text{ab}} = (-20i)(25) = -500i.$$

The current i is

$$i = \frac{5 - 3v}{2000} = \frac{5 - 3V_{\text{Th}}}{2000}.$$

In writing the equation for i, we recognize that the Thévenin voltage is identical to the control voltage. When we combine these two equations, we obtain

$$V_{\text{Th}} = -5 \text{ V}.$$

To calculate the short-circuit current, we place a short circuit across a,b. When the terminals a,b are shorted together, the control voltage v is reduced to zero. Therefore, with the short in place, the circuit shown in Fig. 3.37 becomes the one shown in Fig. 3.38. With the short circuit shunting the 25 Ω resistor, all the current from the dependent current source appears in the short, so

$$i_{\text{sc}} = -20i.$$

As the voltage controlling the dependent voltage source has been reduced to zero, the current controlling the dependent current source is

$$i = \frac{5}{2000} = 2.5 \text{ mA}.$$

Combining these two equations yields a short-circuit current of

$$i_{\text{sc}} = -20(2.5) = -50 \text{ mA}.$$

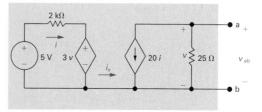

Figure 3.37 A circuit used to illustrate a Thévenin equivalent when the circuit contains dependent sources.

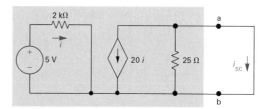

Figure 3.38 The circuit shown in Fig. 3.37 with terminals a and b short-circuited.

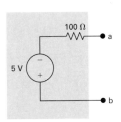

100 Ω
a
5 V
b

Figure 3.39 The Thévenin equivalent for the circuit shown in Fig. 3.37.

From i_{sc} and V_{Th} we get

$$R_{Th} = \frac{V_{Th}}{i_{sc}} = \frac{-5}{-50} \times 10^3 = 100\ \Omega.$$

Figure 3.39 illustrates the Thévenin equivalent for the circuit shown in Fig. 3.37. Note that the reference polarity marks on the Thévenin voltage source in Fig. 3.39 agree with the preceding equation for V_{Th}.

DRILL EXERCISES

3.19 Find the Thévenin equivalent circuit with respect to the terminals a,b for the circuit shown.

ANSWER: $V_{ab} = V_{Th} = 64.8$ V, $R_{Th} = 6\ \Omega$.

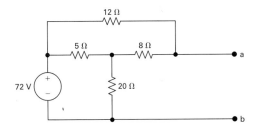

3.20 Find the Norton equivalent circuit with respect to the terminals a,b for the circuit shown.

ANSWER: $I_N = 6$ A (directed toward a), $R_N = 7.5\ \Omega$.

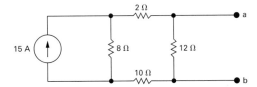

3.21 A 100 kΩ resistor is connected across the terminals AB in the circuit shown. What is the voltage v_{AB}?

ANSWER: 120 V.

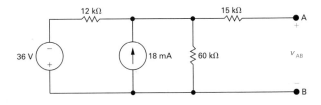

3.10 MORE ON DERIVING A THÉVENIN EQUIVALENT

The technique for determining R_{Th} that we discussed and illustrated in Section 3.9 is not always the easiest method available. Two other methods generally are simpler to use. The first is useful if the network contains only independent sources. To calculate R_{Th} for such a network, we first deactivate all independent sources and then calculate the resistance seen looking into the network at the designated terminal pair. A voltage source is deactivated by replacing it with a short circuit. A current source is deactivated by replacing it with an open circuit. For example, consider the circuit shown in Fig. 3.40. Deactivating the independent sources simplifies the circuit to the one shown in Fig. 3.41. The resistance seen looking into the terminals a,b is denoted R_{ab}, which consists of the 4 Ω resistor in series with the parallel combinations of the 5 and 20 Ω resistors. Thus

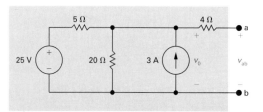

Figure 3.40 A circuit used to illustrate a Thévenin equivalent.

$$R_{ab} = R_{Th} = 4 + \frac{5 \times 20}{25} = 8 \ \Omega. \qquad (3.46)$$

Note that the derivation of R_{Th} with Eq. (3.46) is much simpler than the same derivation with Eqs. (3.42)–(3.45).

If the circuit or network contains dependent sources, an alternative procedure for finding the Thévenin resistance R_{Th} is as follows. We first deactivate all independent sources, and we then apply either a test voltage source or a test current source to the Thévenin terminals a,b. The Thévenin resistance equals the ratio of the voltage across the test source to the current delivered by the test source. Example 3.9 illustrates this alternative procedure for finding R_{Th}, using the same circuit as Example 3.8.

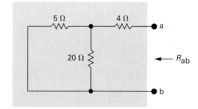

Figure 3.41 The circuit shown in Fig. 3.40 after deactivation of the independent sources.

EXAMPLE 3.9

Find the Thévenin resistance R_{Th} for the circuit in Fig. 3.37, using the alternative method described.

SOLUTION

We first deactivate the independent voltage source from the circuit and then excite the circuit from the terminals a,b with either a test voltage source or a test current source. If we apply a test voltage source, we will know the voltage of the dependent voltage source and hence the controlling current i. Therefore we opt for the test voltage source. Figure 3.42 shows the circuit for computing the Thévenin resistance.

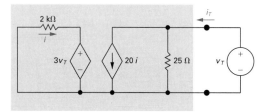

Figure 3.42 An alternative method for computing the Thévenin resistance.

The externally applied test voltage source is denoted v_T, and the current that it delivers to the circuit is labeled i_T. To find the Thévenin resistance, we simply solve the circuit shown in Fig. 3.42 for the ratio of the voltage to the current at the test source; that is, $R_{Th} = v_T / i_T$. From Fig. 3.42,

$$i_T = \frac{v_T}{25} + 20i, \qquad (3.47)$$

$$i = \frac{-3v_T}{2000} \qquad (3.48)$$

We then substitute Eq. (3.48) into Eq. (3.47) and solve the resulting equation for the ratio v_T / i_T:

$$i_T = \frac{v_T}{25} - \frac{60v_T}{2000}, \qquad (3.49)$$

$$\frac{i_T}{v_T} = \frac{1}{25} - \frac{6}{200} = \frac{50}{5000} = \frac{1}{100}. \qquad (3.50)$$

From Eqs. (3.49) and (3.50),

$$R_{Th} = \frac{v_T}{i_T} = 100 \ \Omega. \qquad (3.51)$$

In general, these computations are easier than those involved in computing the short-circuit current. Moreover, in a network containing only resistors and dependent sources, you must use the alternative method, because the ratio of the Thévenin voltage to the short-circuit current is indeterminate. That is, it is the ratio 0/0. (See Problems 3.58 and 3.59.)

DRILL EXERCISES

3.22 Find the Thévenin equivalent circuit with respect to the terminals a,b for the circuit shown.

ANSWER: $V_{Th} = v_{ab} = 8$ V, $R_{Th} = 1 \ \Omega$.

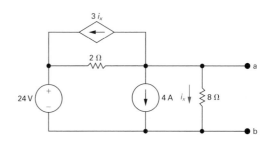

3.23 Find the Thévenin equivalent circuit with respect to the terminals a,b for the circuit shown. (**Hint:** Define the voltage at the leftmost node as v, and write two nodal equations with V_{Th} as the right node voltage.)

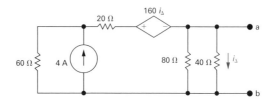

ANSWER: $V_{\text{Th}} = v_{\text{ab}} = 30$ V, $R_{\text{Th}} = 10\ \Omega$.

3.11 MAXIMUM POWER TRANSFER

Circuit analysis plays an important role in the analysis of systems designed to transfer power from a source to a load. We discuss power transfer in terms of two basic types of systems. The first emphasizes the efficiency of the power transfer. Power utility systems are a good example of this type because they are concerned with the generation, transmission, and distribution of large quantities of electric power. If a power utility system is inefficient, a large percentage of the power generated is lost in the transmission and distribution processes, and thus wasted.

The second basic type of system emphasizes the amount of power transferred. Communication and instrumentation systems are good examples because in the transmission of information, or data, via electric signals, the power available at the transmitter or detector is limited. Thus, transmitting as much of this power as possible to the receiver, or load, is desirable. In such applications the amount of power being transferred is small, so the efficiency of transfer is not a primary concern. We now consider maximum power transfer in systems that can be modeled by a purely resistive circuit.

Maximum power transfer can best be described with the aid of the circuit shown in Fig. 3.43. We assume a resistive network containing independent and dependent sources and a designated pair of terminals, a,b, to which a load, R_L, is to be connected. The problem is to determine the value of R_L that permits maximum power delivery to R_L. The first step in this process is to recognize that a resistive network can always be replaced by its Thévenin equivalent. Therefore, we redraw the circuit shown in Fig. 3.43 as the one shown in Fig. 3.44. Replacing the original network by its Thévenin equivalent greatly simplifies the task of finding R_L. Derivation of R_L requires expressing the power dissipated in R_L

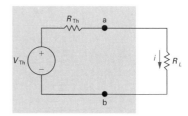

Figure 3.43 A circuit describing maximum power transfer.

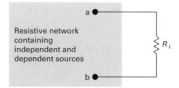

Figure 3.44 A circuit used to determine the value of R_L for maximum power transfer.

as a function of the three circuit parameters V_{Th}, R_{Th}, and R_L. Thus

$$p = i^2 R_L = \left(\frac{V_{Th}}{R_{Th} + R_L} \right)^2 R_L. \tag{3.52}$$

Next, we recognize that for a given circuit, V_{Th} and R_{Th} will be fixed. Therefore the power dissipated is a function of the single variable R_L. To find the value of R_L that maximizes the power, we use elementary calculus. We begin by writing an equation for the derivative of p with respect to R_L:

$$\frac{dp}{dR_L} = V_{Th}^2 \left[\frac{(R_{Th} + R_L)^2 - R_L \cdot 2(R_{Th} + R_L)}{(R_{Th} + R_L)^4} \right]. \tag{3.53}$$

The derivative is zero and p is maximized when

$$(R_{Th} + R_L)^2 = 2R_L(R_{Th} + R_L). \tag{3.54}$$

Solving Eq. (3.54) yields

$$R_L = R_{Th}. \tag{3.55}$$

Thus maximum power transfer occurs when the load resistance R_L equals the Thévenin resistance R_{Th}. To find the maximum power delivered to R_L, we simply substitute Eq. (3.55) into Eq. (3.52):

$$p_{max} = \frac{V_{Th}^2 R_L}{(2R_L)^2} = \frac{V_{Th}^2}{4R_L}. \tag{3.56}$$

The analysis of a circuit when the load resistor is adjusted for maximum power transfer is illustrated in Example 3.10.

EXAMPLE 3.10

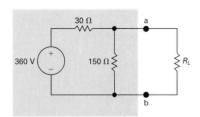

Figure 3.45 The circuit for Example 3.10.

(a) For the circuit shown in Fig. 3.45, find the value of R_L that results in maximum power being transferred to R_L.

(b) Calculate the maximum power that can be delivered to R_L.

(c) When R_L is adjusted for maximum power transfer, what percentage of the power delivered by the 360 V source reaches R_L?

SOLUTION

(a) The Thévenin voltage for the circuit to the left of the terminals a,b is

$$V_{Th} = \frac{150}{180} \times 360$$

$$= 300 \text{ V.}$$

The Thévenin resistance is

$$R_{Th} = \frac{(150)(30)}{180}$$

$$= 25 \ \Omega.$$

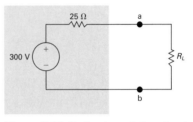

Replacing the circuit to the left of the terminals a,b with its Thévenin equivalent gives us the circuit shown in Fig. 3.46, which indicates that R_L must equal 25 Ω for maximum power transfer.

Figure 3.46 Reduction of the circuit shown in Fig. 3.45 by means of a Thévenin equivalent.

(b) The maximum power that can be delivered to R_L is

$$p_{max} = \left(\frac{300}{50}\right)^2 (25)$$

$$= 900 \text{ W.}$$

(c) When R_L equals 25 Ω, the voltage v_{ab} is

$$v_{ab} = \left(\frac{300}{50}\right) (25)$$

$$= 150 \text{ V.}$$

From Fig. 3.45, when v_{ab} equals 150 V, the current in the voltage source in the direction of the voltage rise across the source is

$$i_s = \frac{360 - 150}{30} = \frac{210}{30}$$

$$= 7 \text{ A.}$$

Therefore, the source is delivering 2520 W to the circuit, or

$$p_s = -i_s(360)$$

$$= -2520 \text{ W.}$$

The percentage of the source power delivered to the load is

$$\frac{900}{2520} \times 100 = 35.71\%.$$

DRILL EXERCISES

3.24 **(a)** Find the value of R that enables the circuit shown to deliver maximum power to the terminals a,b.

(b) Find the maximum power delivered to R.

ANSWER: (a) 3 Ω; (b) 1.2 kW.

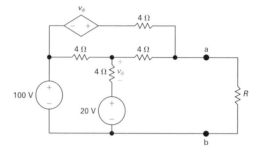

3.25 Assume that the circuit in Drill Exercise 3.24 is delivering maximum power to the load resistor R.

(a) How much power is the 100 V source delivering to the network?

(b) Repeat (a) for the dependent voltage source.

(c) What percentage of the total power generated by these two sources is delivered to the load resistor R?

ANSWER: (a) 3000 W; (b) 800 W; (c) 31.58%.

Practical Perspective

Circuits with Realistic Resistors

It is not possible to fabricate identical electrical components. For example, resistors produced from the same manufacturing process can vary in value by as much as 20%. Therefore, in creating an electrical system the designer must consider the impact that component variation will have on the performance of the system. One way to evaluate this impact is by performing sensitivity analysis. Sensitivity analysis permits the designer to calculate the impact of variations in the component values on the output of the system. We will see how this information enables a designer to specify an acceptable component value tolerance for each of the system's components.

Consider the circuit shown in Fig. 3.47. To illustrate sensitivity analysis, we will investigate the sensitivity of the node voltages v_1 and v_2 to changes in the resistor R_1. Using nodal analysis we can derive the expressions for v_1 and v_2 as functions of the circuit resistors and source currents. The results are given in Eqs. (3.57) and (3.58):

$$v_1 = \frac{R_1\{R_3 R_4 I_{g2} - [R_2(R_3 + R_4) + R_3 R_4]I_{g1}\}}{(R_1 + R_2)(R_3 + R_4) + R_3 R_4}, \quad (3.57)$$

$$v_2 = \frac{R_3 R_4[(R_1 + R_2)I_{g2} - R_1 I_{g1}]}{(R_1 + R_2)(R_3 + R_4) + R_3 R_4}. \quad (3.58)$$

The sensitivity of v_1 with respect to R_1 is found by differentiating Eq. (3.57) with respect to R_1, and similarly the sensitivity of v_2 with respect to R_1 is found by differentiating Eq. (3.58) with respect to R_1. We get

$$\frac{dv_1}{dR_1} = \frac{[R_3 R_4 + R_2(R_3 + R_4)]\{R_3 R_4 I_{g2} - [R_3 R_4 + R_2(R_3 + R_4)]I_{g1}\}}{[(R_1 + R_2)(R_3 + R_4) + R_3 R_4]^2}, \quad (3.59)$$

$$\frac{dv_2}{dR_1} = \frac{R_3 R_4\{R_3 R_4 I_{g2} - [R_2(R_3 + R_4) + R_3 R_4]I_{g1}\}}{[(R_1 + R_2)(R_3 + R_4) + R_3 R_4]^2}. \quad (3.60)$$

We now consider an example with actual component values to illustrate the use of Eqs. (3.59) and (3.60).

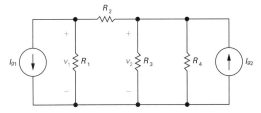

Figure 3.47 Circuit used to introduce sensitivity analysis.

Example

Assume the nominal values of the components in the circuit in Fig. 3.47 are: $R_1 = 25\ \Omega$; $R_2 = 5\ \Omega$; $R_3 = 50\ \Omega$; $R_4 = 75\ \Omega$; $I_{g1} = 12$ A, and $I_{g2} = 16$ A. Use sensitivity analysis to predict the values of v_1 and v_2 if the value of R_1 is different by 10% from its nominal value.

Solution

From Eqs. (3.57) and (3.58) we find the nominal values of v_1 and v_2. Thus

$$v_1 = \frac{25\{2750(16) - [5(125) + 3750]12\}}{30(125) + 3750}, \quad (3.61)$$

$$= 25 \text{ V},$$

and

$$v_2 = \frac{3750[30(16) - 25(12)]}{30(125) + 3750},$$

$$= 90 \text{ V}. \tag{3.62}$$

Now from Eqs. (3.59) and (3.60) we can find the sensitivity of v_1 and v_2 to changes in R_1. Hence

$$\frac{dv_1}{dR_1} = \frac{[3750 + 5(125)] - \{3750(16) - [3750 + 5(125)]12\}}{[(30)(125) + 3750]^2},$$

$$= \frac{7}{12} \text{ V}/\Omega \tag{3.63}$$

and

$$\frac{dv_2}{dR_1} = \frac{3750\{3750(16) - [5(125) + 3750]12\}}{(7500)^2},$$

$$= 0.5 \text{ V}/\Omega. \tag{3.64}$$

How do we use the results given by Eqs. (3.63) and (3.64)? Assume that R_1 is 10% less than its nominal value, that is, $R_1 = 22.5 \ \Omega$. Then $\Delta R_1 = -2.5 \ \Omega$, and Eq. (3.63) predicts Δv_1 will be

$$\Delta v_1 = \left(\frac{7}{12}\right)(-2.5) = -1.4583 \text{ V}.$$

Therefore, if R_1 is 10% less than its nominal value, our analysis predicts that v_1 will be

$$v_1 = 25 - 1.4583 = 23.5417 \text{ V}. \tag{3.65}$$

Similarly for Eq. (3.64) we have

$$\Delta v_2 = 0.5(-2.5) = -1.25 \text{ V}$$

$$v_2 = 90 - 1.25 = 88.75 \text{ V}. \tag{3.66}$$

We attempt to confirm the results in Eqs. (3.65) and (3.66) by substituting the value $R_1 = 22.5 \ \Omega$ into Eqs. (3.57) and (3.58). When we do, the results are

$$v_1 = 23.4780 \text{ V}, \tag{3.67}$$

$$v_2 = 88.6960 \text{ V}. \tag{3.68}$$

Why is there a difference between the values predicted from the sensitivity analysis and the exact values computed by substituting for R_1 in the equations for v_1 and v_2? We can see from Eqs. (3.59) and (3.60) that the sensitivity of v_1 and v_2 with respect to R_1 is a function of R_1, because R_1 appears in the denominator of both Eqs. (3.59) and (3.60). This means that as R_1 changes, the sensitivities change and hence we cannot expect Eqs. (3.59) and (3.60) to give exact results for large changes in R_1. Note that for a 10% change in R_1, the percent error between the predicted and exact values of v_1 and v_2 is small. Specifically, the percent error in $v_1 = 0.2713\%$ and the percent error in $v_2 = 0.0676\%$.

From this example, we can see that a tremendous amount of work is involved if we are to determine the sensitivity of v_1 and v_2 to changes in the remaining component values, namely R_2, R_3, R_4, I_{g1}, and I_{g2}. Fortunately, PSpice has a sensitivity function that will perform sensitivity analysis for us. It calculates two types of sensitivity. The first is known as the one-unit sensitivity, and the second as the 1% sensitivity. In the example circuit, a one-unit change in a resistor would change its value by $1\ \Omega$ and a one-unit change in a current source would change its value by 1 A. In contrast, 1% sensitivity analysis determines the effect of changing resistors or sources by 1% of their nominal values.

The result of PSpice sensitivity analysis of the circuit in Fig. 3.47 is shown in Table 3.2. Because we are analyzing a linear

TABLE 3.2   PSpice Sensitivity Analysis Results

ELEMENT NAME	ELEMENT VALUE	ELEMENT SENSITIVITY (VOLTS/UNIT)	NORMALIZED SENSITIVITY (VOLTS/PERCENT)
(a) DC Sensitivities of Node Voltage V1			
R1	25	0.5833	0.1458
R2	5	−5.417	−0.2708
R3	50	0.45	0.225
R4	75	0.2	0.15
IG1	12	−14.58	−1.75
IG2	16	12.5	2
(b) Sensitivities of Output V2			
R1	25	0.5	0.125
R2	5	6.5	0.325
R3	50	0.54	0.27
R4	75	0.24	0.18
IG1	12	−12.5	−1.5
IG2	16	15	2.4

circuit, we can use superposition to predict values of v_1 and v_2 if more than one component's value changes. For example, let us assume R_1 decreases to 24 Ω and R_2 decreases to 4 Ω. From Table 3.2 we can combine the unit sensitivity of v_1 to changes in R_1 and R_2 to get

$$\frac{\Delta v_1}{\Delta R_1} + \frac{\Delta v_1}{\Delta R_2} = 0.5833 - 5.417 = -4.8337 \text{ V}/\Omega.$$

Similarly,

$$\frac{\Delta v_2}{\Delta R_1} + \frac{\Delta v_2}{\Delta R_2} = 0.5 + 6.5 = 7.0 \text{ V}/\Omega.$$

Thus if both R_1 and R_2 decreased by 1 Ω we would predict

$$v_1 = 25 + 4.8227 = 29.8337 \text{ V},$$

$$v_2 = 90 - 7 = 83 \text{ V}.$$

If we substitute $R_1 = 24$ Ω and $R_2 = 4$ Ω into Eqs. (3.57) and (3.58) we get

$$v_1 = 29.793 \text{ V},$$

$$v_2 = 82.759 \text{ V}.$$

In both cases our predictions are within a fraction of a volt of the actual node voltage values.

Circuit designers use the results of sensitivity analysis to determine which component value variation has the greatest impact on the output of the circuit. As we can see from the PSpice sensitivity analysis in Table 3.2, the node voltages v_1 and v_2 are much more sensitive to changes in R_2 than to changes in R_1. Specifically, v_1 is (5.417/0.5833) or approximately 9 times more sensitive to changes in R_2 than to changes in R_1 and v_2 is (6.5/0.5) or 13 times more sensitive to changes in R_2 than to changes in R_1. Hence in the example circuit, the tolerance on R_2 must be more stringent than the tolerance on R_1 if it is important to keep v_1 and v_2 close to their nominal values.

SUMMARY

- For the topics in this chapter, mastery of some basic terms and the concepts they represent is necessary. Those terms are **node, essential node, path, branch, essential branch, mesh,** and **planar circuit**. Table 3.1 provides definitions and examples of these terms.

- Two new circuit analysis techniques were introduced in this chapter:

    - The **node-voltage method** works with both planar and nonplanar circuits. A reference node is chosen from among the essential nodes. Voltage variables are assigned at the remaining essential nodes, and Kirchhoff's current law is used to write one equation per voltage variable. The number of equations is $n_e - 1$, where n_e is the number of essential nodes.

    - The **mesh-current method** works only with planar circuits. Mesh currents are assigned to each mesh, and Kirchhoff's voltage law is used to write one equation per mesh. The number of equations is $b - (n - 1)$, where b is the number of branches in which the current is unknown, and n is the number of nodes. The mesh currents are used to find the branch currents.

- A circuit simplification technique was introduced in this chapter:

    - A **Thévenin equivalent** or **Norton equivalent** allows us to simplify a circuit comprised of sources and resistors into an equivalent circuit consisting of a voltage source and a series resistor (Thévenin) or a current source and a parallel resistor (Norton). The simplified circuit and the original circuit must be equivalent in terms of their terminal voltage and current. Thus keep in mind that (1) the Thévenin voltage (V_{Th}) is the open-circuit voltage across the terminals of the original circuit, (2) the Thévenin resistance (R_{Th}) is the ratio of the Thévenin voltage to the short-circuit current across the terminals of the original circuit; and (3) the Norton equivalent is obtained by performing a source transformation on a Thévenin equivalent.

- **Maximum power transfer** is a technique for calculating the maximum value of p that can be delivered to a load, R_L. Maximum power transfer occurs when $R_L = R_{Th}$, the Thévenin resistance as seen from the resistor R_L. The equation for the maximum power transferred is

$$p = \frac{V_{Th}^2}{4R_L}.$$

PROBLEMS

3.1. Assume the current i_g in the circuit in Fig. P3.1 is known. The resistors R_1–R_5 are also known.

   (a) How many unknown currents are there?

   (b) How many independent equations can be written using Kirchhoff's current law (KCL)?

   (c) Write an independent set of KCL equations.

   (d) How many independent equations can be derived from Kirchhoff's voltage law (KVL)?

   (e) Write a set of independent KVL equations.

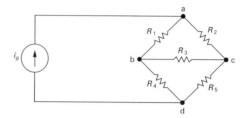

Figure P3.1

3.2. Use the node-voltage method to find how much power the 2 A source extracts from the circuit in Fig. P3.2.

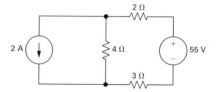

Figure P3.2

3.3. Use the node-voltage method to find v_1 and the power delivered by the 60 V voltage source in the circuit in Fig. P3.3.

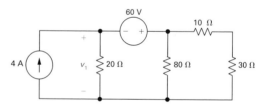

Figure P3.3

3.4. **(a)** Use the node-voltage method to find the branch currents i_a–i_e in the circuit shown in Fig. P3.4.

   (b) Find the total power developed in the circuit.

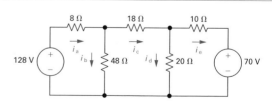

Figure P3.4

3.5. Use the node-voltage method to find v_o in the circuit in Fig. P3.5.

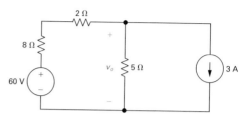

Figure P3.5

3.6. The circuit shown in Fig. P3.6 is a dc model of a residential power distribution circuit.

(a) Use the node-voltage method to find the branch currents i_1-i_6.

(b) Test your solution for the branch currents by showing that the total power dissipated equals the total power developed.

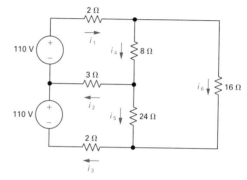

Figure P3.6

3.7. Use the node-voltage method to find v_1 and v_2 in the circuit shown in Fig. P3.7.

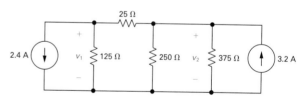

Figure P3.7

3.8. (a) Use the node-voltage method to find v_1, v_2, and v_3 in the circuit in Fig. P3.8.

(b) How much power does the 640 V voltage source deliver to the circuit?

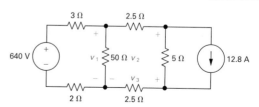

Figure P3.8

3.9. Use the node-voltage method to find v_1 and v_2 in the circuit in Fig. P3.9.

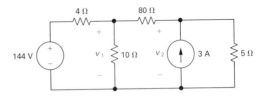

Figure P3.9

3.10. Use the node-voltage method to find the value of v_o in the circuit in Fig. P3.10.

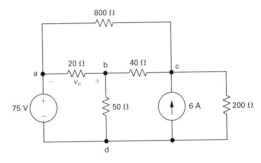

Figure P3.10

3.11. Use the node-voltage method to find the value of v_o in the circuit in Fig. P3.11.

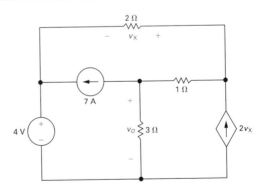

Figure P3.11

3.12. Use the node-voltage method to find the total power dissipated in the circuit in Fig. P3.12.

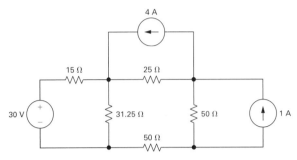

Figure P3.12

3.13. Use the node-voltage method to find i_o in the circuit in Fig. P3.13.

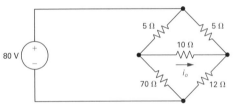

Figure P3.13

3.14. A 100 Ω resistor is connected in series with the 3 A current source in the circuit in Fig. P3.5.

   (a) Find v_o.

   (b) Find the power developed by the 3 A current source.

   (c) Find the power developed by the 60 V voltage source.

   (d) Verify that the total power developed equals the total power dissipated.

   (e) What effect will any finite resistance connected in series with the 3 A current source have on the value of v_o?

3.15. **(a)** Find the power developed by the 3 A current source in the circuit in Fig. P3.5.

   (b) Find the power developed by the 60 V voltage source in the circuit in Fig. P3.5.

   (c) Verify that the total power developed equals the total power dissipated.

3.16. **(a)** Use the node-voltage method to find the branch currents i_1, i_2, and i_3 in the circuit in Fig. P3.16.

(b) Check your solution for i_1, i_2, and i_3 by showing that the power dissipated in the circuit equals the power developed.

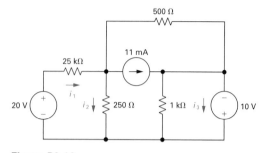

Figure P3.16

3.17. **(a)** Use the node-voltage method to find the power dissipated in the 5 Ω resistor in the circuit in Fig. P3.17.

(b) Find the power supplied by the 500 V source.

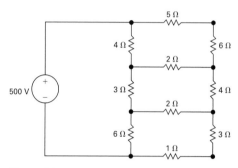

Figure P3.17

3.18. **(a)** Use the node-voltage method to show that the output voltage v_o in the circuit in Fig. P3.18 is equal to the average value of the source voltages.

(b) Find v_o if $v_1 = 150$ V, $v_2 = 200$ V, and $v_3 = -50$ V.

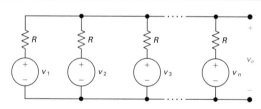

Figure P3.18

3.19. Use the node-voltage method to find v_o in the circuit in Fig. P3.19.

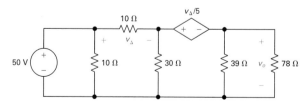

Figure P3.19

3.20. Use the node-voltage method to calculate the power delivered by the dependent voltage source in the circuit in Fig. P3.20.

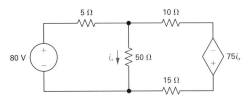

Figure P3.20

3.21. (a) Use the node-voltage method to find v_o in the circuit in Fig. P3.21.

(b) Find the power absorbed by the dependent source.

(c) Find the total power developed by the independent sources.

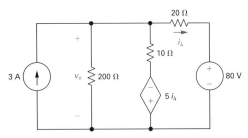

Figure P3.21

3.22. (a) Find the node voltages v_1, v_2, and v_3 in the circuit in Fig. P3.22.

(b) Find the total power dissipated in the circuit.

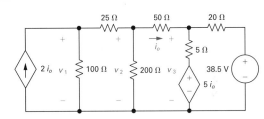

Figure P3.22

3.23. (a) Use the node-voltage method to find the total power developed in the circuit in Fig. P3.23.

(b) Check your answer by finding the total power absorbed in the circuit.

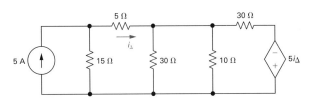

Figure P3.23

3.24. Assume you are a project engineer and one of your staff is assigned to analyze the circuit shown in Fig. P3.24. The reference node and node numbers given on the figure were assigned by the analyst. Her solution gives the values of v_3 and v_4 as 235 V and 222 V, respectively.

　　　Test these values by checking the total power developed in the circuit against the total power dissipated. Do you agree with the solution submitted by the analyst?

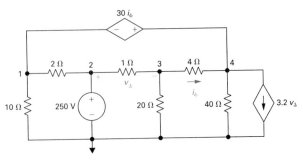

Figure P3.24

3.25. Use the node-voltage method to find the power developed by the 20 V source in the circuit in Fig. P3.25.

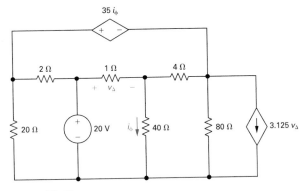

Figure P3.25

3.26. Use the mesh-current method to find the total power dissipated in the circuit in Fig. P3.26.

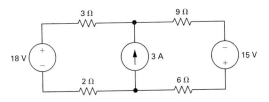

Figure P3.26

3.27. Assume the 18 V source in the circuit in Fig. P3.26 is increased to 100 V. Find the total power dissipated in the circuit.

3.28. **(a)** Assume the 18 V source in the circuit in Fig. P3.26 is changed to −10 V. Find the total power dissipated in the circuit.

(b) Repeat (a) if the 3 A current source is re-placed by a short circuit.

(c) Explain why the answers to (a) and (b) are the same.

3.29. **(a)** Use the mesh-current method to find the branch currents i_a, i_b, and i_c in the circuit in Fig. P3.29.

(b) Repeat (a) if the polarity of the 64 V source is reversed.

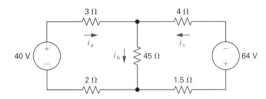

Figure P3.29

3.30. **(a)** Use the mesh-current method to find how much power the 30 A current source delivers to the circuit in Fig. P3.30.

(b) Find the total power delivered to the circuit.

(c) Check your calculations by showing that the total power developed in the circuit equals the total power dissipated.

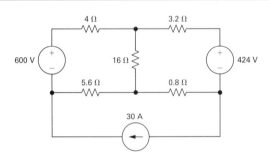

Figure P3.30

3.31. **(a)** Use the mesh-current method to find the total power developed in the circuit in Fig. P3.31.

(b) Check your answer by showing that the to-tal power developed equals the total power dissipated.

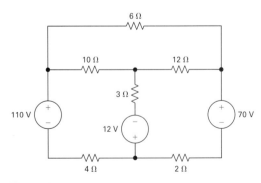

Figure P3.31

3.32. **(a)** Use the mesh-current method to solve for i_Δ in the circuit in Fig. P3.32.

(b) Find the power delivered by the independent current source.

(c) Find the power delivered by the dependent voltage source.

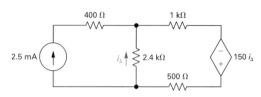

Figure P3.32

3.33. Use the mesh-current method to find the total power developed in the circuit in Fig. P3.33.

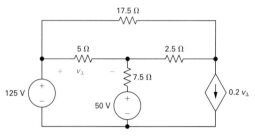

Figure P3.33

3.34. Use the mesh-current method to find the power dissipated in the 8 Ω resistor in the circuit in Fig. P3.34.

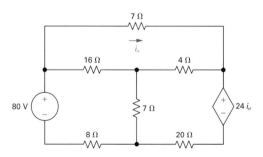

Figure P3.34

3.35. **(a)** Use the mesh-current method to find the branch currents in $i_a - i_e$ in the circuit in Fig. P3.35.

(b) Check your solution by showing that the total power developed in the circuit equals the total power dissipated.

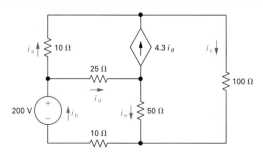

Figure P3.35

3.36. Use the mesh-current method to find the power delivered by the dependent voltage source in the circuit seen in Fig. P3.36.

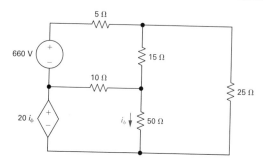

Figure P3.36

3.37. Use the mesh-current method to find the total power developed in the circuit in Fig. P3.37.

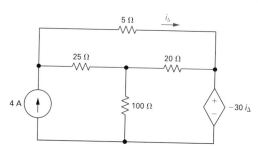

Figure P3.37

3.38. (a) Use the mesh-current method to find v_o in the circuit in Fig. P3.38.

(b) Find the power delivered by the dependent source.

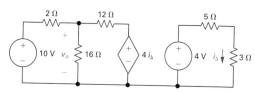

Figure P3.38

3.39. Use the mesh-current method to find the power developed in the dependent voltage source in the circuit in Fig. P3.39.

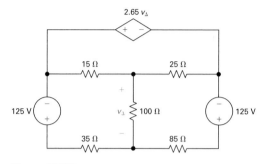

Figure P3.39

3.40. **(a)** Use the mesh-current method to determine which sources in the circuit in Fig. P3.40 are generating power.

(b) Find the total power dissipated in the circuit.

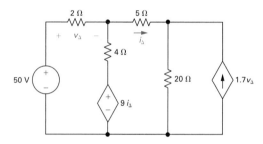

Figure P3.40

3.41. **(a)** Use the mesh-current method to find the power delivered to the 2 Ω resistor in the circuit in Fig. P3.41.

(b) What percentage of the total power developed in the circuit is delivered to the 2 Ω resistor?

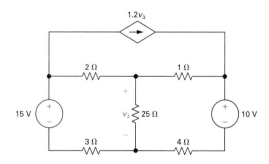

Figure P3.41

3.42. The variable dc voltage source in the circuit in Fig. P3.42 is adjusted so that i_o is zero.

(a) Find the value of V_{dc}.

(b) Check your solution by showing the power developed equals the power dissipated.

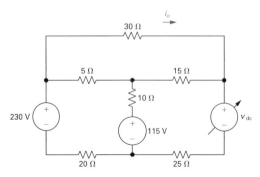

Figure P3.42

3.43. The variable dc current source in the circuit in Fig. P3.43 is adjusted so that the power developed by the 15 A current source is 3750 W. Find the value of i_{dc}.

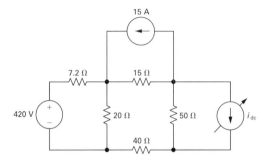

Figure P3.43

3.44. The circuit in Fig. P3.44 is a direct-current version of a typical three-wire distribution system. The resistors R_a, R_b, and R_c represent the resistances of the three conductors that connect the three loads R_1, R_2, and R_3 to the 125/250 V voltage supply. The resistors R_1 and R_2 represent loads connected to the 125 V circuits, and R_3 represents a load connected to the 250 V circuit.

(a) Calculate v_1, v_2, and v_3.

(b) Calculate the power delivered to R_1, R_2, and R_3.

(c) What percentage of the total power developed by the sources is delivered to the loads?

(d) The R_b branch represents the neutral conductor in the distribution circuit. What

adverse effect occurs if the neutral conductor is opened? (*Hint:* Calculate v_1 and v_2 and note that appliances or loads designed for use in this circuit would have a nominal voltage rating of 125 V.)

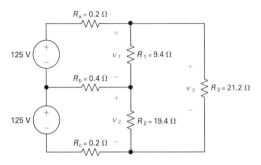

Figure P3.44

3.45. Show that whenever $R_1 = R_2$ in the circuit in Fig. P3.44, the current in the neutral conductor is zero. (*Hint:* Solve for the neutral conductor current as a function of R_1 and R_2).

3.46. **(a)** Find the branch currents i_a–i_e for the circuit shown in Fig. P3.46.

(b) Check your answers by showing that the total power generated equals the total power dissipated.

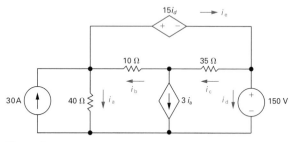

Figure P3.46

3.47. **(a)** Would you use the node-voltage or mesh-current method to find the power absorbed by the 20 V source in the circuit in Fig. P3.47? Explain your choice.

(b) Use the method you selected in (a) to find the power absorbed by the 20 V source.

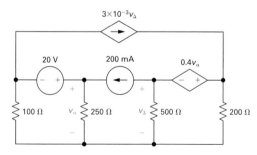

Figure P3.47

3.48. Assume you have been asked to find the power dissipated in the 10 Ω resistor in the circuit in Fig. P3.48.

(a) Which method of circuit analysis would you recommend? Explain why.

(b) Use your recommended method of analysis to find the power dissipated in the 10 Ω resistor.

(c) Would you change your recommendation if the problem had been to find the power developed by the 19.6 A current source? Explain.

(d) Find the power delivered by the 19.6 A current source.

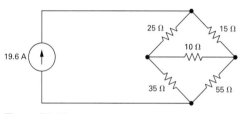

Figure P3.48

3.49. A 25 Ω resistor is placed in parallel with the 19.6 A current source in the circuit in Fig. P3.48. Assume you have been asked to calculate the power developed by the current source.

(a) Which method of circuit analysis would you recommend? Explain why.

(b) Find the power developed by the current source.

 3.50. Find the Thévenin equivalent with respect to the terminals a,b for the circuit in Fig. P3.50.

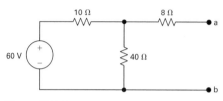

Figure P3.50

 3.51. **(a)** Find the Thévenin equivalent with respect to the terminals a,b for the circuit in Fig. P3.51 by finding the open-circuit voltage and the short-circuit current.

(b) Solve for the Thévenin resistance by removing the independent sources. Compare your result to the Thévenin resistance found in (a).

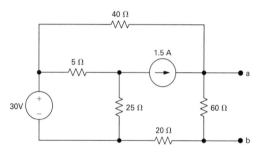

Figure P3.51

 3.52. Find the Thévenin equivalent with respect to the terminals a,b for the circuit in Fig. P3.52.

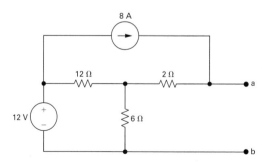

Figure P3.52

3.53. Determine the Thévenin equivalent with respect to the terminals a,b for the circuit shown in Fig. P3.53.

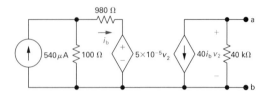

Figure P3.53

3.54. Find the Thévenin equivalent with respect to the terminals a,b for the circuit in Fig. P3.54.

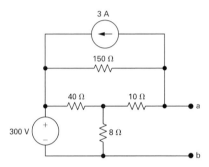

Figure P3.54

3.55. Find the Norton equivalent with respect to the terminals a,b in the circuit in Fig. P3.55.

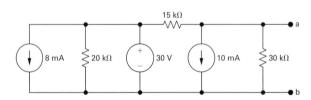

Figure P3.55

3.56. Find the Thévenin equivalent with respect to the terminals a,b for the circuit seen in Fig. P3.56.

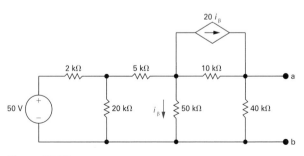

Figure P3.56

3.57. A Thévenin equivalent can also be determined from measurements made at the pair of terminals of interest. Assume the following measurements were made at the terminals a,b in the circuit in Fig. P3.57.

When a 15 kΩ resistor is connected to the terminals a,b, the voltage v_{ab} is measured and found to be 45 V.

When a 5 kΩ resistor is connected to the terminals a,b, the voltage is measured and found to be 25 V.

Find the Thévenin equivalent of the network with respect to the terminals a,b.

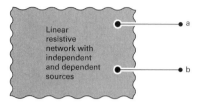

Figure P3.57

3.58. Find the Thévenin equivalent with respect to the terminals a,b in the circuit in Fig. P3.58.

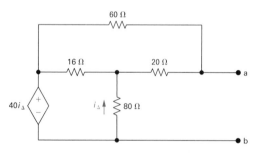

Figure P3.58

3.59. Find the Thévenin equivalent with respect to the terminals a,b for the circuit seen in Fig. P3.59.

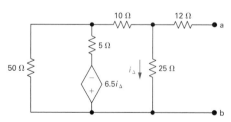

Figure P3.59

3.60. **(a)** Find the value of the variable resistor R_o in the circuit in Fig. P3.60 that will result in maximum power dissipation in the 6 Ω resistor. (**Hint:** Hasty conclusions could be hazardous to your career.)

(b) What is the maximum power that can be delivered to the 6 Ω resistor?

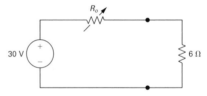

Figure P3.60

3.61. A variable resistor R_o is connected across the terminals a,b in the circuit in Fig. P3.56. The variable resistor is adjusted until maximum power is transferred to R_o.

(a) Find the value of R_o.

(b) Find the maximum power delivered to R_o.

(c) Find the percentage of the total power developed in the circuit that is delivered to R_o.

3.62. **(a)** Calculate the power delivered for each value of R_o used in Problem 2.25.

(b) Plot the power delivered to R_o versus R_o.

(c) At what value of R_o is the power delivered to R_o a maximum?

3.63. The variable resistor (R_o) in the circuit in Fig. P3.63 is adjusted until the power dissipated in the resistor is 250 W. Find the values of R_o that satisfy this condition.

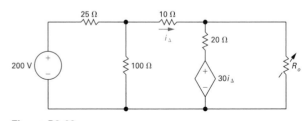

Figure P3.63

3.64. The variable resistor in the circuit in Fig. P3.64 is adjusted for maximum power transfer to R_o.

(a) Find the value of R_o.

(b) Find the maximum power that can be delivered to R_o.

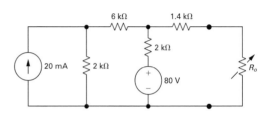

Figure P3.64

3.65. What percentage of the total power developed in the circuit in Fig. P3.64 is delivered to R_o when R_o is set for maximum power transfer?

 3.66. The variable resistor (R_o) in the circuit in Fig. P3.66 is adjusted for maximum power transfer to R_o.

(a) Find the value of R_o.

(b) Find the maximum power that can be delivered to R_o.

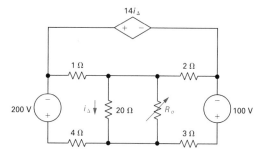

Figure P3.66

 3.67. What percentage of the total power developed in the circuit in Fig. P3.66 is delivered to R_o?

 3.68. The variable resistor (R_L) in the circuit in Fig. P3.68 is adjusted for maximum power transfer to R_L.

(a) Find the numerical value of R_L.

(b) Find the maximum power transferred to R_L.

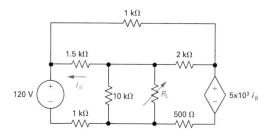

Figure P3.68

 3.69. The variable resistor (R_o) in the circuit in Fig. P3.69 is adjusted for maximum power transfer to R_o. What percentage of the total power developed in the circuit is delivered to R_o?

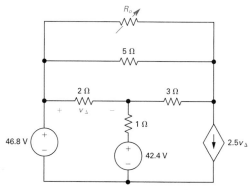

Figure P3.69

3.70. The variable resistor (R_o) in the circuit in
Fig. P3.70 is adjusted until it absorbs maximum
power from the circuit.

(a) Find the value of R_o.

(b) Find the maximum power.

(c) Find the percentage of the total power developed in the circuit that is delivered to R_o.

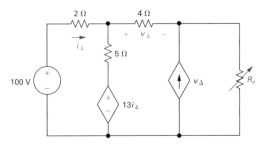

Figure P3.70

3.71. The variable resistor in the circuit in Fig. P3.71 is
adjusted for maximum power transfer to R_o.

(a) Find the numerical value of R_o.

(b) Find the maximum power delivered to R_o.

(c) How much power does the 280 V source deliver to the circuit when R_o is adjusted to the
value found in (a)?

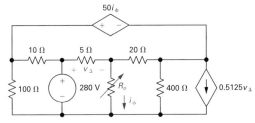

Figure P3.71

3.72. Laboratory measurements on a dc voltage source
yield a terminal voltage of 75 V with no load connected to the source and 60 V when loaded with a
20 Ω resistor.

(a) What is the Thévenin equivalent with respect
to the terminals of the dc voltage source?

(b) Show that the Thévenin resistance of the
source is given by the expression

$$R_{Th} = \left(\frac{V_{Th}}{v_o} - 1 \right) R_L,$$

where

$V_{Th} = $ the Thévenin voltage

$v_o = $ the terminal voltage corresponding

to the load resistance R_L.

3.73. Two ideal dc voltage sources are connected by electrical conductors that have a resistance of r Ω/m, as shown in Fig. P3.73. A load having a resistance of R Ω moves between the two voltage sources. Let x equal the distance between the load and the source v_1, and let L equal the distance between the sources.

(a) Show that

$$v = \frac{v_1 RL + R(v_2 - v_1)x}{RL + 2rLx - 2rx^2}.$$

(b) Show that the voltage v will be minimum when

$$x = \frac{L}{v_2 - v_1}\left[-v_1 \pm \sqrt{v_1 v_2 - \frac{R}{2rL}(v_1 - v_2)^2}\right].$$

(c) Find x when $L = 16$ km, $v_1 = 1000$ V, $v_2 = 1200$ V, $R = 3.9$ Ω, and $r = 5 \times 10^{-5}$ Ω/m.

(d) What is the minimum value of v for the circuit of part (c)?

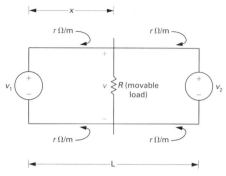

Figure P3.73

◆ 3.74. For the circuit in Fig. 3.47 derive the expressions for the sensitivity of v_1 and v_2 to changes in the source currents I_{g1} and I_{g2}.

◆ 3.75. Assume the nominal values for the components in the circuit in Fig. 3.47 are: $R_1 = 25$ Ω; $R_2 = 5$ Ω; $R_3 = 50$ Ω; $R_4 = 75$ Ω; $I_{g1} = 12$ A; and $I_{g2} = 16$ A.
Predict the values of v_1 and v_2 if I_{g1} decreases to 11 A and all other components stay at their nominal values. Check your predictions using a tool like PSpice or Matlab.

◆ 3.76. Repeat Problem 3.75 if I_{g2} increases to 17 A, and all other components stay at their nominal values. Check your predictions using a tool like PSpice or Matlab.

◆ 3.77. Repeat Problem 3.75 if I_{g1} decreases to 11 A and I_{g2} increases to 17 A. Check your predictions using a tool like PSpice or Matlab.

◆ **3.78.** Use the results given in Table 3.2 to predict the values of v_1 and v_2 if R_1 and R_3 increase to 10% above their nominal values and R_2 and R_4 decrease to 10% below their nominal values. I_{g1} and I_{g2} remain at their nominal values. Compare your predicted values of v_1 and v_2 with their actual values.

The Flash Converter

In Chapter 2 we discussed a resistive ladder circuit used to convert a digital signal into an analog signal. In this chapter we will discuss the operation of a high-speed analog-to-digital converter known as a flash converter. The basic structure of a flash converter is shown in the accompanying figure. We will discuss the operation of this analog-to-digital converter at the end of the chapter after we have discussed the operation of comparators, the triangular elements in the accompanying figure.

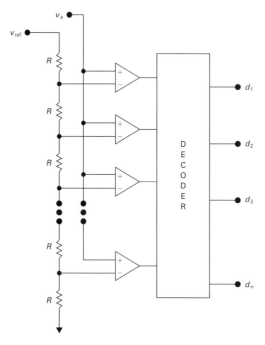

Basic structure of an analog-to-digital flash converter.

4 The Operational Amplifier

Chapter Contents

The electronic circuit known as an operational amplifier has become increasingly important. However, a detailed analysis of this circuit requires an understanding of electronic devices such as diodes and transistors. You may wonder, then, why we are introducing the circuit before discussing the circuit's electronic components. There are two reasons. First, you can develop an appreciation for how the operational amplifier can be used as a circuit building block by focusing on its terminal behavior. At an introductory level, you need not fully understand the operation of the electronic components that govern terminal behavior. Second, you can combine the operational amplifier with resistors to perform some very useful functions, such as scaling, summing, sign changing, and subtracting. After introducing inductors and capacitors in Chapter 5, we will show you how to use the operational amplifier to design integrating and differentiating circuits.

Our focus on the terminal behavior of the operational amplifier implies taking a black box approach to its operation; that is, we are not interested in the internal structure of the amplifier or in the currents and voltages that exist in this structure. The important thing to remember is that the internal behavior of the amplifier accounts for the voltage and current constraints imposed at the terminals. (For now, we ask that you accept these constraints on faith.)

The operational amplifier circuit first came into existence as a basic building block in analog computers. It was referred to as *operational* because it was used to implement the mathematical operations of integration, differentiation, addition, sign changing, and scaling. In recent years, the range of application has broadened beyond implementing mathematical operations; however, the original name for the circuit persists. Engineers and technicians have a penchant for creating technical jargon; hence the operational amplifier is widely known as the **op amp.**

4.1 OPERATIONAL AMPLIFIER TERMINALS

Because we are stressing the terminal behavior of the operational amplifier (op amp), we begin by discussing the terminals on a commercially available device. In 1968, Fairchild Semiconductor introduced an op amp that has found widespread acceptance: the μA741. (The μA prefix is used by Fairchild to indicate a microcircuit fabrication of the amplifier.) This amplifier is available in several different packages. For our discussion, we assume an eight-lead DIP.[1] Figure 4.1 shows a top view of the package, with the terminal designations given alongside the terminals. The terminals of primary interest are

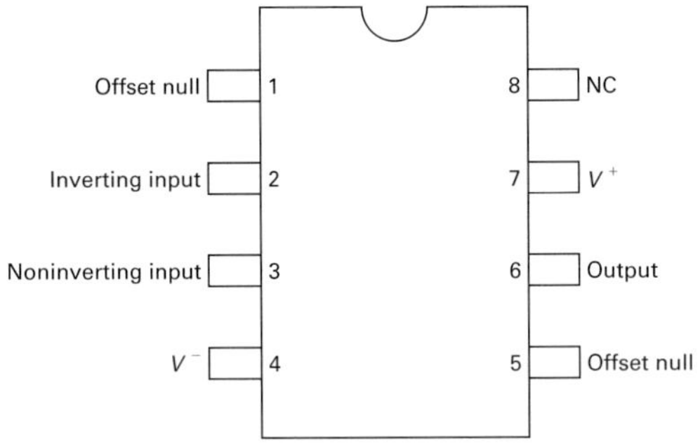

- inverting input
- noninverting input
- output
- positive power supply (V^+)
- negative power supply (V^-)

The remaining three terminals are of little or no concern. The offset null terminals may be used in an auxiliary circuit to compensate for a degradation in performance because of aging and imperfections. However, the degradation in most cases is negligible, so the offset terminals often are unused and play a secondary role in circuit analysis. Terminal 8 is of no interest simply because it is an unused terminal; NC stands for no connection, which means that the terminal is not connected to the amplifier circuit.

Figure 4.1 The eight-lead DIP package (top view).

- - -

[1] DIP is an abbreviation for *dual in-line package*. This means that the terminals on each side of the package are in line, and that the terminals on opposite sides of the package also line up.

Figure 4.2 shows a widely used circuit symbol for an op amp that contains the five terminals of primary interest. Using word labels for the terminals is inconvenient in circuit diagrams, so we simplify the terminal designations in the following way. The noninverting input terminal is labeled plus (+), and the inverting input terminal is labeled minus (−). The power supply terminals, which are always drawn outside the triangle, are marked V^+ and V^-. The terminal at the apex of the triangular box is always understood to be the output terminal. Figure 4.3 summarizes these simplified designations.

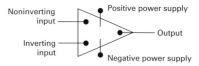

Figure 4.2 The circuit symbol for an operational amplifier (op amp).

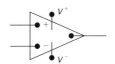

Figure 4.3 A simplified circuit symbol for an op amp.

4.2 TERMINAL VOLTAGES AND CURRENTS

We are now ready to introduce the terminal voltages and currents used to describe the behavior of the op amp. The voltage variables are measured from a common reference node.[2] Figure 4.4 shows the voltage variables with their reference polarities.

All voltages are considered as voltage rises from the common node. This convention is the same as that used in the node-voltage method of analysis. A positive supply voltage (V_{CC}) is connected between V^+ and the common node. A negative supply voltage ($-V_{CC}$) is connected between V^- and the common node. The voltage between the inverting input terminal and the common node is denoted v_n. The voltage between the noninverting input terminal and the common node is designated as v_p. The voltage between the output terminal and the common node is denoted v_o.

Figure 4.5 shows the current variables with their reference directions. Note that all the current reference directions are into the terminals of the operational amplifier: i_n is the current into the inverting input terminal; i_p is the current into the noninverting input terminal; i_o is the current into the output terminal; i_{c+} is the current into the positive power supply terminal; and i_{c-} is the current into the negative power supply terminal.

The terminal behavior of the op amp as a linear circuit element is characterized by constraints on the input voltages and the input currents. The voltage constraint is derived from the voltage transfer characteristic of the op amp integrated circuit and is pictured in Fig. 4.6.

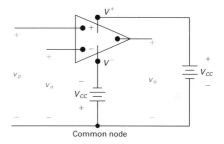

Figure 4.4 Terminal voltage variables.

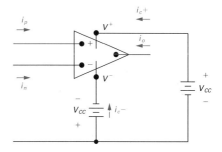

Figure 4.5 Terminal current variables.

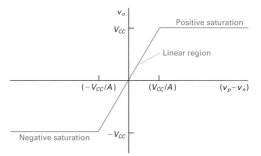

Figure 4.6 The voltage transfer characteristic of an op amp.

---

[2] The common node is external to the op amp. It is the reference terminal of the circuit in which the op amp is embedded.

The voltage transfer characteristic describes how the output voltage varies as a function of the input voltages—that is, how voltage is transferred from the input to the output. Note that for the op amp, the output voltage is a function of the difference between the input voltages, $v_p - v_n$. The equation for the voltage transfer characteristic is

$$v_o = \begin{cases} -V_{CC}, & A(v_p - v_n) < -V_{CC}, \\ A(v_p - v_n), & -V_{CC} \leq A(v_p - v_n) \leq +V_{CC}, \\ +V_{CC}, & A(v_p - v_n) > +V_{CC}. \end{cases} \quad (4.1)$$

We see from Fig. 4.6 and Eq. (4.1) that the op amp has three distinct regions of operation. When the magnitude of the input voltage difference ($|v_p - v_n|$) is small, the op amp behaves as a linear device, as the output voltage is a linear function of the input voltages. Outside this linear region, the output of the op amp saturates, and the op amp behaves as a nonlinear device, because the output voltage is no longer a linear function of the input voltages. When it is operating linearly, the op amp's output voltage is equal to the difference in its input voltages times the multiplying constant, or **gain**, A.

When we confine the op amp to its linear operating region, a constraint is imposed on the input voltages, v_p and v_n. The constraint is based on typical numerical values for V_{CC} and A in Eq. (4.1). For most op amps, the recommended dc power supply voltages seldom exceed 20 V, and the gain, A, is rarely less than 10,000, or 10^4. We see from both Fig. 4.6 and Eq. (4.1) that in the linear region, the magnitude of the input voltage difference ($|v_p - v_n|$) must be less than $20/10^4$, or 2 mV.

Typically, node voltages in the circuits we study are much larger than 2 mV, so a voltage difference of less than 2 mV means the two voltages are essentially equal. Thus, when an op amp is constrained to its linear operating region and the node voltages are much larger than 2 mV, the constraint on the input voltages of the op amp is

$$v_p = v_n. \quad (4.2)$$

Note that Eq. (4.2) characterizes the relationship between the input voltages for an ideal op amp—that is, an op amp whose value of A is infinite.

The input voltage constraint in Eq. (4.2) is called the *virtual short* condition at the input of the op amp. It is natural to ask how the virtual short is maintained at the input of the op amp when the op amp is embedded in a circuit, thus ensuring linear operation. The answer is that a signal is fed back from the output terminal to the inverting input terminal. This configuration is known as **negative feedback**, because the signal fed back from the output subtracts from the input signal. The negative feedback causes the input voltage difference to decrease. Because the output voltage is proportional to the input voltage difference, the output voltage is also decreased, and the op amp operates in its linear region.

If a circuit containing an op amp does not provide a negative feedback path from the op amp output to the inverting input, then the op amp will normally saturate. The difference in the input signals must be extremely small to prevent saturation with no negative feedback. But even if the circuit provides a negative feedback path for the op amp, linear operation is not ensured. So how do we know whether the op amp is operating in its linear region?

The answer is, we don't! We deal with this dilemma by assuming linear operation, performing the circuit analysis, and then checking our results for contradictions. For example, suppose we assume that an op amp in a circuit is operating in its linear region, and we compute the output voltage of the op amp to be 10 V. On examining the circuit, we discover that V_{CC} is 6 V, resulting in a contradiction, because the op amp's output voltage can be no larger than V_{CC}. Thus our assumption of linear operation was invalid, and the op amp output must be saturated at 6 V.

We have identified a constraint on the input voltages that is based on the voltage transfer characteristic of the op amp integrated circuit, the assumption that the op amp is restricted to its linear operating region and to typical values for V_{CC} and A. Equation (4.2) represents the voltage constraint for an ideal op amp, that is, having a value of A that is infinite.

We now turn our attention to the constraint on the input currents. Analysis of the op amp integrated circuit reveals that the equivalent resistance seen by the input terminals of the op amp is very large, typically 1 MΩ or more. Ideally, the equivalent input resistance is infinite, resulting in the current constraint

$$i_p = i_n = 0. \qquad (4.3)$$

Note that the current constraint is not based on assuming the op amp is confined to its linear operating region as was the voltage constraint. Together, Eqs. (4.2) and (4.3) form the constraints on terminal behavior that define our ideal op amp model.

From Kirchhoff's current law we know that the sum of the currents entering the operational amplifier is zero, or

$$i_p + i_n + i_o + i_{c+} + i_{c-} = 0. \qquad (4.4)$$

Substituting the constraint given by Eq. (4.3) into Eq. (4.4) gives

$$i_o = -(i_{c+} + i_{c-}). \qquad (4.5)$$

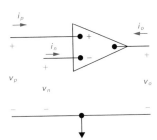

Figure 4.7 The op amp symbol with the power supply terminals removed.

The significance of Eq. (4.5) is that, even though the current at the input terminals is negligible, there may still be appreciable current at the output terminal.

Before we start analyzing circuits containing op amps, let's further simplify the circuit symbol. When we know that the amplifier is operating within its linear region, the dc voltages $\pm V_{CC}$ do not enter into the circuit equations. In this case, we can remove the power supply terminals from the symbol and the dc power supplies from the circuit, as shown in Fig. 4.7. A word of caution: Because the power supply terminals have been omitted, there is a danger of inferring from the symbol that $i_p + i_n + i_o = 0$. We have already noted that such is not the case; that is, $i_p + i_n + i_o + i_{c+} + i_{c-} = 0$. In other words, the ideal op amp model constraint that $i_p = i_n = 0$ does not imply that $i_o = 0$.

Note that the positive and negative power supply voltages do not have to be equal in magnitude. In the linear operating region, v_o must lie between the two supply voltages. For example, if $V^+ = 15$ V and $V^- = -10$ V, then -10 V $\leq v_o \leq 15$ V. Be aware also that the value of A is not constant under all operating conditions. For now, however, we assume that it is. A discussion of how and why the value of A can change must be delayed until after you have studied the electronic devices and components used to fabricate an amplifier.

Example 4.1 illustrates the judicious application of Eqs. (4.2) and (4.3). When we use these equations to predict the behavior of a circuit containing an op amp, in effect we are using an ideal model of the device.

EXAMPLE 4.1

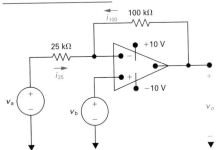

Figure 4.8 The circuit for Example 4.1.

The op amp in the circuit shown in Fig. 4.8 is ideal.

(a) Calculate v_o if $v_a = 1$ V and $v_b = 0$ V.

(b) Repeat (a) for $v_a = 1$ V and $v_b = 2$ V.

(c) If $v_a = 1.5$ V, specify the range of v_b that avoids amplifier saturation.

SOLUTION

(a) Because a negative feedback path exists from the op amp's output to its inverting input through the 100 kΩ resistor, let's assume the op amp is confined to its linear operating region. We can write a node-voltage equation at the inverting input terminal. The voltage at the inverting input terminal is 0, as $v_p = v_b = 0$ from the connected voltage source, and $v_n = v_p$ from the voltage constraint Eq. (4.2). The node-voltage equation at v_n is thus

$$i_{25} + i_{100} = i_n.$$

From Ohm's law,

$$i_{25} = (v_a - v_n)/25 = \frac{1}{25} \text{ mA},$$

$$i_{100} = (v_o - v_n)/100 = v_o/100 \text{ mA}.$$

The current constraint requires $i_n = 0$. Substituting the values for the three currents into the node-voltage equation, we obtain

$$\frac{1}{25} + \frac{v_o}{100} = 0.$$

Hence, v_o is -4 V. Note that because v_o lies between ± 10 V, the op amp is in its linear region of operation.

(b) Using the same process as in (a), we get

$$v_p = v_b = v_n = 2 \text{ V},$$

$$i_{25} = \frac{v_a - v_n}{25} = \frac{1 - 2}{25} = -\frac{1}{25} \text{ mA},$$

$$i_{100} = \frac{v_o - v_n}{100} = \frac{v_o - 2}{100} \text{ mA};$$

$$i_{25} = -i_{100}.$$

Therefore, $v_o = 6$ V. Again, v_o lies within ± 10 V.

(c) As before, $v_n = v_p = v_b$, and $i_{25} = -i_{100}$. Because $v_a = 1.5$ V,

$$\frac{1.5 - v_b}{25} = -\frac{v_o - v_b}{100}.$$

Solving for v_b as a function of v_o gives

$$v_b = \frac{1}{5}(6 + v_o).$$

Now, if the amplifier is to be within the linear region of operation, $-10 \text{ V} \le v_o \le 10$ V. Substituting these limits on v_o into the expression for v_b, we see that v_b is limited to

$$-0.8 \text{ V} \le v_b \le 3.2 \text{ V}.$$

DRILL EXERCISE

4.1 Assume that the op amp in the circuit shown is ideal.

    (a) Calculate v_o for the following values of v_s: 0.4, 2.0, 3.5, −0.6, −1.6, and −2.4 V.

    (b) Specify the range of v_s required to avoid amplifier saturation.

    ANSWER: (a) −2, −10, −15, 3, 8, and 10 V;
(b) −2 V $\leq v_s \leq$ 3 V.

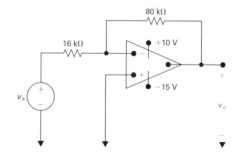

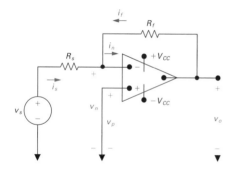

Figure 4.9 An inverting-amplifier circuit.

4.3 THE INVERTING-AMPLIFIER CIRCUIT

We are now ready to discuss the operation of some important op amp circuits, using Eqs. (4.2) and (4.3) to model the behavior of the device itself. Figure 4.9 shows an inverting-amplifier circuit. We assume that the op amp is operating in its linear region. Note that, in addition to the op amp, the circuit consists of two resistors (R_f and R_s), a voltage signal source (v_s), and a short circuit connected between the noninverting input terminal and the common node.

    We now analyze this circuit, assuming an ideal op amp. The goal is to obtain an expression for the output voltage, v_o, as a function of the source voltage, v_s. We employ a single node-voltage equation at the inverting terminal of the op amp, given as

$$i_s + i_f = i_n. \tag{4.6}$$

The voltage constraint of Eq. (4.2) sets the voltage at $v_n = 0$, because the voltage at $v_p = 0$. Therefore,

$$i_s = \frac{v_s}{R_s}, \tag{4.7}$$

$$i_f = \frac{v_o}{R_f}. \tag{4.8}$$

Now we invoke the constraint stated in Eq. (4.3), namely,

$$i_n = 0. \tag{4.9}$$

Substituting Eqs. (4.7)–(4.9) into Eq. (4.6) yields the sought-after result:

$$v_o = \frac{-R_f}{R_s} v_s. \tag{4.10}$$

Note that the output voltage is an inverted, scaled replica of the input. The sign reversal from input to output is, of course, the reason for referring to the circuit as an *inverting* amplifier. The scaling factor, or gain, is the ratio R_f/R_s.

The result given by Eq. (4.10) is valid only if the op amp shown in the circuit in Fig. 4.9 is ideal—that is, if A is infinite and the input resistance is infinite. For a practical op amp, Eq. (4.10) is an approximation, usually a good one. Equation (4.10) is important because it tells us that if the op amp gain A is large, we can specify the gain of the inverting amplifier with the external resistors R_f and R_s. The upper limit on the gain, R_f/R_s, is determined by the power supply voltages and the value of the signal voltage v_s. If we assume equal power supply voltages, that is, $V^+ = -V^- = V_{CC}$, we get

$$|v_o| < V_{CC}, \qquad \left|\frac{R_f}{R_s} v_s\right| < V_{CC}, \qquad \frac{R_f}{R_s} < \left|\frac{V_{CC}}{v_s}\right|. \tag{4.11}$$

For example, if $V_{CC} = 15$ V and $v_s = 10$ mV, the ratio R_f/R_s must be less than 1500.

In the inverting amplifier circuit shown in Fig. 4.9, the resistor R_f provides the negative feedback connection. That is, it connects the output terminal to the inverting input terminal. If R_f is removed, the feedback path is opened and the amplifier is said to be operating *open loop*. Figure 4.10 shows the open-loop operation.

Opening the feedback path drastically changes the behavior of the circuit. First, the output voltage is now

$$v_o = -A v_n, \tag{4.12}$$

assuming as before that $V^+ = -V^- = V_{CC}$; then $|v_n| < V_{CC}/A$ for linear operation. Because the inverting input current is almost zero, the voltage drop across R_s is almost zero, and the inverting input voltage nearly equals the signal voltage, v_s; that is, $v_n \approx v_s$. Hence, the op amp can operate open loop in the linear mode only if $|v_s| < V_{CC}/A$. If $|v_s| > V_{CC}/A$, the op amp simply saturates. In particular, if $v_s < -V_{CC}/A$, the op amp saturates at $+V_{CC}$, and if $v_s > V_{CC}/A$, the op amp saturates at $-V_{CC}$. Because the relationship shown in Eq. (4.12) occurs when there is no feedback path, the value of A is often called the **open-loop gain** of the op amp. We will have more to say about open-loop operation when we discuss the comparator in Section 4.7.

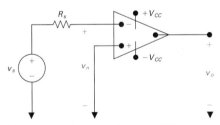

Figure 4.10 An inverting amplifier operating open loop.

DRILL EXERCISE

4.2 The source voltage v_s in the circuit in Drill Exercise 4.1 is −640 mV. The 80 kΩ feedback resistor is replaced by a variable resistor R_x. What range of R_x allows the inverting amplifier to operate in its linear region?

ANSWER: $0 \leq R_x \leq 250$ kΩ.

4.4 THE SUMMING-AMPLIFIER CIRCUIT

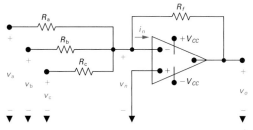

Figure 4.11 A summing amplifier.

The output voltage of a summing amplifier is an inverted, scaled sum of the voltages applied to the input of the amplifier. Figure 4.11 shows a summing amplifier with three input voltages.

We obtain the relationship between the output voltage v_o and the three input voltages, v_a, v_b, and v_c, by summing the currents away from the inverting input terminal:

$$\frac{v_n - v_a}{R_a} + \frac{v_n - v_b}{R_b} + \frac{v_n - v_c}{R_c} + \frac{v_n - v_o}{R_f} + i_n = 0. \quad (4.13)$$

Assuming an ideal op amp, we can use the voltage and current constraints together with the ground imposed at v_p by the circuit to see that $v_n = v_p = 0$ and $i_n = 0$. This reduces Eq. (4.13) to

$$v_o = -\left(\frac{R_f}{R_a} v_a + \frac{R_f}{R_b} v_b + \frac{R_f}{R_c} v_c \right). \quad (4.14)$$

Equation (4.14) states that the output voltage is an inverted, scaled sum of the three input voltages.

If $R_a = R_b = R_c = R_s$, then Eq. (4.14) reduces to

$$v_o = -\frac{R_f}{R_s}(v_a + v_b + v_c). \quad (4.15)$$

Finally, if we make $R_f = R_s$, the output voltage is just the inverted sum of the input voltages. That is,

$$v_o = -(v_a + v_b + v_c). \quad (4.16)$$

Although we illustrated the summing amplifier with just three input signals, the number of input voltages can be increased as needed. For example, you might wish to sum 16 individually recorded audio signals to form a single audio signal. The summing amplifier configuration in Fig. 4.11 could include 16 different input resistor values so that each of the input audio tracks appears in the output signal with a different amplification factor. The summing amplifier thus plays the role of an audio mixer. As with inverting-amplifier circuits, the scaling factors in summing-amplifier circuits are determined by the external resistors $R_f, R_a, R_b, R_c, \ldots, R_n$.

DRILL EXERCISE

4.3   **(a)** Find v_o in the circuit shown if $v_a = 0.1$ V and $v_b = 0.25$ V.

   (b) If $v_b = 0.25$ V, how large can v_a be before the op amp saturates?

   (c) If $v_a = 0.10$ V, how large can v_b be before the op amp saturates?

   (d) Repeat (a), (b), and (c) with the polarity of v_b reversed.

   ANSWER: (a) -7.5 V; (b) 0.15 V; (c) 0.5 V; (d) -2.5, 0.25, and 2 V.

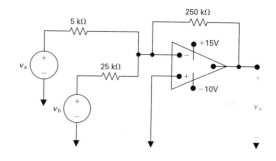

4.5 THE NONINVERTING-AMPLIFIER CIRCUIT

Figure 4.12 depicts a noninverting-amplifier circuit. The signal source is represented by v_g in series with the resistor R_g. In deriving the expression for the output voltage as a function of the source voltage, we assume an ideal op amp operating within its linear region. Thus, as before, we use Eqs. (4.2) and (4.3) as the basis for the derivation. Because the op amp input current is zero, we can write $v_p = v_g$ and, from Eq. (4.2), $v_n = v_g$ as well. Now, because the input current is zero ($i_n = i_p = 0$), the resistors R_f and R_s form an unloaded voltage divider across v_o. Therefore,

$$v_n = v_g = \frac{v_o R_s}{R_s + R_f}. \qquad (4.17)$$

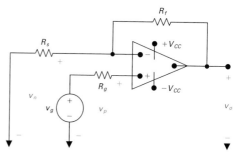

Figure 4.12 A noninverting amplifier.

Solving Eq. (4.17) for v_o gives us the sought-after expression:

$$v_o = \frac{R_s + R_f}{R_s} v_g. \tag{4.18}$$

Operation in the linear region requires that

$$\frac{R_s + R_f}{R_s} < \left| \frac{V_{CC}}{v_g} \right|.$$

Note again that, because of the ideal op amp assumption, we can express the output voltage as a function of the input voltage and the external resistors—in this case, R_s and R_f.

DRILL EXERCISE

4.4 Assume that the op amp in the circuit shown is ideal.

(a) Find the output voltage when the variable resistor is set to 60 kΩ.

(b) How large can R_x be before the amplifier saturates?

ANSWER: (a) 4.8 V; (b) 75 kΩ.

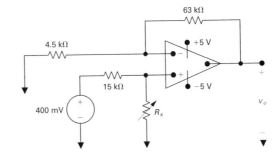

4.6 THE DIFFERENCE-AMPLIFIER CIRCUIT

The output voltage of a difference amplifier is proportional to the difference between the two input voltages. To demonstrate, we analyze the difference-amplifier circuit shown in Fig. 4.13, assuming an ideal op amp operating in its linear region. We derive the relationship between v_o and the two input voltages v_a and v_b by summing the currents away from the inverting input node:

$$\frac{v_n - v_a}{R_a} + \frac{v_n - v_o}{R_b} + i_n = 0. \tag{4.19}$$

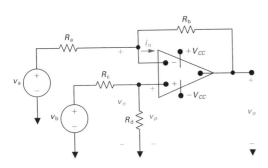

Figure 4.13 A difference amplifier.

Because the op amp is ideal, we use the voltage and current constraints to see that

$$i_n = i_p = 0, \tag{4.20}$$

$$v_n = v_p = \frac{R_d}{R_c + R_d} v_b. \tag{4.21}$$

Combining Eqs. (4.19), (4.20), and (4.21) gives the desired relationship:

$$v_o = \frac{R_d(R_a + R_b)}{R_a(R_c + R_d)} v_b - \frac{R_b}{R_a} v_a. \tag{4.22}$$

Equation (4.22) shows that the output voltage is proportional to the difference between a scaled replica of v_b and a scaled replica of v_a. In general the scaling factor applied to v_b is not the same as that applied to v_a. However, the scaling factor applied to each input voltage can be made equal by setting

$$\frac{R_a}{R_b} = \frac{R_c}{R_d}. \tag{4.23}$$

When Eq. (4.23) is satisfied, the expression for the output voltage reduces to

$$v_o = \frac{R_b}{R_a} (v_b - v_a). \tag{4.24}$$

Equation (4.24) indicates that the output voltage can be made a scaled replica of the difference between the input voltages v_b and v_a. As in the previous ideal amplifier circuits, the scaling is controlled by the external resistors. Furthermore, the relationship between the output voltage and the input voltages is not affected by connecting a nonzero load resistance across the output of the amplifier.

DRILL EXERCISES

4.5   **(a)** Use the principle of superposition to derive Eq. (4.22).

   (b) Derive Eqs. (4.23) and (4.24).

ANSWER: (a) $v_o' = -\dfrac{R_b}{R_a} v_a$, $v_o'' = \dfrac{R_d(R_a + R_b)}{R_a(R_c + R_d)} v_b$, $v_o = v_o' + v_o''$; (b) derivation.

4.6   **(a)** In the difference amplifier shown, $v_b = 4.0$ V. What range of values for v_a will result in linear operation?

(b) Repeat (a) with the 20 kΩ resistor decreased to 8 kΩ.

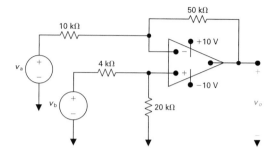

ANSWER: (a) 2 V $\leq v_a \leq$ 6 V;
(b) 1.2 V $\leq v_a \leq$ 5.2 V.

The Difference Amplifier—Another Perspective

We can examine the behavior of a difference amplifier more closely if we redefine its inputs in terms of two other voltages. The first is the **differential mode** input, which is the difference between the two input voltages in Fig. 4.13:

$$v_{dm} = v_b - v_a. \qquad (4.25)$$

The second is the **common mode** input, which is the average of the two input voltages in Fig. 4.13:

$$v_{cm} = (v_a + v_b)/2. \qquad (4.26)$$

Using Eqs. (4.25) and (4.26), we can now represent the original input voltages, v_a and v_b, in terms of the differential mode and common mode voltages, v_{dm} and v_{cm}:

$$v_a = v_{cm} - \frac{1}{2}v_{dm} \qquad (4.27)$$

$$v_b = v_{cm} + \frac{1}{2}v_{dm}. \qquad (4.28)$$

Substituting Eqs. (4.27) and (4.28) into Eq. (4.22) gives the output of the difference amplifier in terms of the differential mode and common mode voltages:

$$v_o = \left[\frac{R_a R_d - R_b R_c}{R_a (R_c + R_d)} \right] v_{cm}$$

$$+ \left[\frac{R_d (R_a + R_b) + R_b (R_c + R_d)}{2 R_a (R_c + R_d)} \right] v_{dm}, \qquad (4.29)$$

$$= A_{cm} v_{cm} + A_{dm} v_{dm}, \qquad (4.30)$$

where A_{cm} is the common mode gain and A_{dm} is the differential mode gain. Now, substitute $R_c = R_a$ and $R_d = R_b$, which are the values for R_c and R_d that satisfy Eq. (4.23), into Eq. (4.29):

$$v_o = (0)v_{cm} + \left(\frac{R_b}{R_a}\right)v_{dm}. \qquad (4.31)$$

Thus, an ideal difference amplifier has $A_{cm} = 0$, amplifies only the differential mode portion of the input voltage, and eliminates the common mode portion of the input voltage. Figure 4.14 shows a difference-amplifier circuit with differential mode and common mode input voltages in place of v_a and v_b.

Equation (4.30) provides an important perspective on the function of the difference amplifier, since in many applications it is the differential mode signal that contains the information of interest, whereas the common mode signal is the noise found in all electric signals. For example, an electrocardiograph electrode measures the voltages produced by your body to regulate your heartbeat. These voltages have very small magnitudes compared with the electrical noise that the electrode picks up from sources such as lights and electrical equipment. The noise appears as the common mode portion of the measured voltage, whereas the heart-rate voltages comprise the differential mode portion. Thus an ideal difference amplifier would amplify only the voltage of interest and would suppress the noise.

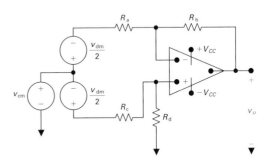

Figure 4.14 A difference amplifier with common mode and differential mode input voltages.

Measuring Difference-Amplifier Performance— The Common Mode Rejection Ratio

An ideal difference amplifier has zero common mode gain and nonzero (and usually large) differential mode gain. Two factors have an influence on the ideal common mode gain—resistance mismatches (that is, Eq. [4.23] is not satisfied) or a nonideal op amp (that is, Eq. [4.20] is not satisfied). We focus here on the effect of resistance mismatches on the performance of a difference amplifier.

Suppose that resistor values are chosen that do not precisely satisfy Eq. (4.23). Instead, the relationship among the resistors R_a, R_b, R_c, and R_d is

$$\frac{R_a}{R_b} = (1 - \epsilon)\frac{R_c}{R_d},$$

so

$$R_a = (1 - \epsilon)R_c \quad \text{and} \quad R_b = R_d, \qquad (4.32)$$

or

$$R_d = (1 - \epsilon) R_b \quad \text{and} \quad R_a = R_c, \tag{4.33}$$

where ϵ is a very small number. We can see the effect of this resistance mismatch on the common mode gain of the difference amplifier by substituting Eq. (4.33) into Eq. (4.29) and simplifying the expression for A_{cm}:

$$A_{cm} = \frac{R_a(1 - \epsilon)R_b - R_a R_b}{R_a[R_a + (1 - \epsilon)R_b]} \tag{4.34}$$

$$= \frac{-\epsilon R_b}{R_a + (1 - \epsilon)R_b} \tag{4.35}$$

$$\approx \frac{-\epsilon R_b}{R_a + R_b}. \tag{4.36}$$

We can make the approximation to give Eq. (4.36) because ϵ is very small, and therefore $(1 - \epsilon)$ is approximately 1 in the denominator of Eq. (4.35). Note that, when the resistors in the difference amplifier satisfy Eq. (4.23), $\epsilon = 0$ and Eq. (4.36) gives $A_{cm} = 0$.

Now calculate the effect of the resistance mismatch on the differential mode gain by substituting Eq. (4.33) into Eq. (4.29) and simplifying the expression for A_{dm}:

$$A_{dm} = \frac{(1 - \epsilon)R_b(R_a + R_b) + R_b[R_a + (1 - \epsilon)R_b]}{2R_a[R_a + (1 - \epsilon)R_b]} \tag{4.37}$$

$$= \frac{R_b}{R_a}\left[1 - \frac{(\epsilon/2)R_a}{R_a + (1 - \epsilon)R_b}\right] \tag{4.38}$$

$$\approx \frac{R_b}{R_a}\left[1 - \frac{(\epsilon/2)R_a}{R_a + R_b}\right] \tag{4.39}$$

We use the same rationale for the approximation in Eq. (4.39) as in the computation of A_{cm}. When the resistors in the difference amplifier satisfy Eq. (4.23), $\epsilon = 0$ and Eq. (4.39) gives $A_{dm} = R_b/R_a$.

The **common mode rejection ratio (CMRR)** can be used to measure how nearly ideal a difference amplifier is. It is defined as the ratio of the differential mode gain to the common mode gain:

$$\text{CMRR} = \left|\frac{A_{dm}}{A_{cm}}\right| \tag{4.40}$$

The higher the CMRR, the more nearly ideal the difference amplifier. We can see the effect of resistance mismatch on the

CMRR by substituting Eqs. (4.36) and (4.39) into Eq. (4.40):

$$\text{CMRR} \approx \left| \frac{\dfrac{R_b}{R_a}[1 - (R_a\epsilon/2)/(R_a + R_b)]}{-\epsilon R_b/(R_a + R_b)} \right| \qquad (4.41)$$

$$\approx \frac{R_a(1 - \epsilon/2) + R_b}{\epsilon R_a} \qquad (4.42)$$

$$\approx \frac{1 + R_b/R_a}{\epsilon}. \qquad (4.43)$$

From Eq. (4.43), if the resistors in the difference amplifier are matched, $\epsilon = 0$ and $\text{CMRR} = \infty$. Even if the resistors are mismatched, we can minimize the impact of the mismatch by making the differential mode gain (R_b/R_a) very large, thereby making the CMRR large.

We said at the outset that another reason for nonzero common mode gain is a nonideal op amp. Note that the op amp is itself a difference amplifier, because in the linear operating region, its output is proportional to the difference of its inputs; that is, $v_o = A(v_p - v_n)$. The output of a nonideal op amp is not strictly proportional to the difference between the inputs (the differential mode input) but also is comprised of a common mode signal. Internal mismatches in the components of the integrated circuit make the behavior of the op amp nonideal, in the same way that the resistor mismatches in the difference-amplifier circuit make its behavior nonideal. Even though a discussion of nonideal op amps is beyond the scope of this text, you may note that the CMRR is often used in assessing how nearly ideal an op amp's behavior is. In fact, it is one of the main ways of rating op amps in practice.

DRILL EXERCISES

4.7 In the difference amplifier shown, compute (a) the differential mode gain, (b) the common mode gain, and (c) the CMRR.

ANSWER: (a) 24.98; (b) −0.04; (c) 624.5.

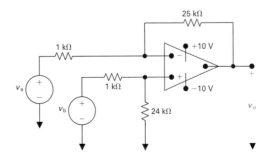

4.8   In the difference amplifier shown, what value of R_x
yields a CMRR ≥ 1000?

ANSWER: 19.93 kΩ or 20.07 kΩ.

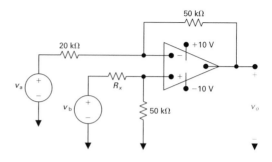

4.7 THE COMPARATOR

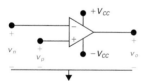

Figure 4.15 The op amp comparator circuit.

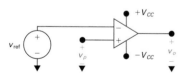

Figure 4.16 A comparator with one input, a
reference voltage.

An operational amplifier can be used as a comparator by removing the feedback (R_f) and input (R_i) resistors of the inverting
amplifier circuit in Fig. 4.9. The op amp comparator circuit is
shown in Fig. 4.15. Since we are assuming an ideal op amp (infinite gain), the output of the comparator will be either plus V_{CC}
or minus V_{CC}, depending on whether v_p is greater than or less
than v_n. Specifically, v_o will equal $+V_{CC}$ if $v_p > v_n$ and $-V_{CC}$ if
$v_p < v_n$.

The comparator can be thought of as an analog-to-digital
converter in that it converts the difference of two analog signals
to a digital decision, i.e., high if the difference is positive and low
if the difference is negative.

Note that one of the inputs to the comparator can be a reference voltage as shown in Fig. 4.16. In this application v_o will be
high or low depending on whether $v_p > v_{\text{ref}}$ (high) or $v_p < v_{\text{ref}}$
(low).

EXAMPLE 4.2

(a) For the comparator circuit in Fig. 4.17 make a sketch of the
output voltage v_o vs. the signal voltage v_s when $v_{\text{ref}} = 5$ V;
$R_1 = 10$ kΩ; and $R_2 = 40$ kΩ.

(b) Repeat part (a) if $v_{\text{ref}} = -10$ V.

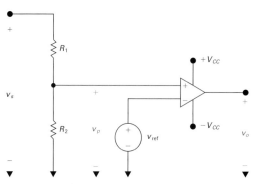

Figure 4.17 Comparator circuit for Example 4.2.

SOLUTION

(a) For the given values of R_1 and R_2 the voltage at the noninverting input terminal of the comparator will be

$$v_p = \frac{v_s(40)}{(10+40)} = 0.8v_s.$$

The output voltage will be $+V_{CC}$ when

$$0.8v_s > v_{\text{ref}}$$

or

$$v_s > (1.25)(5)$$

$$> 6.25 \text{ V}.$$

The output voltage will be $-V_{CC}$ when

$$0.8v_s < v_{\text{ref}}$$

or

$$< 6.25 \text{ V}.$$

A sketch of v_o vs. v_s is shown in Fig. 4.18.

(b) When v_{ref} equals -10 V, the output voltage will be $+V_{CC}$ when

$$v_s > 1.25(-10)$$

$$> -12.5 \text{ V}.$$

The output voltage will be $-V_{CC}$ when

$$v_s < -12.5 \text{ V}.$$

A sketch of v_o vs. v_s is shown in Fig. 4.19.

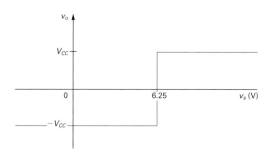

Figure 4.18 v_o vs. v_s for Example 4.2(a).

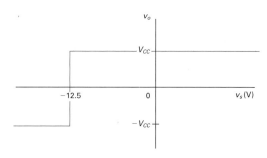

Figure 4.19 v_o vs. v_s for Example 4.2(b).

Practical Perspective

The Flash Converter

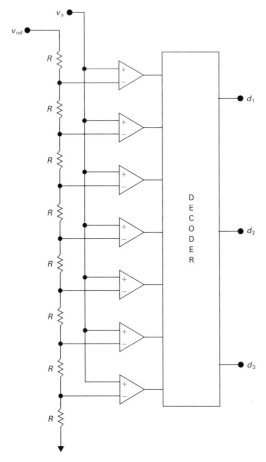

Figure 4.20 3-bit resolution flash converter.

The basic idea behind the flash converter introduced at the beginning of the chapter is to convert an analog signal v_s to a digital array which indicates the value of v_s. To achieve an n-bit resolution of v_s, $2^n - 1$ comparators are needed in the flash converter structure. For purposes of discussion we will assume a 3-bit resolution, therefore we need $2^3 - 1$ or 7 comparators. Our 3-bit resolution converter is shown in Fig. 4.20. In drawing Fig 4.20 we have omitted the power supply voltages on the comparators. It is understood that the output voltage of a comparator is either high or low depending on the voltages applied to its noninverting and inverting input terminals. We will also assume a high voltage represents the binary digit 1 and a low voltage represents the binary digit 0.

Beginning with the lower comparator, the voltage applied to the inverting input terminal starts at $(1/8)v_{ref}$ and then increases by $(1/8)v_{ref}$ for each comparator as we move up the string of comparators. Now assume v_s is less than $(1/8)v_{ref}$. In this instance the output voltage of each comparator will be low. Thus for $v_s < (1/8)v_{ref}$ the logic code in the decoder will be 0000000. When v_s lies between $(1/8)v_{ref}$ and $(2/8)v_{ref}$, i.e., $(1/8)v_{ref} < v_s < (2/8)v_{ref}$, the output of the lower comparator will be high and that of the remaining comparators low. Hence the logic code becomes 0000001.

As v_s continues to increase, the logic code adds a 1 each time v_s exceeds the reference value of an inverting input voltage. Hence the logic code into the decoder as v_s increases is

$$
\begin{array}{ccccccc}
0 & 0 & 0 & 0 & 0 & 0 & 0 \\
0 & 0 & 0 & 0 & 0 & 0 & 1 \\
0 & 0 & 0 & 0 & 0 & 1 & 1 \\
0 & 0 & 0 & 0 & 1 & 1 & 1 \\
0 & 0 & 0 & 1 & 1 & 1 & 1 \\
0 & 0 & 1 & 1 & 1 & 1 & 1 \\
0 & 1 & 1 & 1 & 1 & 1 & 1 \\
1 & 1 & 1 & 1 & 1 & 1 & 1 \\
\end{array}
$$

This array of ones and zeros is known as a thermometer code because as v_s increases, the number of ones in the array increases.

The decoder will convert these eight unique patterns of ones and zeros into the eight unique patterns representing the decimal values from zero to seven as shown in Table 4.1.

TABLE 4.1 Conversion of the Thermometer Code to a Binary Code.

THERMOMETER CODE							BINARY CODE			DECIMAL VALUE
0	0	0	0	0	0	0	0	0	0	0
0	0	0	0	0	0	1	0	0	1	1
0	0	0	0	0	1	1	0	1	0	2
0	0	0	0	1	1	1	0	1	1	3
0	0	0	1	1	1	1	1	0	0	4
0	0	1	1	1	1	1	1	0	1	5
0	1	1	1	1	1	1	1	1	0	6
1	1	1	1	1	1	1	1	1	1	7

SUMMARY

- The equation that defines the voltage transfer characteristic of an ideal op amp is

$$v_o = \begin{cases} -V_{CC}, & A(v_p - v_n) < -V_{CC}, \\ A(v_p - v_n), & -V_{CC} \leq A(v_p - v_n) \leq +V_{CC}, \\ +V_{CC}, & A(v_p - v_n) > +V_{CC}, \end{cases}$$

  where A is a proportionality constant known as the open-loop gain, and V_{CC} represents the power supply voltages.

- A feedback path between an op amp's output and its inverting input can constrain the op amp to its linear operating region where $v_o = A(v_p - v_n)$.

- A voltage constraint exists when the op amp is confined to its linear operating region due to typical values of V_{CC} and A. If the ideal modeling assumptions are made—meaning A is assumed to be infinite—the ideal op amp model is characterized by the voltage constraint

$$v_p = v_n.$$

- A current constraint further characterizes the ideal op amp model, because the ideal input resistance of the op amp integrated circuit is infinite. This current constraint is given by

$$i_p = i_n = 0.$$

- An inverting amplifier is an op amp circuit producing an output voltage that is an inverted, scaled replica of the input.

- A summing amplifier is an op amp circuit producing an output voltage that is a scaled sum of the input voltages.
- A noninverting amplifier is an op amp circuit producing an output voltage that is a scaled replica of the input voltage.
- A difference amplifier is an op amp circuit producing an output voltage that is a scaled replica of the input voltage difference.
- The two voltage inputs to a difference amplifier can be used to calculate the common mode and difference mode voltage inputs, v_{cm} and v_{dm}. The output from the difference amplifier can be written in the form

$$v_o = A_{cm} v_{cm} + A_{dm} v_{dm},$$

where A_{cm} is the common mode gain, and A_{dm} is the differential mode gain.

- In an ideal difference amplifier, $A_{cm} = 0$. To measure how nearly ideal a difference amplifier is, we use the common mode rejection ratio:

$$\text{CMRR} = \left| \frac{A_{dm}}{A_{cm}} \right|.$$

An ideal difference amplifier has an infinite CMRR.

- An operational amplifier can be used as a comparator. The output of a comparator will be $\pm V_{CC}$. If $v_p > v_n$, $v_o = +V_{CC}$. If $v_p < v_n$, $v_o = -V_{CC}$.

PROBLEMS

4.1. A voltmeter with a full-scale reading of 10 V is used to measure the output voltage in the circuit in Fig. P4.1. What is the reading of the voltmeter? Assume the op amp is ideal.

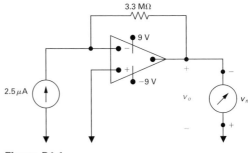

Figure P4.1

4.2. Find i_L (in microamperes) in the circuit in Fig. P4.2.

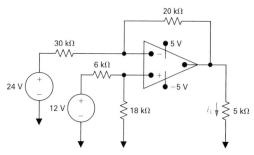

Figure P4.2

4.3. The op amp in the circuit in Fig. P4.3 is ideal.
  (a) Calculate v_o if $v_a = 1.5$ V and $v_b = 0$ V.
  (b) Calculate v_o if $v_a = 3.0$ V and $v_b = 0$ V.
  (c) Calculate v_o if $v_a = 1.0$ V and $v_b = 2$ V.
  (d) Calculate v_o if $v_a = 4.0$ V and $v_b = 2$ V.
  (e) Calculate v_o if $v_a = 6.0$ V and $v_b = 8$ V.
  (f) If $v_b = 4.5$ V, specify the range of v_a such that the amplifier does not saturate.

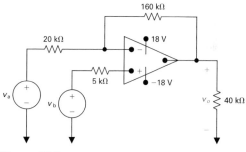

Figure P4.3

4.4. The op amp in the circuit in Fig. P4.4 is ideal. Calculate the following:
  (a) v_a
  (b) v_o
  (c) i_a
  (d) i_o

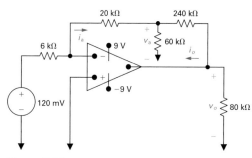

Figure P4.4

4.5. Find i_o in the circuit in Fig. P4.5 if the op amp is ideal.

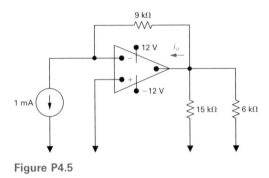

Figure P4.5

4.6. A circuit designer claims the circuit in Fig. P4.6 will produce an output voltage that will vary between ±5 as v_g varies between 0 and 5 V. Assume the op amp is ideal.

(a) Draw a graph of the output voltage v_o as a function of the input voltage v_g for $0 \le v_g \le 5$ V.

(b) Do you agree with the designer's claim?

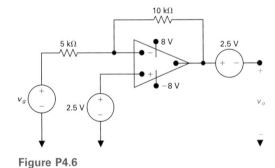

Figure P4.6

4.7. The op amp in the circuit in Fig. P4.7 is ideal.

(a) Find the range of values for σ in which the op amp does not saturate.

(b) Find i_o (in microamperes) when $\sigma = 0.12$.

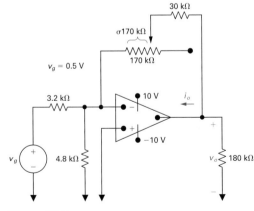

Figure P4.7

4.8. **(a)** The op amp in the circuit shown in Fig. P4.8 is ideal. The adjustable resistor R_Δ has a maximum value of 150 kΩ, and α is restricted to the range of $0.3 \leq \alpha \leq 0.75$. Calculate the range of v_o if $v_g = 50$ mV.

(b) If α is not restricted, at what value of α will the op amp saturate?

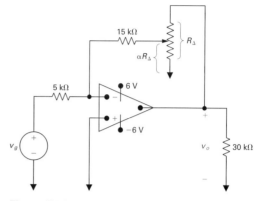

Figure P4.8

4.9. **(a)** The op amp in Fig. P4.9 is ideal. Find v_o if $v_a = 18$ V, $v_b = 6$ V, $v_c = -15$ V, and $v_d = 8$ V.

(b) Assume v_a, v_b, and v_d retain their values as given in (a). Specify the range of v_c such that the op amp operates within its linear region.

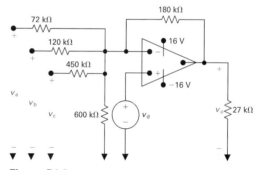

Figure P4.9

4.10. The 180 kΩ feedback resistor in the circuit in Fig. P4.9 is replaced by a variable resistor R_f. The voltages v_a–v_d have the same values as given in Problem 4.9(a).

(a) What value of R_f will cause the op amp to saturate? Note that $0 \leq R_f \leq \infty$.

(b) When R_f has the value found in (a), what is the current (in microamperes) into the output terminal of the op amp?

4.11. The op amp in Fig. P4.11 is ideal.

   (a) Find v_o if $v_a = 1.2$ V, $v_b = -1.5$ V, and $v_c = 4$ V.

   (b) The voltages v_a and v_c remain at 1.2 V and 4 V, respectively. What are the limits on v_b if the op amp operates within its linear region?

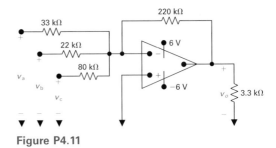

Figure P4.11

4.12. Refer to the circuit in Fig. 4.11, where the op amp is assumed to be ideal. Given that $R_a = 3$ kΩ, $R_b = 5$ kΩ, $R_c = 25$ kΩ, $v_a = 150$ mV, $v_b = 100$ mV, $v_c = 250$ mV, and $V_{CC} = \pm 6$ V, specify the range of R_f for which the op amp operates within its linear region.

❖ **4.13.** Design an inverting summing amplifier so that

$$v_o = -(6v_a + 9v_b + 4v_c + 3v_d).$$

If the feedback resistor (R_f) is chosen to be 72 kΩ, draw a circuit diagram of the amplifier and specify the values of R_a, R_b, R_c, and R_d.

4.14. **(a)** Show that when the ideal op amp in Fig. P4.14 is operating in its linear region,

$$i_a = \frac{3v_g}{R}.$$

   (b) Show that the ideal op amp will saturate when

$$R_a = \frac{R(\pm V_{CC} - 2v_g)}{3v_g}.$$

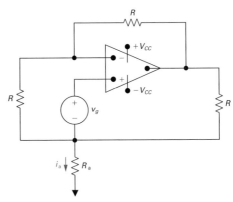

Figure P4.14

4.15. The circuit inside the shaded area in Fig. P4.15 is a constant current source for a limited range of values of R_L.

(a) Find the value of i_L for $R_L = 2.5$ kΩ.

(b) Find the maximum value for R_L for which i_L will have the value in (a).

(c) Assume that $R_L = 6.5$ kΩ. Explain the operation of the circuit. You can assume that $i_n = i_p \approx 0$ under all operating conditions.

(d) Sketch i_L versus R_L for $0 \le R_L \le 6.5$ kΩ.

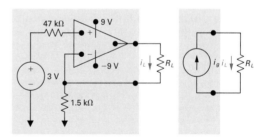

Figure P4.15

4.16. The variable resistor R_o in the circuit in Fig. P4.16 is adjusted until the source current i_g is zero. The op amps are ideal, and $0 \le v_g \le 1.2$ V.

(a) What is the value of R_o?

(b) If $v_g = 1.0$ V, how much power (in microwatts) is dissipated in R_o?

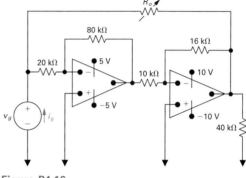

Figure P4.16

4.17. Find i_a in the circuit in Fig. P4.17.

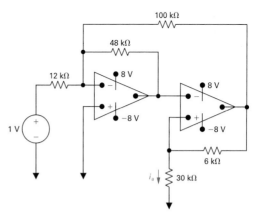

Figure P4.17

4.18. The op amps in the circuit in Fig. P4.18 are ideal.

    (a) Find i_a.

    (b) Find the value of the left source voltage for which $i_a = 0$.

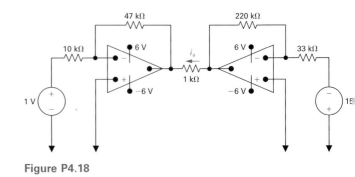

Figure P4.18

4.19. Assume that the ideal op amp in the circuit in Fig. P4.19 is operating in its linear region.

    (a) Calculate the power delivered to the 16 kΩ resistor.

    (b) Repeat (a) with the op amp removed from the circuit, that is, with the 16 kΩ resistor connected in the series with the voltage source and the 48 kΩ resistor.

    (c) Find the ratio of the power found in (a) to that found in (b).

    (d) Does the insertion of the op amp between the source and the load serve a useful purpose? Explain.

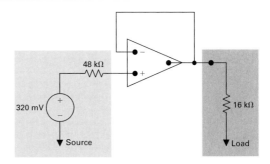

Figure P4.19

4.20. Assume that the ideal op amp in the circuit seen in Fig. P4.20 is operating in its linear region.

    (a) Show that $v_o = [(R_1 + R_2)/R_1]v_s$.

    (b) What happens if $R_1 \to \infty$ and $R_2 \to 0$?

    (c) Explain why this circuit is referred to as a voltage follower when $R_1 = \infty$ and $R_2 = 0$.

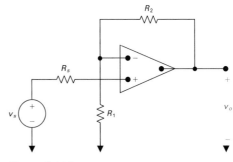

Figure P4.20

4.21. The op amp in the circuit shown in Fig. P4.21 is ideal.

(a) Calculate v_o when v_g equals 3 V.

(b) Specify the range of values of v_g so that the op amp operates in a linear mode.

(c) Assume that v_g equals 5 V and that the 48 kΩ resistor is replaced with a variable resistor. What value of the variable resistor will cause the op amp to saturate?

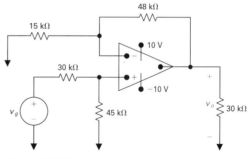

Figure P4.21

4.22. The op amp in the noninverting summing amplifier of Fig. P4.22 is ideal.

(a) Specify the values of R_f, R_b, and R_c so that

$$v_o = 3v_a + 2v_b + v_c.$$

(b) Find (in microamperes) i_a, i_b, i_c, i_g, and i_s when $v_a = 0.80$ V, $v_b = 1.5$ V, and $v_c = 2.1$ V.

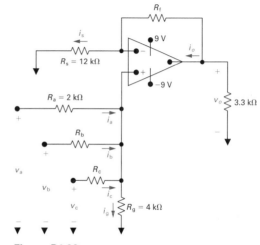

Figure P4.22

4.23. The op amp in the noninverting amplifier shown in Fig. P4.23 is ideal. The signal voltages v_a and v_b are 500 mV and 1200 mV, respectively.

(a) Calculate v_o in volts.

(b) Find i_a and i_b in microamperes.

(c) What are the weighting factors associated with v_a and v_b?

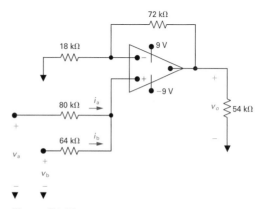

Figure P4.23

4.24. The circuit in Fig. P4.24 is a noninverting summing amplifier. Assume the op amp is ideal. Design the circuit so that

$$v_o = 5v_a + 4v_b + v_c.$$

(a) Specify the numerical values of R_b, R_c, and R_f.

(b) Calculate (in microamperes) i_a, i_b, and i_c when $v_a = 0.5$ V, $v_b = 1.0$ V, and $v_c = 1.5$ V.

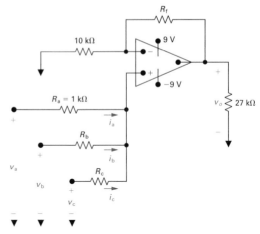

Figure P4.24

4.25. The resistors in the difference amplifier shown in Fig. 4.13 are $R_a = 20$ kΩ, $R_b = 80$ kΩ, $R_c = 47$ kΩ, and $R_d = 33$ kΩ. The signal voltages v_a and v_b are 0.45 and 0.9 V, respectively, and $V_{CC} = \pm 9$ V.

(a) Find v_o.

(b) What is the resistance seen by the signal source v_a?

(c) What is the resistance seen by the signal source v_b?

4.26. The op amp in the adder-subtracter circuit shown in Fig. P4.26 is ideal.

(a) Find v_o when $v_a = 0.5$ V, $v_b = 0.3$ V, $v_c = 0.6$ V, and $v_d = 0.8$ V.

(b) If v_a, v_b, and v_d are held constant, what values of v_c will not saturate the op amp?

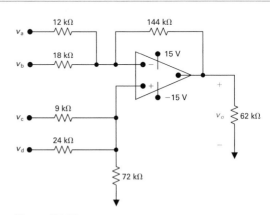

Figure P4.26

4.27. The op amp in the circuit of Fig. P4.27 is ideal.

(a) Plot v_o versus α when $R_f = 4R_1$ and $v_g = 2$ V. Use increments of 0.1 and note by hypothesis that $0 \leq \alpha \leq 1.0$.

(b) Write an equation for the straight line you plotted in (a). How are the slope and intercept of the line related to v_g and the ratio R_f/R_1?

(c) Using the results from (b), choose values for v_g and the ratio R_f/R_1 such that $v_o = -6\alpha + 4$.

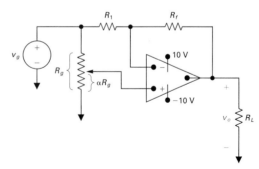

Figure P4.27

 4.28. Select the values of R_b and R_f in the circuit in Fig. P4.28 so that

$$v_o = 2000(i_b - i_a).$$

The op amp is ideal.

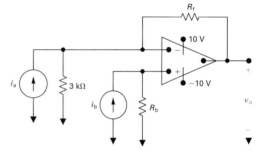

Figure P4.28

 4.29. Design the difference-amplifier circuit in Fig. P4.29 so that $v_o = 10(v_b - v_a)$, and the voltage source v_b sees an input resistance of 220 kΩ. Specify the values of R_a, R_b, and R_f. Use the ideal model for the op amp.

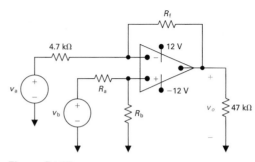

Figure P4.29

❖ **4.30.** Design a difference amplifier (Fig. 4.13) to meet the following criteria: $v_o = 2v_b - 5v_a$. The resistance seen by the signal source v_b is 600 kΩ, and the resistance seen by the signal source v_a is 18 kΩ when the output voltage v_o is zero. Specify the values of R_a, R_b, R_c, and R_d.

4.31. The two op amps in the circuit in Fig. P4.31 are ideal. Calculate v_{o1} and v_{o2}.

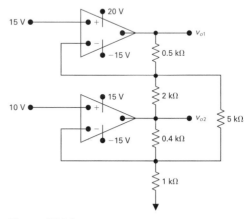

Figure P4.31

4.32. Find v_o and i_o in the circuit shown in Fig. P4.32, assuming the op amps are ideal.

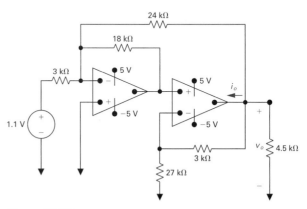

Figure P4.32

4.33. The op amps in the circuit shown in Fig. P4.33 are ideal.

(a) Find v_o as a function of α, σ, v_{g1}, and v_{g2} when the op amps operate within their linear regions.

(b) Describe the behavior of the circuit when $\alpha = \sigma = 1.0$.

(c) Describe the behavior of the circuit when $\alpha = \sigma = 0$.

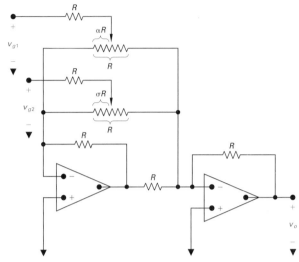

Figure P4.33

4.34. The signal voltage v_g in the circuit shown in Fig. P4.34 is described by the following equations:

$$v_g = 0, \quad t \leq 0,$$

$$v_g = 10\sin(\pi/3)t \text{ V}, \quad 0 \leq t \leq \infty.$$

Sketch v_o versus t, assuming the op amp is ideal.

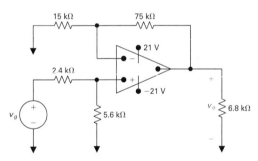

Figure P4.34

4.35. The voltage v_g shown in Fig. P4.35(a) is applied to the inverting amplifier shown in Fig. P4.35(b). Sketch v_o versus t, assuming the op amp is ideal.

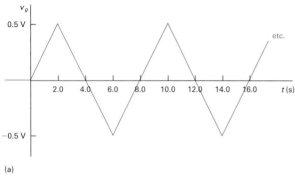

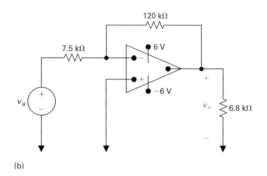

(a) (b)

Figure P4.35

4.36. **(a)** For the comparator circuit in Fig. P4.36 make a sketch of the output voltage v_o vs. the signal voltage v_s when $v_{ref} = 5$ V; $R_1 = 10$ kΩ; and $R_2 = 40$ kΩ.

(b) Repeat part (a) if $v_{ref} = -10$ V.

(c) Compare your plots with those obtained in Example 4.2.

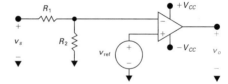

Figure P4.36

4.37. **(a)** Show that the threshold value of v_s in the single-ended comparator circuit shown in Fig. P4.37 is

$$v_s = -\frac{R_1}{R_2} v_{ref}.$$

(b) Assume $v_{ref} = -10$ V; $R_1 = 10$ kΩ; and $R_2 = 20$ kΩ. Sketch v_o vs. v_s for -10 V $\le v_s \le 10$ V.

(c) Repeat (b) if $v_{ref} = 10$ V.

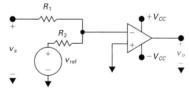

Figure P4.37

◆ **4.38.** Assume the reference voltage in the analog-to-digital structure in Fig. 4.20 is 7 V. Write the thermometer codes for the following values of v_s: 1 V; 3 V; 5 V; and 7 V.

◆ **4.39.** Assume the last resistor in the string of resistors connected to v_{ref} is connected to $-v_{ref}$ instead of ground as shown in Fig. P4.39. Calculate the inverting input reference voltages v_1 through v_7 as a fraction of v_{ref}.

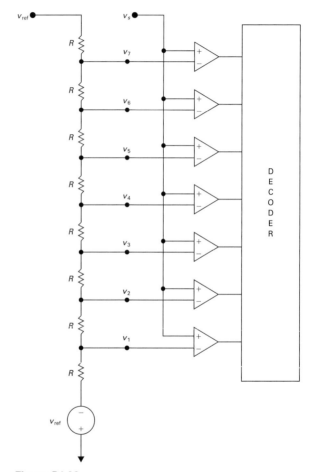

Figure P4.39

◆ **4.40.** Assume in Problem 4.39 that $v_{ref} = 7$ V. Determine the thermometer code for each of the following values of v_s: -7 V; -5 V; -3 V; -1 V; 1 V; 3 V; 5 V; and 7 V.

Dual Slope Analog-to-Digital Converter

In the Practical Perspective introduced in Chapter 4 we showed how an operational amplifier, when used as a comparator, could be employed to convert an analog signal to a digital signal. In this chapter we will show how a capacitor can be combined with an operational amplifier to form an integrating amplifier which in turn can be used to convert an analog signal to a digital signal. The circuit is known as a dual-slope, or dual-ramp, converter and is widely used in digital multimeters (like the one shown in the accompanying figure) to convert an analog signal to a digital readout. We will analyze the dual-slope converter after we have introduced the integrating amplifier circuit in Section 5.6.

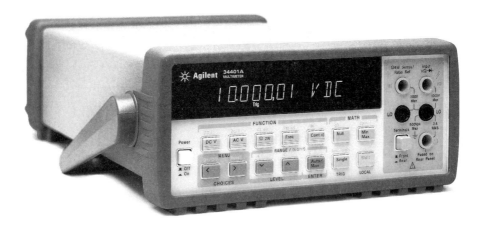

5 The Natural and Step Response of RL and RC Circuits

Chapter Contents

We begin this chapter by introducing the last two ideal circuit elements mentioned in Section 1.3, namely, inductors and capacitors. Be assured that the circuit analysis techniques introduced in Chapters 2 and 3 apply to circuits containing inductors and capacitors. Therefore, once you understand the terminal behavior of these elements in terms of current and voltage, you can use Kirchhoff's laws to describe any interconnections with the other basic elements. Like other components, inductors and capacitors are easier to describe in terms of circuit variables rather than electromagnetic field variables. However, before we focus on the circuit descriptions, a brief review of the field concepts underlying these basic elements is in order.

An inductor is an electrical component that opposes any change in electrical current. It is composed of a coil of wire wound around a supporting core whose material may be magnetic or nonmagnetic. The behavior of inductors is based on phenomena associated with magnetic fields. The source of the magnetic field is charge in motion, or current. If the current is varying with time, the magnetic field is varying with time. A time-varying magnetic field induces a voltage in any conductor linked by the field. The circuit parameter of **inductance** relates the induced voltage to the current. We discuss this quantitative relationship in Section 5.1.

A capacitor is an electrical component that consists of two conductors separated by an insulator or dielectric material. The capacitor is the only device other than a battery that can store electrical charge. The behavior of capacitors is based on phenomena associated with electric fields. The source of the electric field is separation of charge, or voltage. If the voltage is varying with time, the electric field is varying with time. A time-varying electric field produces a displacement current in the space occupied by the field. The circuit parameter of **capacitance** relates the displacement current to the voltage, where the displacement current is equal to the conduction current at the terminals of the capacitor. We discuss this quantitative relationship in Section 5.2. Section 5.3 describes techniques used to simplify circuits with series or parallel combinations of capacitors or inductors.

Energy can be stored in both magnetic and electric fields. Hence you should not be too surprised to learn that inductors and capacitors are capable of storing energy. For example, energy can be stored in an inductor and then released to fire a spark plug. Energy can be stored in a capacitor and then released to fire a flashbulb. In ideal inductors and capacitors, only as much energy can be extracted as has been stored. Because inductors and capacitors cannot generate energy, they are classified as **passive elements**.

5.1 THE INDUCTOR

Inductance is the circuit parameter used to describe an inductor. Inductance is symbolized by the letter L, is measured in henrys (H), and is represented graphically as a coiled wire—a reminder that inductance is a consequence of a conductor linking a magnetic field. Figure 5.1(a) shows an inductor. Assigning the reference direction of the current in the direction of the voltage drop across the terminals of the inductor, as shown in Fig. 5.1(b), yields

$$v = L\frac{di}{dt}, \tag{5.1}$$

where v is measured in volts, L in henrys, i in amperes, and t in seconds. Equation (5.1) reflects the passive sign convention shown in Fig. 5.1(b); that is, the current reference is in the direction of the voltage drop across the inductor. If the current reference is in the direction of the voltage rise, Eq. (5.1) is written with a minus sign.

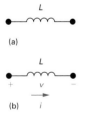

(a)

(b) i

Figure 5.1 (a) The graphic symbol for an inductor with an inductance of L henrys. (b) Assigning reference voltage and current to the inductor, following the passive sign convention.

Note from Eq. (5.1) that the voltage across the terminals of an inductor is proportional to the time rate of change of the current in the inductor. We can make two important observations here. First, if the current is constant, the voltage across the ideal inductor is zero. Thus the inductor behaves as a short circuit in the presence of a constant, or dc, current. Second, current cannot change instantaneously in an inductor; that is, the current cannot change by a finite amount in zero time. Equation (5.1) tells us that this change would require an infinite voltage, and infinite voltages are not possible. For example, when someone opens the switch on an inductive circuit in an actual system, the current initially continues to flow in the air across the switch, a phenomenon called **arcing**. The arc across the switch prevents the current from dropping to zero instantaneously. Switching inductive circuits is an important engineering problem, because arcing and voltage surges must be controlled to prevent equipment damage. The first step to understanding the nature of this problem is to master the introductory material presented in this and the following two chapters. Example 5.1 illustrates the application of Eq. (5.1) to a simple circuit.

EXAMPLE 5.1

The independent current source in the circuit shown in Fig. 5.2 generates zero current for $t < 0$ and a pulse $10te^{-5t}$ A, for $t > 0$.

(a) Sketch the current waveform.

(b) At what instant of time is the current maximum?

(c) Express the voltage across the terminals of the 100 mH inductor as a function of time.

(d) Sketch the voltage waveform.

(e) Are the voltage and the current at a maximum at the same time?

(f) At what instant of time does the voltage change polarity?

(g) Is there ever an instantaneous change in voltage across the inductor? If so, at what time?

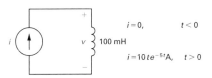

Figure 5.2 The circuit for Example 5.1.

SOLUTION

(a) Figure 5.3 shows the current waveform.

(b) $di/dt = 10(-5te^{-5t} + e^{-5t}) = 10e^{-5t}(1 - 5t)$ A/s; $di/dt = 0$ when $t = \frac{1}{5}$ s. (See Fig. 5.3.)

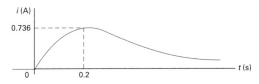

Figure 5.3 The current waveform for Example 5.1.

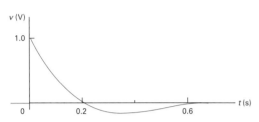

Figure 5.4 The voltage waveform for Example 5.1.

(c) $v = L\,di/dt = (0.1)10e^{-5t}(1 - 5t) = e^{-5t}(1 - 5t)$ V, $t > 0$; $v = 0$, $t < 0$.

(d) Figure 5.4 shows the voltage waveform.

(e) No; the voltage is proportional to di/dt, not i.

(f) At 0.2 s, which corresponds to the moment when di/dt is passing through zero and changing sign.

(g) Yes, at $t = 0$. Note that the voltage can change instantaneously across the terminals of an inductor.

Current in an Inductor in Terms of the Voltage Across the Inductor

Equation (5.1) expresses the voltage across the terminals of an inductor as a function of the current in the inductor. Also desirable is the ability to express the current as a function of the voltage. To find i as a function of v, we start by multiplying both sides of Eq. (5.1) by a differential time dt:

$$v\,dt = L\left(\frac{di}{dt}\right) dt. \tag{5.2}$$

Multiplying the rate at which i varies with t by a differential change in time generates a differential change in i, so we write Eq. (5.2) as

$$v\,dt = L\,di. \tag{5.3}$$

We next integrate both sides of Eq. (5.3). For convenience, we interchange the two sides of the equation and write

$$L \int_{i(t_0)}^{i(t)} dx = \int_{t_0}^{t} v\,d\tau. \tag{5.4}$$

Note that we use x and τ as the variables of integration, whereas i and t become limits on the integrals. Then, from Eq. (5.4),

$$i(t) = \frac{1}{L} \int_{t_0}^{t} v\,d\tau + i(t_0), \tag{5.5}$$

where $i(t)$ is the current corresponding to t, and $i(t_0)$ is the value of the inductor current when we initiate the integration, namely, t_0. In many practical applications, t_0 is zero and Eq. (5.5) becomes

$$i(t) = \frac{1}{L} \int_{0}^{t} v\,d\tau + i(0). \tag{5.6}$$

Equations (5.1) and (5.5) both give the relationship between the voltage and current at the terminals of an inductor. Equation (5.1) expresses the voltage as a function of current, whereas Eq. (5.5) expresses the current as a function of voltage. In both equations the reference direction for the current is in the direction of the voltage drop across the terminals. Note that $i(t_0)$ carries its own algebraic sign. If the initial current is in the same direction as the reference direction for i, it is a positive quantity. If the initial current is in the opposite direction, it is a negative quantity. Example 5.2 illustrates the application of Eq. (5.5).

EXAMPLE 5.2

The voltage pulse applied to the 100 mH inductor shown in Fig. 5.5 is 0 for $t < 0$ and is given by the expression

$$v(t) = 20te^{-10t} \text{ V}$$

for $t > 0$. Also assume $i = 0$ for $t \le 0$.

(a) Sketch the voltage as a function of time.

(b) Find the inductor current as a function of time.

(c) Sketch the current as a function of time.

SOLUTION

(a) The voltage as a function of time is shown in Fig. 5.6.

(b) The current in the inductor is 0 at $t = 0$. Therefore, the current for $t > 0$ is

$$i = \frac{1}{0.1} \int_0^t 20\tau e^{-10\tau} d\tau + 0$$

$$= 200 \left[\frac{-e^{-10\tau}}{100}(10\tau + 1) \right]\Bigg|_0^t$$

$$= 2(1 - 10te^{-10t} - e^{-10t})\text{A}, \qquad t > 0.$$

(c) Figure 5.7 shows the current as a function of time.

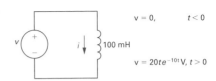

Figure 5.5 The circuit for Example 5.2.

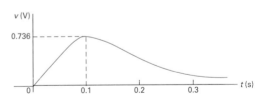

Figure 5.6 The voltage waveform for Example 5.2.

Figure 5.7 The current waveform for Example 5.2.

Note in Example 5.2 that i approaches a constant value of 2 A as t increases. We say more about this result after discussing the energy stored in an inductor.

Power and Energy in the Inductor

The power and energy relationships for an inductor can be derived directly from the current and voltage relationships. If the current reference is in the direction of the voltage drop across the terminals of the inductor, the power is

$$p = vi. \tag{5.7}$$

Remember that power is in watts, voltage is in volts, and current is in amperes. If we express the inductor voltage as a function of the inductor current, Eq. (5.7) becomes

$$p = Li\frac{di}{dt}. \tag{5.8}$$

We can also express the current in terms of the voltage:

$$p = v\left[\frac{1}{L}\int_{t_0}^{t} v\,d\tau + i(t_0)\right]. \tag{5.9}$$

Equation (5.8) is useful in expressing the energy stored in the inductor. Power is the time rate of expending energy, so

$$p = \frac{dw}{dt} = Li\frac{di}{dt}. \tag{5.10}$$

Multiplying both sides of Eq. (5.10) by a differential time gives the differential relationship

$$dw = Li\,di. \tag{5.11}$$

Both sides of Eq. (5.11) are integrated with the understanding that the reference for zero energy corresponds to zero current in the inductor. Thus

$$\int_{0}^{w} dx = L\int_{0}^{i} y\,dy,$$

$$w = \frac{1}{2}Li^2. \tag{5.12}$$

As before, we use different symbols of integration to avoid confusion with the limits placed on the integrals. In Eq. (5.12), the energy is in joules, inductance is in henrys, and current is in amperes. To illustrate the application of Eqs. (5.7) and (5.12), we return to Examples 5.1 and 5.2 by means of Example 5.3.

EXAMPLE 5.3

(a) For Example 5.1, plot i, v, p, and w versus time. Line up the plots vertically to allow easy assessment of each variable's behavior.

(b) In what time interval is energy being stored in the inductor?

(c) In what time interval is energy being extracted from the inductor?

(d) What is the maximum energy stored in the inductor?

(e) Evaluate the integrals

$$\int_0^{0.2} p \, dt \quad \text{and} \quad \int_{0.2}^{\infty} p \, dt$$

and comment on their significance.

(f) Repeat (a)–(c) for Example 5.2.

(g) In Example 5.2, why is there a sustained current in the inductor as the voltage approaches zero?

SOLUTION

(a) The plots of i, v, p, and w follow directly from the expressions for i and v obtained in Example 5.1 and are shown in Fig. 5.8. In particular, $p = vi$, and $w = (\frac{1}{2})Li^2$.

(b) An increasing energy curve indicates that energy is being stored. Thus energy is being stored in the time interval 0– 0.2 s. Note that this corresponds to the interval when $p > 0$.

(c) A decreasing energy curve indicates that energy is being extracted. Thus energy is being extracted in the time interval 0.2 s–∞. Note that this corresponds to the interval when $p < 0$.

(d) From Eq. (5.12) we see that energy is at a maximum when current is at a maximum; glancing at the graphs confirms this. From Example 5.1, maximum current = 0.736 A. Therefore, $\omega_{\text{max}} = 27.07$ mJ.

(e) From Example 5.1,

$$i = 10te^{-5t} \text{ A} \quad \text{and} \quad v = e^{-5t}(1 - 5t) \text{ V}.$$

Therefore

$$p = vi = 10te^{-10t} - 50t^2 e^{-10t} \text{ W}.$$

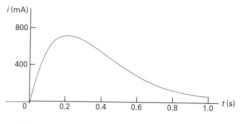

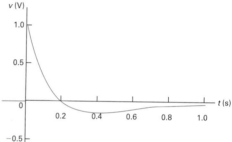

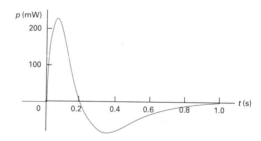

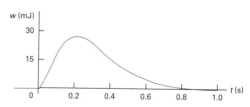

Figure 5.8 The variables i, v, p, and w versus t for Example 5.1.

Thus

$$\int_0^{0.2} p\,dt = 10\left[\frac{e^{-10t}}{100}(-10t-1)\right]_0^{0.2}$$

$$-50\left\{\frac{t^2 e^{-10t}}{-10} + \frac{2}{10}\left[\frac{e^{-10t}}{100}(-10t-1)\right]\right\}_0^{0.2}$$

$$= 0.2e^{-2} = 27.07 \text{ mJ},$$

$$\int_{0.2}^{\infty} p\,dt = 10\left[\frac{e^{-10t}}{100}(-10t-1)\right]_{0.2}^{\infty}$$

$$-50\left\{\frac{t^2 e^{-10t}}{-10} + \frac{2}{10}\left[\frac{e^{-10t}}{100}(-10t-1)\right]\right\}_{0.2}^{\infty}$$

$$= -0.2e^{-2} = -27.07 \text{ mJ}.$$

Based on the definition of p, the area under the plot of p versus t represents the energy expended over the interval of integration. Hence the integration of the power between 0 and 0.2 s represents the energy stored in the inductor during this time interval. The integral of p over the interval 0.2 s–∞ is the energy extracted. Note that in this time interval, all the energy originally stored is removed; that is, after the current peak has passed, no energy is stored in the inductor.

(f) The plots of v, i, p, and w follow directly from the expressions for v and i given in Example 5.2 and are shown in Fig. 5.9. Note that in this case the power is always positive, and hence energy is always being stored during the voltage pulse.

(g) The application of the voltage pulse stores energy in the inductor. Because the inductor is ideal, this energy cannot dissipate after the voltage subsides to zero. Therefore, a sustained current circulates in the circuit. A lossless inductor obviously is an ideal circuit element. Practical inductors require a resistor in the circuit model. (More about this later.)

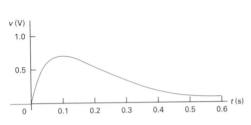

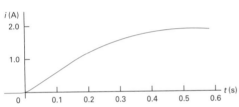

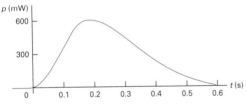

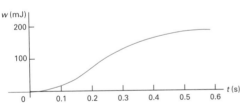

Figure 5.9 The variables v, i, p, and w versus t for Example 5.2.

DRILL EXERCISE

5.1   The current source in the circuit shown generates the current pulse

$$i_g(t) = 0, \qquad t < 0,$$

$$i_g(t) = 8e^{-300t} - 8e^{-1200t} \text{ A}, \qquad t \geq 0.$$

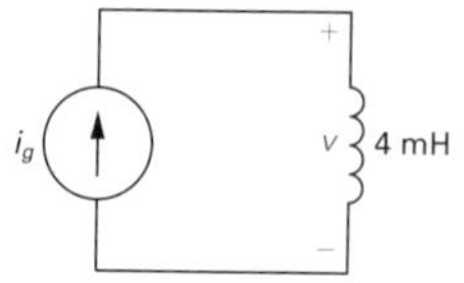

Find (a) $v(0)$; (b) the instant of time, greater than zero, when the voltage v passes through zero; (c) the expression for the power delivered to the inductor; (d) the instant when the power delivered to the inductor is maximum; (e) the maximum power; (f) the maximum energy stored in the inductor; and (g) the instant of time when the stored energy is maximum.

ANSWER: (a) 28.8 V; (b) 1.54 ms;
(c) $384e^{-1500t} - 76.8e^{-600t} - 307.2e^{-2400t}$ W;
(d) 411.05 μs; (e) 32.72 W; (f) 28.57 mJ;
(g) 1.54 ms.

5.2 THE CAPACITOR

The circuit parameter of capacitance is represented by the letter C, is measured in farads (F), and is symbolized graphically by two short parallel conductive plates, as shown in Fig. 5.10(a). Because the farad is an extremely large quantity of capacitance, practical capacitor values usually lie in the picofarad (pF) to microfarad (μF) range.

The graphic symbol for a capacitor is a reminder that capacitance occurs whenever electrical conductors are separated by a dielectric, or insulating, material. This condition implies that electric charge is not transported through the capacitor. Although applying a voltage to the terminals of the capacitor cannot move a charge through the dielectric, it can displace a charge within the dielectric. As the voltage varies with time, the displacement of charge also varies with time, causing what is known as the **displacement current**.

(a)

(b)

Figure 5.10 (a) The circuit symbol for a capacitor. (b) Assigning reference voltage and current to the capacitor, following the passive sign convention.

At the terminals, the displacement current is indistinguishable from a conduction current. The current is proportional to the rate at which the voltage across the capacitor varies with time, or, mathematically,

$$i = C\frac{dv}{dt},\qquad\qquad (5.13)$$

where i is measured in amperes, C in farads, v in volts, and t in seconds.

Equation (5.13) reflects the passive sign convention shown in Fig. 5.10(b); that is, the current reference is in the direction of the voltage drop across the capacitor. If the current reference is in the direction of the voltage rise, Eq. (5.13) is written with a minus sign.

Two important observations follow from Eq. (5.13). First, voltage cannot change instantaneously across the terminals of a capacitor. Equation (5.13) indicates that such a change would produce infinite current, a physical impossibility. Second, if the voltage across the terminals is constant, the capacitor current is zero. The reason is that a conduction current cannot be established in the dielectric material of the capacitor. Only a time-varying voltage can produce a displacement current. Thus a capacitor behaves as an open circuit in the presence of a constant voltage.

Equation (5.13) gives the capacitor current as a function of the capacitor voltage. Expressing the voltage as a function of the current is also useful. To do so, we multiply both sides of Eq. (5.13) by a differential time dt and then integrate the resulting differentials:

$$i\,dt = C\,dv \qquad \text{or} \qquad \int_{v(t_0)}^{v(t)} dx = \frac{1}{C}\int_{t_0}^{t} i\,d\tau.$$

Carrying out the integration of the left-hand side of the second equation gives

$$v(t) = \frac{1}{C}\int_{t_0}^{t} i\,d\tau + v(t_0).\qquad\qquad (5.14)$$

In many practical applications of Eq. (5.14), the initial time is zero; that is, $t_0 = 0$. Thus Eq. (5.14) becomes

$$v(t) = \frac{1}{C}\int_{0}^{t} i\,d\tau + v(0).\qquad\qquad (5.15)$$

We can easily derive the power and energy relationships for the capacitor. From the definition of power,

$$p = vi = Cv\frac{dv}{dt},$$ (5.16)

or

$$p = i\left[\frac{1}{C}\int_{t_0}^{t} i\,d\tau + v(t_0)\right].$$ (5.17)

Combining the definition of energy with Eq. (5.16) yields

$$dw = Cv\,dv,$$

from which

$$\int_0^w dx = C\int_0^v y\,dy,$$

or

$$w = \frac{1}{2}Cv^2.$$ (5.18)

In the derivation of Eq. (5.18), the reference for zero energy corresponds to zero voltage.

Examples 5.4 and 5.5 illustrate the application of the current, voltage, power, and energy relationships for a capacitor.

EXAMPLE 5.4

The voltage pulse described by the following equations is impressed across the terminals of a 0.5 μF capacitor:

$$v(t) = \begin{cases} 0, & t \le 0; \\ 4t \text{ V}, & 0 \le t \le 1; \\ 4e^{-(t-1)} \text{ V}, & 1 \le t \le \infty. \end{cases}$$

(a) Derive the expressions for the capacitor current, power, and energy.

(b) Sketch the voltage, current, power, and energy as functions of time. Line up the plots vertically.

(c) Specify the interval of time when energy is being stored in the capacitor.

(d) Specify the interval of time when energy is being delivered by the capacitor.

(e) Evaluate the integrals

$$\int_0^1 p\, dt \qquad \text{and} \qquad \int_1^\infty p\, dt$$

and comment on their significance.

SOLUTION

(a) From Eq. (5.13),

$$i = \begin{cases} (0.5 \times 10^{-6})(0) = 0, & t < 0; \\ (0.5 \times 10^{-6})(4) = 2\,\mu\text{A}, & 1 < t < 1; \\ (0.5 \times 10^{-6})(-4e^{-(t-1)}) = -2e^{-(t-1)}\,\mu\text{A}, & 1 < t < \infty. \end{cases}$$

The expression for the power is derived from Eq. (5.16):

$$p = \begin{cases} 0, & t < 0; \\ (4t)(2) = 8t\,\mu\text{W}, & 0 \le t < 1; \\ (4e^{-(t-1)})(-2e^{-(t-1)}) = -8e^{-2(t-1)}\,\mu\text{W}, & 1 < t \le \infty. \end{cases}$$

The energy expression follows directly from Eq. (5.18):

$$w = \begin{cases} 0 & t < 0; \\ \frac{1}{2}(0.5)16t^2 = 4t^2\,\mu\text{J}, & 0 \le t < 1; \\ \frac{1}{2}(0.5)16e^{-2(t-1)} = 4e^{-2(t-1)}\,\mu\text{J}, & 1 \le t \le \infty. \end{cases}$$

(b) Figure 5.11 shows the voltage, current, power, and energy as functions of time.

(c) Energy is being stored in the capacitor whenever the power is positive. Hence energy is being stored in the interval 0–1 s.

(d) Energy is being delivered by the capacitor whenever the power is negative. Thus energy is being delivered for all t greater than 1 s.

(e) The integral of $p\, dt$ is the energy associated with the time interval corresponding to the limits on the integral. Thus the first integral represents the energy stored in the capacitor between 0 and 1 s, whereas the second integral represents

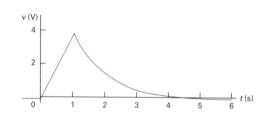

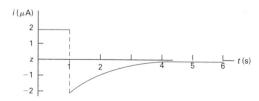

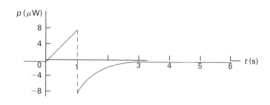

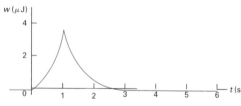

Figure 5.11 The variables v, i, p, and w versus t for Example 5.4.

the energy returned, or delivered, by the capacitor in the interval 1 s to ∞:

$$\int_0^1 p\,dt = \int_0^1 8t\,dt = 4t^2\Big|_0^1 = 4\,\mu\text{J},$$

$$\int_1^\infty p\,dt = \int_1^\infty (-8e^{-2(t-1)})\,dt = (-8)\frac{e^{-2(t-1)}}{-2}\Big|_1^\infty = -4\,\mu\text{J}.$$

The voltage applied to the capacitor returns to zero as time increases without limit, so the energy returned by this ideal capacitor must equal the energy stored.

EXAMPLE 5.5

An uncharged $0.2\ \mu\text{F}$ capacitor is driven by a triangular current pulse. The current pulse is described by

$$i(t) = \begin{cases} 0, & t \le 0; \\ 5000t\ \text{A}, & 0 \le t \le 20\ \mu\text{s}; \\ 0.2 - 5000t\ \text{A}, & 20 \le t \le 40\ \mu\text{s}; \\ 0, & t \ge 40\ \mu\text{s}. \end{cases}$$

(a) Derive the expressions for the capacitor voltage, power, and energy for each of the four time intervals needed to describe the current.

(b) Plot i, v, p, and w versus t. Align the plots as specified in the previous examples.

(c) Why does a voltage remain on the capacitor after the current returns to zero?

SOLUTION

(a) For $t \le 0$, v, p, and w all are zero.
    For $0 \le t \le 20\ \mu\text{s}$,

$$v = 5 \times 10^6 \int_0^t 5000\tau\,d\tau + 0 = 12.5 \times 10^9 t^2\ \text{V},$$

$$p = vi = 62.5 \times 10^{12} t^3\ \text{W},$$

$$w = \frac{1}{2}Cv^2 = 15.625 \times 10^{12} t^4\ \text{J}.$$

For $20\,\mu s \le t \le 40\,\mu s$,

$$v = 5 \times 10^6 \int_{20\mu s}^{t} (0.2 - 5000\tau)\,d\tau + 5.$$

(Note that 5 V is the voltage on the capacitor at the end of the preceding interval.) Then,

$$v = (10^6 t - 12.5 \times 10^9 t^2 - 10)\ \text{V},$$

$$p = vi,$$

$$= (62.5 \times 10^{12} t^3 - 7.5 \times 10^9 t^2 + 2.5 \times 10^5 t - 2)\ \text{W},$$

$$w = \frac{1}{2}Cv^2,$$

$$= (15.625 \times 10^{12} t^4 - 2.5 \times 10^9 t^3 + 0.125 \times 10^6 t^2 - 2t + 10^{-5})\ \text{J}.$$

For $t \ge 40\,\mu s$,

$$v = 10\ \text{V},$$

$$p = vi = 0,$$

$$w = \frac{1}{2}Cv^2 = 10\,\mu\text{J}.$$

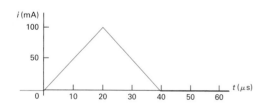

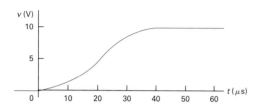

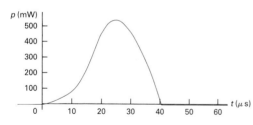

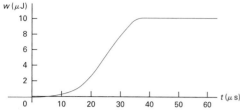

Figure 5.12 The variables $i, v, p,$ and w versus t for Example 5.5.

(b) The excitation current and the resulting voltage, power, and energy are plotted in Fig. 5.12.

(c) Note that the power is always positive for the duration of the current pulse, which means that energy is continuously being stored in the capacitor. When the current returns to zero, the stored energy is trapped because the ideal capacitor offers no means for dissipating energy. Thus a voltage remains on the capacitor after i returns to zero.

DRILL EXERCISES

5.2   The voltage at the terminals of the 0.6 μF capacitor shown in the figure is 0 for $t < 0$ and
$40e^{-15,000t} \sin 30,000t$ V for $t \geq 0$. Find (a) $i(0)$;
(b) the power delivered to the capacitor at
$t = \pi/80$ ms; and (c) the energy stored in the capacitor at $t = \pi/80$ ms.

0.6 μF

ANSWER: (a) 0.72 A; (b) −649.2 mW;
(c) 126.13 μJ.

5.3   The current in the capacitor of Drill Exercise 5.2
is 0 for $t < 0$ and $3 \cos 50,000t$ A for $t \geq 0$. Find
(a) $v(t)$; (b) the maximum power delivered to the
capacitor at any one instant of time; and (c) the
maximum energy stored in the capacitor at any one
instant of time.

ANSWER: (a) $100 \sin 50,000t$ V; (b) 150 W;
(c) 3 mJ.

5.3 SERIES-PARALLEL COMBINATIONS OF INDUCTANCE AND CAPACITANCE

Just as series-parallel combinations of resistors can be reduced
to a single equivalent resistor, series-parallel combinations of inductors or capacitors can be reduced to a single inductor or capacitor. Figure 5.13 shows inductors in series. Here, the inductors
are forced to carry the same current; thus we define only one
current for the series combination. The voltage drops across the
individual inductors are

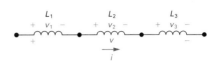

Figure 5.13 Inductors in series.

$$v_1 = L_1 \frac{di}{dt}, \qquad v_2 = L_2 \frac{di}{dt}, \qquad \text{and} \qquad v_3 = L_3 \frac{di}{dt}.$$

The voltage across the series connection is

$$v = v_1 + v_2 + v_3 = (L_1 + L_2 + L_3) \frac{di}{dt},$$

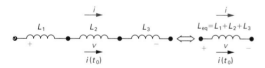

Figure 5.14 An equivalent circuit for inductors in series carrying an initial current $i(t_0)$.

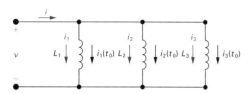

Figure 5.15 Three inductors in parallel.

from which it should be apparent that the equivalent inductance of series-connected inductors is the sum of the individual inductances. For n inductors in series,

$$L_{\text{eq}} = L_1 + L_2 + L_3 + \cdots + L_n. \tag{5.19}$$

If the original inductors carry an initial current of $i(t_0)$, the equivalent inductor carries the same initial current. Figure 5.14 shows the equivalent circuit for series inductors carrying an initial current.

Inductors in parallel have the same terminal voltage. In the equivalent circuit, the current in each inductor is a function of the terminal voltage and the initial current in the inductor. For the three inductors in parallel shown in Fig. 5.15, the currents for the individual inductors are

$$i_1 = \frac{1}{L_1} \int_{t_0}^{t} v \, d\tau + i_1(t_0),$$

$$i_2 = \frac{1}{L_2} \int_{t_0}^{t} v \, d\tau + i_2(t_0),$$

$$i_3 = \frac{1}{L_3} \int_{t_0}^{t} v \, d\tau + i_3(t_0). \tag{5.20}$$

The current at the terminals of the three parallel inductors is the sum of the inductor currents:

$$i = i_1 + i_2 + i_3. \tag{5.21}$$

Substituting Eq. (5.20) into Eq. (5.21) yields

$$i = \left(\frac{1}{L_1} + \frac{1}{L_2} + \frac{1}{L_3} \right) \int_{t_0}^{t} v \, d\tau + i_1(t_0) + i_2(t_0) + i_3(t_0). \tag{5.22}$$

Now we can interpret Eq. (5.22) in terms of a single inductor; that is,

$$i = \frac{1}{L_{\text{eq}}} \int_{t_0}^{t} v \, d\tau + i(t_0). \tag{5.23}$$

Comparing Eq. (5.23) with (5.22) yields

$$\frac{1}{L_{\text{eq}}} = \frac{1}{L_1} + \frac{1}{L_2} + \frac{1}{L_3} \tag{5.24}$$

$$i(t_0) = i_1(t_0) + i_2(t_0) + i_3(t_0). \tag{5.25}$$

Figure 5.16 shows the equivalent circuit for the three parallel inductors in Fig. 5.15.

The results expressed in Eqs. (5.24) and (5.25) can be extended to n inductors in parallel:

$$\frac{1}{L_{eq}} = \frac{1}{L_1} + \frac{1}{L_2} + \cdots + \frac{1}{L_n} \qquad (5.26)$$

$$i(t_0) = i_1(t_0) + i_2(t_0) + \cdots + i_n(t_0). \qquad (5.27)$$

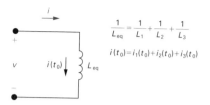

Figure 5.16 An equivalent circuit for three inductors in parallel.

Capacitors connected in series can be reduced to a single equivalent capacitor. The reciprocal of the equivalent capacitance is equal to the sum of the reciprocals of the individual capacitances. If each capacitor carries its own initial voltage, the initial voltage on the equivalent capacitor is the algebraic sum of the initial voltages on the individual capacitors. Figure 5.17 and the following equations summarize these observations:

$$\frac{1}{C_{eq}} = \frac{1}{C_1} + \frac{1}{C_2} + \cdots + \frac{1}{C_n}, \qquad (5.28)$$

$$v(t_0) = v_1(t_0) + v_2(t_0) + \cdots + v_n(t_0). \qquad (5.29)$$

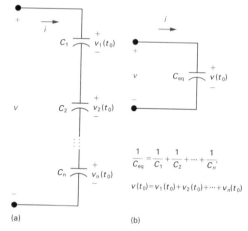

Figure 5.17 An equivalent circuit for capacitors connected in series. (a) The series capacitors. (b) The equivalent circuit.

We leave the derivation of the equivalent circuit for series-connected capacitors as an exercise. (See Problem 5.21.)

The equivalent capacitance of capacitors connected in parallel is simply the sum of the capacitances of the individual capacitors, as Fig. 5.18 and the following equation show:

$$C_{eq} = C_1 + C_2 + \cdots + C_n. \qquad (5.30)$$

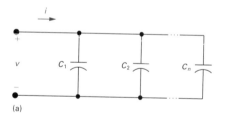

Capacitors connected in parallel must carry the same voltage. Therefore, if there is an initial voltage across the original parallel capacitors, this same initial voltage appears across the equivalent capacitance C_{eq}. The derivation of the equivalent circuit for parallel capacitors is left as an exercise. (See Problem 5.22.)

We say more about series-parallel equivalent circuits of inductors and capacitors later in the chapter, where we interpret results based on their use.

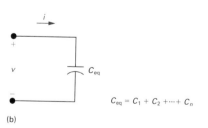

Figure 5.18 An equivalent circuit for capacitors connected in parallel. (a) Capacitors in parallel. (b) The equivalent circuit.

DRILL EXERCISES

5.4 The initial values of i_1 and i_2 in the circuit shown are +3 and −5 A, respectively. The voltage at the terminals of the parallel inductors for $t \geq 0$ is $-30e^{-5t}$ mV.

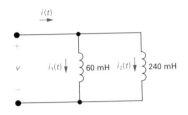

    (a) If the parallel inductors are replaced by a single inductor, what is its inductance?

    (b) What is the initial current and its reference direction in the equivalent inductor?

    (c) Use the equivalent inductor to find $i(t)$.

    (d) Find $i_1(t)$ and $i_2(t)$. Verify that the solutions for $i_1(t)$, $i_2(t)$, and $i(t)$ satisfy Kirchhoff's current law.

ANSWER: (a) 48 mH; (b) 2 A, up;
(c) $0.125e^{-5t} - 2.125$ A; (d) $i_1(t) = 0.1e^{-5t} + 2.9$ A,
$i_2(t) = 0.025e^{-5t} - 5.025$ A.

5.5 The current at the terminals of the two capacitors shown is $240e^{-10t}\ \mu$A for $t \geq 0$. The initial values of v_1 and v_2 are −10 and −5 V, respectively. Calculate the total energy trapped in the capacitors as $t \to \infty$. (*Hint:* Don't combine the capacitors in series—find the energy trapped in each, and then add.)

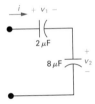

ANSWER: $20\ \mu$J.

5.4 NATURAL RESPONSE OF RL AND RC CIRCUITS

In our introduction to this chapter, we noted that an important attribute of inductors and capacitors is their ability to store energy. We are now in a position to determine the currents and voltages that arise when energy is either released or acquired by an inductor or capacitor in response to an abrupt change in a dc voltage or current source. We will focus on circuits that consist only of sources, resistors, and either (but not both) inductors or capacitors. For brevity, such configurations are called **RL** (resistor-inductor) **circuits** and **RC** (resistor-capacitor).

Our analysis of RL and RC circuits will be divided into three phases. In the first phase, we consider the currents and voltages that arise when stored energy in an inductor or capacitor is suddenly released to a resistive network. This happens when the inductor or capacitor is abruptly disconnected from its dc source. Thus we can reduce the circuit to one of the two equivalent forms shown in Fig. 5.19. The currents and voltages that arise in this configuration are referred to as the **natural response** of the circuit, to emphasize that the nature of the circuit itself, not external sources of excitation, determines its behavior.

In the second phase of our analysis, we consider the currents and voltages that arise when energy is being acquired by an inductor or capacitor due to the sudden application of a dc voltage or current source. This response is referred to as the **step response**. The process for finding both the natural and step responses is the same; thus, in the third phase of our analysis, we develop a general method that can be used to find the response of RL and RC circuits to any abrupt change in a dc voltage or current source.

Figure 5.20 shows the four possibilities for the general configuration of RL and RC circuits. Note that when there are no independent sources in the circuit, the Thévenin voltage or Norton current is zero, and the circuit reduces to one of those shown in Fig. 5.19; that is, we have a natural-response problem.

RL and RC circuits are also known as **first-order circuits**, because their voltages and currents are described by first-order differential equations. No matter how complex a circuit may appear, if it can be reduced to a Thévenin or Norton equivalent connected to the terminals of an equivalent inductor or capacitor, it is a first-order circuit. (Note that if multiple inductors or capacitors exist in the original circuit, they must be interconnected so that they can be replaced by a single equivalent element.)

The natural response of an RL circuit can best be described in terms of the circuit shown in Fig. 5.21. We assume that the independent current source generates a constant current of I_s A, and that the switch has been in a closed position for a long time. We define the phrase *a long time* more accurately later in this section. For now it means that all currents and voltages have reached a constant value. Thus only constant, or dc, currents can exist in the circuit just prior to the switch's being opened, and therefore the inductor appears as a short circuit ($L\,di/dt = 0$) prior to the release of the stored energy.

Because the inductor appears as a short circuit, the voltage across the inductive branch is zero, and there can be no current

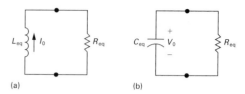

Figure 5.19 The two forms of the circuits for natural response. (a) RL circuit. (b) RC circuit.

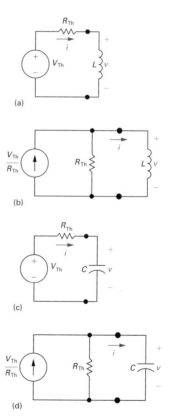

Figure 5.20 Four possible first-order circuits. (a) An inductor connected to a Thévenin equivalent. (b) An inductor connected to a Norton equivalent. (c) A capacitor connected to a Thévenin equivalent. (d) A capacitor connected to a Norton equivalent.

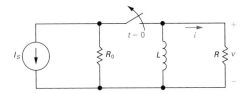

Figure 5.21 An *RL* circuit.

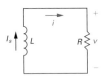

Figure 5.22 The circuit shown in Fig. 5.21, for $t \geq 0$.

in either R_o or R. Therefore, all the source current I_s appears in the inductive branch. Finding the natural response requires finding the voltage and current at the terminals of the resistor after the switch has been opened, that is, after the source has been disconnected and the inductor begins releasing energy. If we let $t = 0$ denote the instant when the switch is opened, the problem becomes one of finding $v(t)$ and $i(t)$ for $t \geq 0$. For $t \geq 0$, the circuit shown in Fig. 5.21 reduces to the one shown in Fig. 5.22.

Deriving the Expression for the Current

To find $i(t)$, we use Kirchhoff's voltage law to obtain an expression involving i, R, and L. Summing the voltages around the closed loop gives

$$L \frac{di}{dt} + Ri = 0, \tag{5.31}$$

where we use the passive sign convention. Equation (5.31) is known as a first-order ordinary differential equation, because it contains terms involving the ordinary derivative of the unknown, that is, di/dt. The highest order derivative appearing in the equation is 1; hence the term **first-order**.

We can go one step further in describing this equation. The coefficients in the equation, R and L, are constants; that is, they are not functions of either the dependent variable i or the independent variable t. Thus the equation can also be described as an ordinary differential equation with constant coefficients.

To solve Eq. (5.31), we divide by L, transpose the term involving i to the right-hand side, and then multiply both sides by a differential time dt. The result is

$$\frac{di}{dt} dt = -\frac{R}{L} i \, dt. \tag{5.32}$$

Next, we recognize the left-hand side of Eq. (5.32) as a differential change in the current i, that is, di. We now divide through by i, getting

$$\frac{di}{i} = -\frac{R}{L} dt. \tag{5.33}$$

We obtain an explicit expression for i as a function of t by integrating both sides of Eq. (5.33). Using x and y as variables of integration yields

$$\int_{i(t_0)}^{i(t)} \frac{dx}{x} = -\frac{R}{L} \int_{t_0}^{t} dy, \tag{5.34}$$

in which $i(t_0)$ is the current corresponding to time t_0, and $i(t)$ is the current corresponding to time t. Here, $t_0 = 0$. Therefore, carrying out the indicated integration gives

$$\ln \frac{i(t)}{i(0)} = -\frac{R}{L}t. \qquad (5.35)$$

Based on the definition of the natural logarithm,

$$i(t) = i(0)e^{-(R/L)t}. \qquad (5.36)$$

Remember that an instantaneous change of current cannot occur in an inductor. Therefore, in the first instant after the switch has been opened, the current in the inductor remains unchanged. If we use 0^- to denote the time just prior to switching, and 0^+ for the time immediately following switching, then

$$i(0^-) = i(0^+) = I_0,$$

where, as in Fig. 5.19, I_0 denotes the initial current in the inductor. The initial current in the inductor is oriented in the same direction as the reference direction of i. Hence Eq. (5.36) becomes

$$i(t) = I_0 e^{-(R/L)t}, \quad t \geq 0, \qquad (5.37)$$

which shows that the current starts from an initial value I_0 and decreases exponentially toward zero as t increases. Figure 5.23 shows this response.

We derive the voltage across the resistor in Fig. 5.22 from a direct application of Ohm's law:

$$v = iR = I_0 R e^{-(R/L)t}, \quad t \geq 0^+. \qquad (5.38)$$

Note that in contrast to the expression for the current shown in Eq. (5.37), the voltage is defined only for $t > 0$, not at $t = 0$. The reason is that a step change occurs in the voltage at zero. Note that for $t < 0$, the derivative of the current is zero, so the voltage is also zero. (This result follows from $v = L\,di/dt = 0$.) Thus

$$v(0^-) = 0, \qquad (5.39)$$

$$v(0^+) = I_0 R, \qquad (5.40)$$

where $v(0^+)$ is obtained from Eq. (5.38) with $t = 0^+$.[1] With this step change at an instant in time, the value of the voltage at $t = 0$

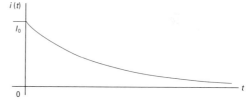

Figure 5.23 The current response for the circuit shown in Fig. 5.22.

---

[1] We can define the expressions 0^- and 0^+ more formally. The expression $x(0^-)$ refers to the limit of the variable x as $t \to 0$ from the left, or from negative time. The expression $x(0^+)$ refers to the limit of the variable x as $t \to 0$ from the right, or from positive time.

is unknown. Thus we use $t \geq 0^+$ in defining the region of validity for these solutions.

We derive the power dissipated in the resistor from any of the following expressions:

$$p = vi, \quad p = i^2 R, \quad \text{or} \quad p = \frac{v^2}{R}. \tag{5.41}$$

Whichever form is used, the resulting expression can be reduced to

$$p = I_0^2 R e^{-2(R/L)t}, \quad t \geq 0^+. \tag{5.42}$$

The energy delivered to the resistor during any interval of time after the switch has been opened is

$$w = \int_0^t p \, dx = \int_0^t I_0^2 R e^{-2(R/L)x} \, dx$$

$$= \frac{1}{2(R/L)} I_0^2 R (1 - e^{-2(R/L)t})$$

$$= \frac{1}{2} L I_0^2 (1 - e^{-2(R/L)t}), \quad t \geq 0. \tag{5.43}$$

Note from Eq. (5.43) that as t becomes infinite, the energy dissipated in the resistor approaches the initial energy stored in the inductor.

The Significance of the Time Constant

The expressions for $i(t)$ [Eq. (5.37)] and $v(t)$ [Eq. (5.38)] include a term of the form $e^{-(R/L)t}$. The coefficient of t—namely, R/L—determines the rate at which the current or voltage approaches zero. The reciprocal of this ratio is the **time constant** of the circuit, denoted

$$\tau = \text{time constant} = \frac{L}{R}. \tag{5.44}$$

Using the time-constant concept, we write the expressions for current, voltage, power, and energy as

$$i(t) = I_0 e^{-t/\tau}, \quad t \geq 0, \tag{5.45}$$

$$v(t) = I_0 R e^{-t/\tau}, \quad t \geq 0^+, \tag{5.46}$$

$$p = I_0^2 R e^{-2t/\tau}, \quad t \geq 0^+, \tag{5.47}$$

$$w = \frac{1}{2} L I_0^2 (1 - e^{-2t/\tau}), \quad t \geq 0. \tag{5.48}$$

TABLE 5.1   Value of $e^{-t/\tau}$ for t equal to integral multiples of τ

t	$e^{-t/\tau}$	t	$e^{-t/\tau}$
τ	3.6788×10^{-1}	6τ	2.4788×10^{-3}
2τ	1.3534×10^{-1}	7τ	9.1188×10^{-4}
3τ	4.9787×10^{-2}	8τ	3.3546×10^{-4}
4τ	1.8316×10^{-2}	9τ	1.2341×10^{-4}
5τ	6.7379×10^{-3}	10τ	4.5400×10^{-5}

The time constant is an important parameter for first-order circuits, so mentioning several of its characteristics is worthwhile. First, it is convenient to think of the time elapsed after switching in terms of integral multiples of τ. Thus one time constant after the inductor has begun to release its stored energy to the resistor, the current has been reduced to e^{-1}, or approximately 0.37 of its initial value.

Table 5.1 gives the value of $e^{-t/\tau}$ for integral multiples of τ from 1 to 10. Note that when the elapsed time exceeds five time constants, the current is less than 1% of its initial value. Thus we sometimes say that five time constants after switching has occurred, the currents and voltages have, for most practical purposes, reached their final values. For single time-constant circuits (first-order circuits) with 1% accuracy, the phrase *a long time* implies that five or more time constants have elapsed. Thus the existence of current in the *RL* circuit shown in Fig. 5.19(a) is a momentary event and is referred to as the **transient response** of the circuit. The response that exists a long time after the switching has taken place is called the **steady-state response**. The phrase *a long time* then also means the time it takes the circuit to reach its steady-state value.

Any first-order circuit is characterized, in part, by the value of its time constant. If we have no method for calculating the time constant of such a circuit (perhaps because we don't know the values of its components), we can determine its value from a plot of the circuit's natural response. That's because another important characteristic of the time constant is that it gives the time required for the current to reach its final value if the current continues to change at its initial rate. To illustrate, we evaluate di/dt at 0^+ and assume that the current continues to change at this rate:

$$\frac{di}{dt}(0^+) = -\frac{R}{L}I_0 = -\frac{I_0}{\tau}. \tag{5.49}$$

Now, if i starts as I_0 and decreases at a constant rate of I_0/τ amperes per second, the expression for i becomes

$$i = I_0 - \frac{I_0}{\tau}t. \qquad (5.50)$$

Equation (5.50) indicates that i would reach its final value of zero in τ seconds. Figure 5.24 shows how this graphic interpretation is useful in estimating the time constant of a circuit from a plot of its natural response. Such a plot could be generated on an oscilloscope measuring output current. Drawing the tangent to the natural response plot at $t = 0$ and reading the value at which the tangent intersects the time axis gives the value of τ.

Calculating the natural response of an RL circuit can be summarized as follows:

1. Find the initial current, $i(0)$, through the inductor.

2. Find the time constant of the circuit.

3. Use Eq. (5.45) to generate $i(t)$ from $i(0)$ and τ.

All other calculations of interest follow from knowing $i(t)$. Examples 5.6 and 5.7 illustrate the numerical calculations associated with the natural response of an RL circuit.

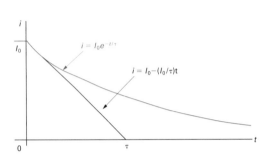

Figure 5.24 A graphic interpretation of the time constant of the RL circuit shown in Fig. 5.22.

EXAMPLE 5.6

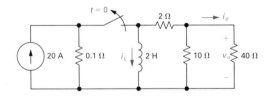

Figure 5.25 The circuit for Example 5.6.

The switch in the circuit shown in Fig. 5.25 has been closed for a long time before it is opened at $t = 0$. Find

(a) $i_L(t)$ for $t \geq 0$

(b) $i_o(t)$ for $t \geq 0^+$

(c) $v_o(t)$ for $t \geq 0^+$

(d) the percentage of the total energy stored in the 2 H inductor that is dissipated in the 10 Ω resistor

SOLUTION

(a) The switch has been closed for a long time prior to $t = 0$, so we know the voltage across the inductor must be zero at $t = 0^-$. Therefore the initial current in the inductor is 20 A at $t = 0^-$. Hence, $i_L(0^+)$ also is 20 A, because an instantaneous change in the current cannot occur in an inductor.

We replace the resistive circuit connected to the terminals of the inductor with a single resistor of 10 Ω:

$$R_{eq} = 2 + (40 \parallel 10) = 10\,\Omega.$$

The time constant of the circuit is L/R_{eq}, or 0.2 s, giving the expression for the inductor current as

$$i_L(t) = 20e^{-5t}\ \text{A}, \quad t \geq 0.$$

(b) We find the current in the 40 Ω resistor most easily by using current division; that is,

$$i_o = -i_L \frac{10}{10 + 40}.$$

Note that this expression is valid for $t \geq 0^+$ because $i_o = 0$ at $t = 0^-$. The inductor behaves as a short circuit prior to the switch being opened, producing an instantaneous change in the current i_o. Then,

$$i_o(t) = -4e^{-5t}\ \text{A}, \quad t \geq 0^+.$$

(c) We find the voltage v_o by direct application of Ohm's law:

$$v_o(t) = 40i_o = -160e^{-5t}\ \text{V}, \quad t \geq 0^+.$$

(d) The power dissipated in the 10 Ω resistor is

$$p_{10\Omega}(t) = \frac{v_o^2}{10} = 2560e^{-10t}\ \text{W}, \quad t \geq 0^+.$$

The total energy dissipated in the 10 Ω resistor is

$$w_{10\Omega}(t) = \int_0^\infty 2560e^{-10t}\, dt = 256\ \text{J}.$$

The initial energy stored in the 2 H inductor is

$$w(0) = \frac{1}{2}Li^2(0) = \frac{1}{2}(2)(400) = 400\ \text{J}.$$

Therefore the percentage of energy dissipated in the 10 Ω resistor is

$$\frac{256}{400}(100) = 64\%.$$

EXAMPLE 5.7

In the circuit shown in Fig. 5.26, the initial currents in inductors L_1 and L_2 have been established by sources not shown. The switch is opened at $t = 0$.

(a) Find i_1, i_2, and i_3 for $t \geq 0$.

(b) Calculate the initial energy stored in the parallel inductors.

(c) Determine how much energy is stored in the inductors as $t \to \infty$.

(d) Show that the total energy delivered to the resistive network equals the difference between the results obtained in (b) and (c).

SOLUTION

(a) The key to finding currents i_1, i_2, and i_3 lies in knowing the voltage $v(t)$. We can easily find $v(t)$ if we reduce the circuit shown in Fig. 5.26 to the equivalent form shown in Fig. 5.27. The parallel inductors simplify to an equivalent inductance of 4 H, carrying an initial current of 12 A. The resistive network reduces to a single resistance of 8 Ω. Hence the initial value of $i(t)$ is 12 A and the time constant is 4/8, or 0.5 s. Therefore

$$i(t) = 12e^{-2t} \text{ A}, \quad t \geq 0.$$

Now $v(t)$ is simply the product $8i$, so

$$v(t) = 96e^{-2t} \text{ V}, \quad t \geq 0^+.$$

The circuit shows that $v(t) = 0$ at $t = 0^-$, so the expression for $v(t)$ is valid for $t \geq 0^+$. After obtaining $v(t)$, we can calculate i_1, i_2, and i_3:

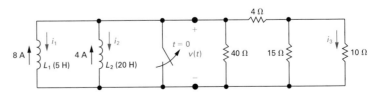

Figure 5.26 The circuit for Example 5.7.

$$i_1 = \frac{1}{5} \int_0^t 96e^{-2x} \, dx - 8$$

$$= 1.6 - 9.6e^{-2t} \text{ A}, \quad t \geq 0,$$

$$i_2 = \frac{1}{20} \int_0^t 96e^{-2x} \, dx - 4$$

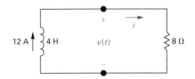

Figure 5.27 A simplification of the circuit shown in Fig. 5.26.

$$= -1.6 - 2.4e^{-2t} \text{ A}, \quad t \geq 0,$$

$$i_3 = \frac{v(t)}{10} \frac{15}{25} = 5.76e^{-2t} \text{ A}, \quad t \geq 0^+.$$

Note that the expressions for the inductor currents i_1 and i_2 are valid for $t \geq 0$, whereas the expression for the resistor current i_3 is valid for $t \geq 0^+$.

(b) The initial energy stored in the inductors is

$$w = \frac{1}{2}(5)(64) + \frac{1}{2}(20)(16) = 320 \text{ J}.$$

(c) As $t \to \infty$, $i_1 \to 1.6$ A and $i_2 \to -1.6$ A. Therefore, a long time after the switch has been opened, the energy stored in the two inductors is

$$w = \frac{1}{2}(5)(1.6)^2 + \frac{1}{2}(20)(-1.6)^2 = 32 \text{ J}.$$

(d) We obtain the total energy delivered to the resistive network by integrating the expression for the instantaneous power from zero to infinity:

$$w = \int_0^\infty p \, dt = \int_0^\infty 1152e^{-4t} \, dt$$

$$= 1152 \left. \frac{e^{-4t}}{-4} \right|_0^\infty = 288 \text{ J}.$$

This result is the difference between the initially stored energy (320 J) and the energy trapped in the parallel inductors (32 J). The equivalent inductor for the parallel inductors (which predicts the terminal behavior of the parallel combination) has an initial energy of 288 J; that is, the energy stored in the equivalent inductor represents the amount of energy that will be delivered to the resistive network at the terminals of the original inductors.

DRILL EXERCISES

5.6 The switch in the circuit shown has been closed for a long time and is opened at $t = 0$.

(a) Calculate the initial value of i.

(b) Calculate the initial energy stored in the inductor.

(c) What is the time constant of the circuit for $t > 0$?

(d) What is the numerical expression for $i(t)$ for $t \geq 0$?

(e) What percentage of the initial energy stored has been dissipated in the 2 Ω resistor 5 ms after the switch has been opened?

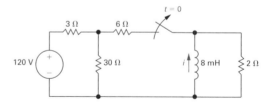

ANSWER: (a) -12.5 A; (b) 625 mJ; (c) 4 ms; (d) $-12.5e^{-250t}$ A; (e) 91.8%.

5.7 At $t = 0$, the switch in the circuit shown moves instantaneously from position a to position b.

(a) Calculate v_o for $t \geq 0^+$.

(b) What percentage of the initial energy stored in the inductor is eventually dissipated in the 4 Ω resistor?

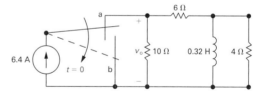

ANSWER: (a) $-8e^{-10t}$ V; (b) 80%.

Natural Response of an RC Circuit

As mentioned earlier, the natural response of an RC circuit is analogous to that of an RL circuit. Consequently, we don't treat the RC circuit in the same detail as we did the RL circuit.

The natural response of an RC circuit is developed from the circuit shown in Fig. 5.28. We begin by assuming that the switch has been in position a for a long time, allowing the loop made up of the dc voltage source V_g, the resistor R_1, and the capacitor C to reach a steady-state condition. Remember that a capacitor behaves as an open circuit in the presence of a constant voltage. Thus the voltage source cannot sustain a current, and

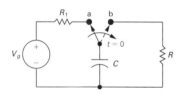

Figure 5.28 An RC circuit.

so the source voltage appears across the capacitor terminals. In Section 5.5, we will discuss how the capacitor voltage actually builds to the steady-state value of the dc voltage source, but for now the important point is that when the switch is moved from position a to position b (at $t = 0$), the voltage on the capacitor is V_g. Because there can be no instantaneous change in the voltage at the terminals of a capacitor, the problem reduces to solving the circuit shown in Fig. 5.29.

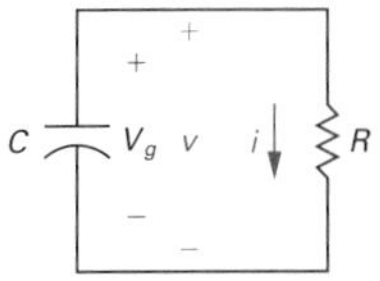

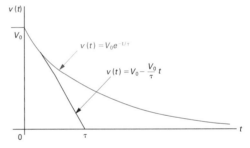

Figure 5.29 The circuit shown in Fig. 5.28, after switching.

Deriving the Expression for the Voltage

We can easily find the voltage $v(t)$ by thinking in terms of node voltages. Using the lower junction between R and C as the reference node and summing the currents away from the upper junction between R and C gives

$$C\frac{dv}{dt} + \frac{v}{R} = 0. \tag{5.51}$$

Comparing Eq. (5.51) with Eq. (5.31) shows that the same mathematical techniques can be used to obtain the solution for $v(t)$. We leave it to you to show that

$$v(t) = v(0)e^{-t/RC}, \quad t \geq 0. \tag{5.52}$$

As we have already noted, the initial voltage on the capacitor equals the voltage source voltage V_g, or

$$v(0^-) = v(0) = v(0^+)$$

$$= V_g = V_0, \tag{5.53}$$

where V_0 denotes the initial voltage on the capacitor. The time constant for the RC circuit equals the product of the resistance and capacitance, namely,

$$\tau = RC. \tag{5.54}$$

Substituting Eqs. (5.53) and (5.54) into Eq. (5.52) yields

$$v(t) = V_0 e^{-t/\tau}, \quad t \geq 0, \tag{5.55}$$

which indicates that the natural response of an RC circuit is an exponential decay of the initial voltage. The time constant RC governs the rate of decay. Figure 5.30 shows the plot of Eq. (5.55) and the graphic interpretation of the time constant.

Figure 5.30 The natural response of an RC circuit.

After determining $v(t)$, we can easily derive the expressions for i, p, and w:

$$i(t) = \frac{v(t)}{R} = \frac{V_0}{R}e^{-t/\tau}, \quad t \geq 0^+, \tag{5.56}$$

$$p = vi = \frac{V_0^2}{R}e^{-2t/\tau}, \quad t \geq 0^+, \tag{5.57}$$

$$w = \int_0^t p\,dx = \int_0^t \frac{V_0^2}{R}e^{-2x/\tau}\,dx$$

$$= \frac{1}{2}CV_0^2(1 - e^{-2t/\tau}), \quad t \geq 0. \tag{5.58}$$

Calculating the natural response of an RC circuit can be summarized as follows:

1. Find the initial voltage, $v(0)$, across the capacitor.

2. Find the time constant of the circuit.

3. Use Eq. (5.55) to generate $v(t)$ from $v(0)$ and τ.

All other calculations of interest follow from knowing $v(t)$. Examples 5.8 and 5.9 illustrate the numerical calculations associated with the natural response of an RC circuit.

EXAMPLE 5.8

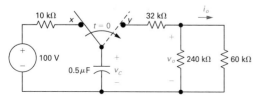

Figure 5.31 The circuit for Example 5.8.

The switch in the circuit shown in Fig. 5.31 has been in position x for a long time. At $t = 0$, the switch moves instantaneously to position y. Find

(a) $v_C(t)$ for $t \geq 0$,

(b) $v_o(t)$ for $t \geq 0^+$,

(c) $i_o(t)$ for $t \geq 0^+$, and

(d) the total energy dissipated in the 60 kΩ resistor.

SOLUTION

(a) Because the switch has been in position x for a long time, the 0.5 μF capacitor will charge to 100 V and be positive at the upper terminal. We can replace the resistive network connected to the capacitor at $t = 0^+$ with an equivalent resistance of 80 kΩ. Hence the time constant of the circuit is $(0.5 \times 10^{-6})(80 \times 10^3)$ or 40 ms. Then,

$$v_C(t) = 100e^{-25t} \text{ V}, \quad t \geq 0.$$

(b) The easiest way to find $v_o(t)$ is to note that the resistive circuit forms a voltage divider across the terminals of the capacitor. Thus

$$v_o(t) = \frac{48}{80}v_C(t) = 60e^{-25t} \text{ V}, \quad t \geq 0^+.$$

This expression for $v_o(t)$ is valid for $t \geq 0^+$ because $v_o(0^-)$ is zero. Thus we have an instantaneous change in the voltage across the 240 kΩ resistor.

(c) We find the current $i_o(t)$ from Ohm's law:

$$i_o(t) = \frac{v_o(t)}{60 \times 10^3} = e^{-25t} \text{ mA}, \quad t \geq 0^+.$$

(d) The power dissipated in the 60 kΩ resistor is

$$p_{60k\Omega}(t) = i_o^2(t)(60 \times 10^3) = 60e^{-50t} \text{ mW}, \quad t \geq 0^+.$$

The total energy dissipated in the 60 kΩ resistor is

$$w_{60k\Omega} = \int_0^\infty i_o^2(t)(60 \times 10^3)\, dt = 1.2 \text{ mJ}.$$

EXAMPLE 5.9

The initial voltages on capacitors C_1 and C_2 in the circuit shown in Fig. 5.32 have been established by sources not shown. The switch is closed at $t = 0$.

(a) Find $v_1(t)$, $v_2(t)$, and $v(t)$ for $t \geq 0$ and $i(t)$ for $t \geq 0^+$.

(b) Calculate the initial energy stored in the capacitors C_1 and C_2.

(c) Determine how much energy is stored in the capacitors as $t \to \infty$.

(d) Show that the total energy delivered to the 250 kΩ resistor is the difference between the results obtained in (b) and (c).

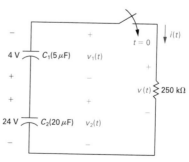

Figure 5.32 The circuit for Example 5.9.

SOLUTION

(a) Once we know $v(t)$, we can obtain the current $i(t)$ from Ohm's law. After determining $i(t)$, we can calculate $v_1(t)$ and $v_2(t)$, because the voltage across a capacitor is a function of the capacitor current. To find $v(t)$, we replace the series-connected capacitors with an equivalent capacitor. It has a capacitance of 4 μF and is charged to a voltage of 20 V. Therefore, the circuit shown in Fig. 5.32 reduces to the one shown in Fig. 5.33, which reveals that the initial value of $v(t)$ is 20 V, and that the time constant of the circuit is $(4)(250) \times 10^{-3}$, or 1 s. Thus the expression for $v(t)$ is

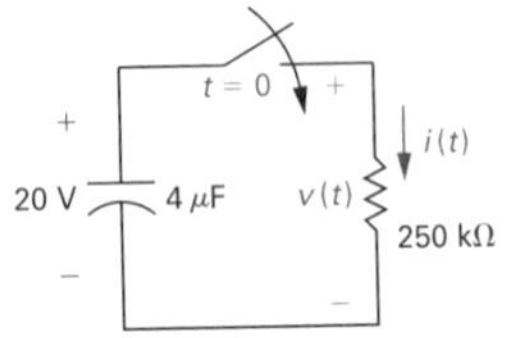

$$v(t) = 20e^{-t} \text{ V}, \quad t \geq 0.$$

The current $i(t)$ is

$$i(t) = \frac{v(t)}{250,000} = 80e^{-t} \ \mu\text{A}, \quad t \geq 0^+.$$

Knowing $i(t)$, we calculate the expressions for $v_1(t)$ and $v_2(t)$:

$$v_1(t) = -\frac{10^6}{5} \int_0^t 80 \times 10^{-6} e^{-x} \, dx - 4$$

$$= (16e^{-t} - 20) \text{ V}, \quad t \geq 0,$$

$$v_2(t) = -\frac{10^6}{20} \int_0^t 80 \times 10^{-6} e^{-x} \, dx + 24$$

$$= (4e^{-t} + 20) \text{ V}, \quad t \geq 0.$$

(b) The initial energy stored in C_1 is

$$w_1 = \frac{1}{2}(5 \times 10^{-6})(16) = 40 \ \mu\text{J}.$$

The initial energy stored in C_2 is

$$w_2 = \frac{1}{2}(20 \times 10^{-6})(576) = 5760 \ \mu\text{J}.$$

The total energy stored in the two capacitors is

$$w_o = 40 + 5760 = 5800 \ \mu\text{J}.$$

Figure 5.33 A simplification of the circuit shown in Fig. 5.32.

(c) As $t \rightarrow \infty$,

$$v_1 \rightarrow -20 \text{ V} \quad \text{and} \quad v_2 \rightarrow +20 \text{ V}.$$

Therefore the energy stored in the two capacitors is

$$w_\infty = \frac{1}{2}(5 + 20) \times 10^{-6}(400) = 5000 \ \mu\text{J}.$$

(d) The total energy delivered to the 250 kΩ resistor is

$$w = \int_0^\infty p \, dt = \int_0^\infty \frac{400e^{-2t}}{250,000} \, dt = 800 \ \mu\text{J}.$$

Comparing the results obtained in (b) and (c) shows that

$$800 \ \mu\text{J} = (5800 - 5000) \ \mu\text{J}.$$

The energy stored in the equivalent capacitor in Fig. 5.33 is $\frac{1}{2}(4 \times 10^{-6})(400)$, or 800 μJ. Because this capacitor predicts the terminal behavior of the original series-connected capacitors, the energy stored in the equivalent capacitor is the energy delivered to the 250 kΩ resistor.

DRILL EXERCISES

5.8   The switch in the circuit shown has been closed for a long time and is opened at $t = 0$. Find

(a) the initial value of $v(t)$

(b) the time constant for $t > 0$

(c) the numerical expression for $v(t)$ after the switch has been opened

(d) the initial energy stored in the capacitor

(e) the length of time required to dissipate 75% of the initially stored energy.

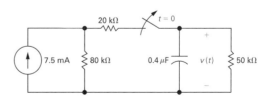

ANSWER: (a) 200 V; (b) 20 ms; (c) $200e^{-50t}$ V; (d) 8 mJ; (e) 13.86 ms.

5.9   The switch in the circuit shown has been closed for a long time before being opened at $t = 0$.

  (a) Find $v_o(t)$ for $t \geq 0$.

  (b) What percentage of the initial energy stored in the circuit has been dissipated after the switch has been open for 60 ms?

  ANSWER: (a) $8e^{-25t} + 4e^{-10t}$ V; (b) 81.05%.

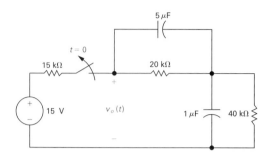

5.5 STEP RESPONSE OF RL AND RC CIRCUITS

In binary digital systems the logic variables are represented by two dc voltage levels. In a positive logic system the higher amplitude represents 1 and the lower amplitude represents 0. (The reverse representation is referred to as negative logic.) A digital logic circuit switches between high and low values as it processes information. The speed at which a voltage changes from low to high (or high to low) governs the number of operations per unit time that can be performed by a digital computer. Studying the step response of RL and RC circuits is a first step in understanding the factors that govern the speed of digital computers. We begin with the step response of an RL circuit.

Step Response of an RL Circuit

To begin, we modify the first-order circuit shown in Fig. 5.20(a) by adding a switch. We use the resulting circuit, shown in Fig. 5.34, in developing the step response of an RL circuit. Energy stored in the inductor at the time the switch is closed is given in terms of a nonzero initial current $i(0)$. The task is to find the expressions for the current in the circuit and for the voltage across the inductor after the switch has been closed. The procedure is the same as that used in Section 5.4; we use circuit analysis to derive the differential equation that describes the circuit in terms of the variable of interest, and then we use elementary calculus to solve the equation.

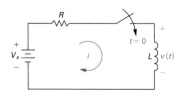

Figure 5.34 A circuit used to illustrate the step response of a first-order RL circuit.

After the switch in Fig. 5.34 has been closed, Kirchhoff's voltage law requires that

$$V_s = Ri + L\frac{di}{dt},$$ (5.59)

which can be solved for the current by separating the variables i and t and then integrating. The first step in this approach is to solve Eq. (5.59) for the derivative di/dt:

$$\frac{di}{dt} = \frac{-Ri + V_s}{L} = \frac{-R}{L}\left(i - \frac{V_s}{R}\right).$$ (5.60)

Next, we multiply both sides of Eq. (5.60) by a differential time dt. This step reduces the left-hand side of the equation to a differential change in the current. Thus

$$\frac{di}{dt}\,dt = \frac{-R}{L}\left(i - \frac{V_s}{R}\right)dt,$$

or

$$di = \frac{-R}{L}\left(i - \frac{V_s}{R}\right)dt.$$ (5.61)

We now separate the variables in Eq. (5.61) to get

$$\frac{di}{i - (V_s/R)} = \frac{-R}{L}\,dt,$$ (5.62)

and then integrate both sides of Eq. (5.62). Using x and y as variables for the integration, we obtain

$$\int_{I_0}^{i(t)} \frac{dx}{x - (V_s/R)} = \frac{-R}{L}\int_0^t dy,$$ (5.63)

where I_0 is the current at $t = 0$ and $i(t)$ is the current at any $t > 0$. Performing the integration called for in Eq. (5.63) generates the expression

$$\ln\frac{i(t) - (V_s/R)}{I_0 - (V_s/R)} = \frac{-R}{L}t,$$ (5.64)

from which

$$\frac{i(t) - (V_s/R)}{I_0 - (V_s/R)} = e^{-(R/L)t},$$

or

$$i(t) = \frac{V_s}{R} + \left(I_0 - \frac{V_s}{R}\right)e^{-(R/L)t}.$$ (5.65)

When the initial energy in the inductor is zero, I_0 is zero. Thus Eq. (5.65) reduces to

$$i(t) = \frac{V_s}{R} - \frac{V_s}{R}e^{-(R/L)t}. \tag{5.66}$$

Equation (5.66) indicates that after the switch has been closed, the current increases exponentially from zero to a final value of V_s/R. The time constant of the circuit, L/R, determines the rate of increase. One time constant after the switch has been closed, the current will have reached approximately 63% of its final value, or

$$i(\tau) = \frac{V_s}{R} - \frac{V_s}{R}e^{-1} \approx 0.6321\frac{V_s}{R}. \tag{5.67}$$

If the current were to continue to increase at its initial rate, it would reach its final value at $t = \tau$; that is, because

$$\frac{di}{dt} = \frac{-V_s}{R}\left(\frac{-1}{\tau}\right)e^{-t/\tau} = \frac{V_s}{L}e^{-t/\tau}, \tag{5.68}$$

the initial rate at which $i(t)$ increases is

$$\frac{di}{dt}(0) = \frac{V_s}{L}. \tag{5.69}$$

If the current were to continue to increase at this rate, the expression for i would be

$$i = \frac{V_s}{L}t, \tag{5.70}$$

from which, at $t = \tau$,

$$i = \frac{V_s}{L}\frac{L}{R} = \frac{V_s}{R}. \tag{5.71}$$

Equations (5.66) and (5.70) are plotted in Fig. 5.35. The values given by Eqs. (5.67) and (5.71) are also shown in this figure.

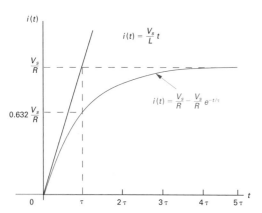

Figure 5.35 The step response of the RL circuit shown in Fig. 5.34 when $I_0 = 0$.

The voltage across an inductor is $L\,di/dt$, so from Eq. (5.65), for $t \geq 0^+$,

$$v = L\left(\frac{-R}{L}\right)\left(I_0 - \frac{V_s}{R}\right)e^{-(R/L)t} = (V_s - I_0R)e^{-(R/L)t}. \quad (5.72)$$

The voltage across the inductor is zero before the switch is closed. Equation (5.72) indicates that the inductor voltage jumps to $V_s - I_0R$ at the instant the switch is closed and then decays exponentially to zero.

Does the value of v at $t = 0^+$ make sense? Because the initial current is I_0 and the inductor prevents an instantaneous change in current, the current is I_0 in the instant after the switch has been closed. The voltage drop across the resistor is I_0R, and the voltage impressed across the inductor is the source voltage minus the voltage drop, that is, $V_s - I_0R$.

When the initial inductor current is zero, Eq. (5.72) simplifies to

$$v = V_s e^{-(R/L)t}. \quad (5.73)$$

If the initial current is zero, the voltage across the inductor jumps to V_s. We also expect the inductor voltage to approach zero as t increases, because the current in the circuit is approaching the constant value of V_s/R. Figure 5.36 shows the plot of Eq. (5.73) and the relationship between the time constant and the initial rate at which the inductor voltage is decreasing.

If there is an initial current in the inductor, Eq. (5.65) gives the solution for it. The algebraic sign of I_0 is positive if the initial current is in the same direction as i; otherwise, I_0 carries a negative sign. Example 5.10 illustrates the application of Eq. (5.65) to a specific circuit.

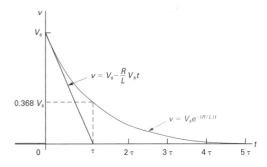

Figure 5.36 Inductor voltage versus time.

EXAMPLE 5.10

The switch in the circuit shown in Fig. 5.37 has been in position a for a long time. At $t = 0$, the switch moves from position a to position b. The switch is a make-before-break type; that is, the connection at position b is established before the connection at position a is broken, so there is no interruption of current through the inductor.

(a) Find the expression for $i(t)$ for $t \geq 0$.

(b) What is the initial voltage across the inductor just after the switch has been moved to position b?

(c) Does this initial voltage make sense in terms of circuit behavior?

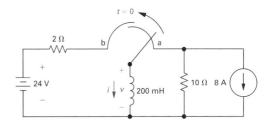

Figure 5.37 The circuit for Example 5.10.

(d) How many milliseconds after the switch has been moved does the inductor voltage equal 24 V?

(e) Plot both $i(t)$ and $v(t)$ versus t.

SOLUTION

(a) The switch has been in position a for a long time, so the 200 mH inductor is a short circuit across the 8 A current source. Therefore the inductor carries an initial current of 8 A. This current is oriented opposite to the reference direction for i; thus I_0 is −8 A. When the switch is in position b, the final value of i will be 24/2, or 12 A. The time constant of the circuit is 200/2, or 100 ms. Substituting these values into Eq. (5.65) gives

$$i = 12 + (-8 - 12)e^{-t/0.1}$$

$$= 12 - 20e^{-10t} \text{ A}, \quad t \geq 0.$$

(b) The voltage across the inductor is

$$v = L\frac{di}{dt} = 0.2(200e^{-10t}) = 40e^{-10t} \text{ V}, \quad t \geq 0^+.$$

The initial inductor voltage is

$$v(0^+) = 40 \text{ V}.$$

(c) Yes; in the instant after the switch has been moved to position b, the inductor sustains a current of 8 A counterclockwise around the newly formed closed path. This current causes a 16 V drop across the 2 Ω resistor. This voltage drop adds to the drop across the source, producing a 40 V drop across the inductor.

(d) We find the time at which the inductor voltage equals 24 V by solving the expression

$$24 = 40e^{-10t}$$

for t:

$$t = \frac{1}{10}\ln\frac{40}{24} = 51.08 \times 10^{-3} = 51.08 \text{ ms.}$$

(e) Figure 5.38 shows the graphs of $i(t)$ and $v(t)$ versus t. Note that the instant of time when the current equals zero corresponds to the instant of time when the inductor voltage equals the source voltage of 24 V, as predicted by Kirchhoff's voltage law.

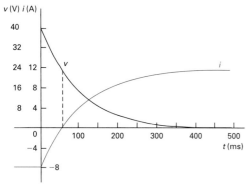

Figure 5.38 The current and voltage waveforms for Example 5.10.

DRILL EXERCISE

5.10 Assume that the switch in the circuit shown in Fig. 5.37 has been in position b for a long time, and at $t = 0$ it moves to position a. Find (a) $i(0^+)$; (b) $v(0^+)$; (c) τ, $t > 0$; (d) $i(t)$, $t \geq 0$; and (e) $v(t)$, $t \geq 0^+$.

ANSWER: (a) 12 A; (b) -200 V; (c) 20 ms; (d) $-8 + 20e^{-50t}$ A, $t \geq 0$; (e) $-200e^{-50t}$ V, $t \geq 0^+$.

We can also describe the voltage $v(t)$ across the inductor in Fig. 5.34 directly, not just in terms of the circuit current. We begin by noting that the voltage across the resistor is the difference between the source voltage and the inductor voltage. We write

$$i(t) = \frac{V_s}{R} - \frac{v(t)}{R}, \qquad (5.74)$$

where V_s is a constant. Differentiating both sides with respect to time yields

$$\frac{di}{dt} = -\frac{1}{R}\frac{dv}{dt}. \qquad (5.75)$$

Then, if we multiply each side of Eq. (5.75) by the inductance L, we get an expression for the voltage across the inductor on the left-hand side, or

$$v = -\frac{L}{R}\frac{dv}{dt}. \qquad (5.76)$$

Putting Eq. (5.76) into standard form yields

$$\frac{dv}{dt} + \frac{R}{L}v = 0. \qquad (5.77)$$

You should verify (in Drill Exercise 5.11) that the solution to Eq. (5.77) is identical to that given in Eq. (5.72).

At this point, a general observation about the step response of an RL circuit is pertinent. (This observation will prove helpful later.) When we derived the differential equation for the inductor current, we obtained Eq. (5.59). We now rewrite Eq. (5.59) as

$$\frac{di}{dt} + \frac{R}{L}i = \frac{V_s}{L}. \qquad (5.78)$$

Observe that Eqs. (5.77) and (5.78) have the same form. Specifically, each equates the sum of the first derivative of the variable and a constant times the variable to a constant value. In

Eq. (5.77), the constant on the right-hand side happens to be zero; hence this equation takes on the same form as the natural response equations in Section 5.4. In both Eq. (5.77) and Eq. (5.78), the constant multiplying the dependent variable is the reciprocal of the time constant, that is, $R/L = 1/\tau$. We encounter a similar situation in the derivations for the step response of an RC circuit. Later we will use these observations to develop a general approach to finding the natural and step responses of RL and RC circuits.

DRILL EXERCISE

5.11 **(a)** Derive Eq. (5.77) by first converting the Thévenin equivalent in Fig. 5.34 to a Norton equivalent and then summing the currents away from the upper node, using the inductor voltage v as the variable of interest.

(b) Use the separation-of-variables technique to find the solution to Eq. (5.77). Verify that your solution agrees with the solution given in Eq. (5.72).

ANSWER: (a) Derivation; (b) verification.

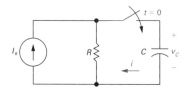

Figure 5.39 A circuit used to illustrate the step response of a first-order RC circuit.

The Step Response of an RC Circuit

We can find the step response of a first-order RC circuit by analyzing the circuit shown in Fig. 5.39. For mathematical convenience, we choose the Norton equivalent of the network connected to the equivalent capacitor. Summing the currents away from the top node in Fig. 5.39 generates the differential equation

$$C\frac{dv_C}{dt} + \frac{v_C}{R} = I_s. \tag{5.79}$$

Division of Eq. (5.49) by C gives

$$\frac{dv_C}{dt} + \frac{v_C}{RC} = \frac{I_s}{C}. \tag{5.80}$$

Comparing Eq. (5.80) with Eq. (5.78) reveals that the form of the solution for v_C is the same as that for the current in the inductive circuit, namely, Eq. (5.65). Therefore, by simply substituting the appropriate variables and coefficients, we can write the solution for v_C directly. The translation requires that I_s replace V_s, C replace L, $1/R$ replace R, and V_0 replace I_0. We get

$$v_C = I_s R + (V_0 - I_s R)e^{-t/RC}, \quad t \geq 0. \qquad (5.81)$$

A similar derivation for the current in the capacitor yields the differential equation

$$\frac{di}{dt} + \frac{1}{RC}i = 0. \qquad (5.82)$$

Equation (5.82) has the same form as Eq. (5.77), hence the solution for i is obtained by using the same translations used for the solution of Eq. (5.80). Thus

$$i = \left(I_s - \frac{V_0}{R}\right)e^{-t/RC}, \quad t \geq 0^+, \qquad (5.83)$$

where V_0 is the initial value of v_C, the voltage across the capacitor.

We obtained Eqs. (5.81) and (5.83) by using a mathematical analogy to the solution for the step response of the inductive circuit. Let's see whether these solutions for the RC circuit make sense in terms of known circuit behavior. From Eq. (5.81), note that the initial voltage across the capacitor is V_0, the final voltage across the capacitor is $I_s R$, and the time constant of the circuit is RC. Also note that the solution for v_C is valid for $t \geq 0$. These observations are consistent with the behavior of a capacitor in parallel with a resistor when driven by a constant current source.

Equation (5.83) predicts that the current in the capacitor at $t = 0^+$ is $I_s - V_0/R$. This prediction makes sense because the capacitor voltage cannot change instantaneously, and therefore the initial current in the resistor is V_0/R. The capacitor branch current changes instantaneously from zero at $t = 0^-$ to $I_s - V_0/R$ at $t = 0^+$. The capacitor current is zero at $t = \infty$. Also note that the final value of $v = I_s R$.

Example 5.11 illustrates how to use Eqs. (5.81) and (5.83) to find the step response of a first-order RC circuit.

EXAMPLE 5.11

The switch in the circuit shown in Fig. 5.40 has been in position 1 for a long time. At $t = 0$, the switch moves to position 2. Find

(a) $v_o(t)$ for $t \geq 0$

(b) $i_o(t)$ for $t \geq 0^+$

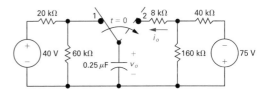

Figure 5.40 The circuit for Example 5.11.

SOLUTION

(a) The switch has been in position 1 for a long time, so the initial value of v_o is 40(60/80), or 30 V. To take advantage of Eqs. (5.81) and (5.83), we find the Norton equivalent with respect to the terminals of the capacitor for $t \geq 0$. To do this, we begin by computing the open-circuit voltage, which is given by the −75 V source divided across the 40 kΩ and 160 kΩ resistors:

$$V_{\text{oc}} = \frac{160 \times 10^3}{(40 + 160) \times 10^3}(-75) = -60 \text{ V}.$$

Next, we calculate the Thévenin resistance, as seen to the right of the capacitor, by shorting the −75 V source and making series and parallel combinations of the resistors:

$$R_{\text{Th}} = 8000 + 40{,}000 \parallel 160{,}000 = 40 \text{ k}\Omega.$$

The value of the Norton current source is the ratio of the open-circuit voltage to the Thévenin resistance, or $-60/(40 \times 10^3) = -1.5$ mA. The resulting Norton equivalent circuit is shown in Fig. 5.41. From Fig. 5.41, $I_s R = -60$ V and $RC = 10$ ms. We have already noted that $v_o(0) = 30$ V, so the solution for v_o is

$$v_o = -60 + [30 - (-60)]e^{-100t}$$

$$= -60 + 90e^{-100t} \text{ V}, \quad t \geq 0.$$

(b) We write the solution for i_o directly from Eq. (5.83) by noting that $I_s = -1.5$ mA and $V_o/R = (30/40) \times 10^{-3}$, or 0.75 mA:

$$i_o = -2.25e^{-100t} \text{ mA}, \quad t \geq 0^+.$$

We check the consistency of the solutions for v_o and i_o by noting that

$$i_o = C\frac{dv_o}{dt} = (0.25 \times 10^{-6})(-9000e^{-100t}) = -2.25e^{-100t} \text{ mA}.$$

Because $dv_o(0^-)/dt = 0$, the expression for i_o clearly is valid only for $t \geq 0^+$.

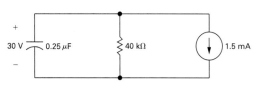

Figure 5.41 The equivalent circuit for $t > 0$ for the circuit shown in Fig. 5.40.

DRILL EXERCISES

5.12 **(a)** Derive Eq. (5.82) by first converting the Norton equivalent circuit shown in Fig. 5.39 to a Thévenin equivalent and then summing the voltages around the closed loop, using the capacitor current i as the relevant variable.

(b) Use the separation of variables technique to find the solution to Eq. (5.82). Verify that your solution agrees with that of Eq. (5.83).

   ANSWER: (a) Derivation; (b) verification.

5.13 **(a)** Find the expression for the voltage across the 160 kΩ resistor in the circuit shown in Fig. 5.40. Let this voltage be denoted v_A, and assume that the reference polarity for the voltage is positive at the upper terminal of the 160 kΩ resistor.

(b) Specify the interval of time for which the expression obtained in (a) is valid.

   ANSWER: (a) $v_A = -60 + 72e^{-100t}$ V; (b) $t \geq 0^+$.

A General Solution for Step and Natural Responses

The general approach to finding either the natural response or the step response of the first-order RL and RC circuits shown in Fig. 5.42 is based on their differential equations being the same (compare Eq. [5.78] and Eq. [5.80]). To generalize the solution of these four possible circuits, we let $x(t)$ represent the unknown quantity, giving $x(t)$ four possible values. It can represent the current or voltage at the terminals of an inductor or the current or voltage at the terminals of a capacitor. From Eqs. (5.77), (5.78), (5.80), and (5.82), we know that the differential equation describing any one of the four circuits in Fig. 5.42 takes the form

$$\frac{dx}{dt} + \frac{x}{\tau} = K, \qquad (5.84)$$

where the value of the constant K can be zero. Because the sources in the circuit are constant voltages and/or currents,

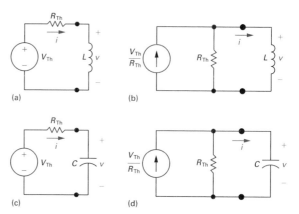

Figure 5.42 Four possible first-order circuits. (a) An inductor connected to a Thévenin equivalent. (b) An inductor connected to a Norton equivalent. (c) A capacitor connected to a Thévenin equivalent. (d) A capacitor connected to a Norton equivalent.

the final value of x will be constant; that is, the final value must satisfy Eq. (5.84), and, when x reaches its final value, the derivative dx/dt must be zero. Hence

$$x_f = K\tau, \qquad (5.85)$$

where x_f represents the final value of the variable.

We solve Eq. (5.84) by separating the variables, beginning by solving for the first derivative:

$$\frac{dx}{dt} = \frac{-x}{\tau} + K = \frac{-(x - K\tau)}{\tau} = \frac{-(x - x_f)}{\tau}. \qquad (5.86)$$

In writing Eq. (5.86), we used Eq. (5.85) to substitute x_f for $K\tau$. We now multiply both sides of Eq. (5.86) by dt and divide by $x - x_f$ to obtain

$$\frac{dx}{x - x_f} = \frac{-1}{\tau} dt. \qquad (5.87)$$

Next, we integrate Eq. (5.87). To obtain as general a solution as possible, we use time t_0 as the lower limit and t as the upper limit. Time t_0 corresponds to the time of the switching or other change. Previously we assumed that $t_0 = 0$, but this change allows the switching to take place at any time. Using u and v as symbols of integration, we get

$$\int_{x(t_0)}^{x(t)} \frac{du}{u - x_f} = -\frac{1}{\tau} \int_{t_0}^{t} dv. \qquad (5.88)$$

Carrying out the integration called for in Eq. (5.88) gives

$$x(t) = x_f + [x(t_0) - x_f]e^{-(t-t_0)/\tau}. \qquad (5.89)$$

The importance of Eq. (5.89) becomes apparent if we write it out in words:

$$\begin{array}{c} \text{the unknown} \\ \text{variable as a} \\ \text{function of time} \end{array} = \begin{array}{c} \text{the final} \\ \text{value of the} \\ \text{variable} \end{array}$$

$$+ \left[\begin{array}{c} \text{the initial} \\ \text{value of the} \\ \text{variable} \end{array} - \begin{array}{c} \text{the final} \\ \text{value of the} \\ \text{variable} \end{array} \right] \times e^{\frac{-[t-(\text{time of switching})]}{(\text{time constant})}}. \quad (5.90)$$

In many cases, the time of switching—that is, t_0—is zero.

When computing the step and natural responses of circuits, it may help to follow these steps:

1. Identify the variable of interest for the circuit. For *RC* circuits, it is most convenient to choose the capacitive voltage; for *RL* circuits, it is best to choose the inductive current.

2. Determine the initial value of the variable, which is its value at t_0. Note that if you choose capacitive voltage or inductive current as your variable of interest, it is not necessary to distinguish between $t = t_0^-$ and $t = t_0^+$.[2] This is because they both are continuous variables. If you choose another variable, you need to remember that its initial value is defined at $t = t_0^+$.

3. Calculate the final value of the variable, which is its value as $t \to \infty$.

4. Calculate the time constant for the circuit.

With these quantities, you can use Eq. (5.90) to produce an equation describing the variable of interest as a function of time. You can then find equations for other circuit variables using the circuit analysis techniques introduced in Chapters 2 and 3 or by repeating the preceding steps for the other variables.

Examples 5.12–5.14 illustrate how to use Eq. (5.90) to find the step response of an *RC* or *RL* circuit.

---

[2] The expressions t_0^- and t_0^+ are analogous to 0^- and 0^+. Thus $x(t_0^-)$ is the limit of $x(t)$ as $t \to t_0$ from the left, and $x(t_0^+)$ is the limit of $x(t)$ as $t \to t_0$ from the right.

EXAMPLE 5.12

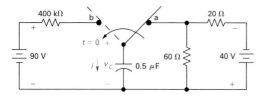

Figure 5.43 The circuit for Example 5.12.

The switch in the circuit shown in Fig. 5.43 has been in position a for a long time. At $t = 0$ the switch is moved to position b.

(a) What is the initial value of v_C?

(b) What is the final value of v_C?

(c) What is the time constant of the circuit when the switch is in position b?

(d) What is the expression for $v_C(t)$ when $t \geq 0$?

(e) What is the expression for $i(t)$ when $t \geq 0^+$?

(f) How long after the switch is in position b does the capacitor voltage equal zero?

(g) Plot $v_C(t)$ and $i(t)$ versus t.

SOLUTION

(a) The switch has been in position a for a long time, so the capacitor looks like an open circuit. Therefore the voltage across the capacitor is the voltage across the 60 Ω resistor. From the voltage-divider rule, the voltage across the 60 Ω resistor is $40 \times [60/(60 + 20)]$, or 30 V. As the reference for v_C is positive at the upper terminal of the capacitor, we have $v_C(0) = -30$ V.

(b) After the switch has been in position b for a long time, the capacitor will look like an open circuit in terms of the 90 V source. Thus the final value of the capacitor voltage is $+90$ V.

(c) The time constant is

$$\tau = RC = (400 \times 10^3)(0.5 \times 10^{-6}) = 0.2 \text{ s.}$$

(d) Substituting the appropriate values for v_f, $v(0)$, and t into Eq. (5.90) yields

$$v_C(t) = 90 + (-30 - 90)e^{-5t} = 90 - 120e^{-5t} \text{ V}, \quad t \geq 0.$$

(e) Here the value for τ doesn't change. Thus we need to find only the initial and final values for the current in the capacitor. When obtaining the initial value, we must get the value of $i(0^+)$, because the current in the capacitor can change instantaneously. This current is equal to the current in the resistor, which from Ohm's law is $[90-(-30)]/(400 \times 10^3) = 300 \, \mu A$. Note that when applying Ohm's law we recognized that the capacitor voltage cannot change instantaneously. The final value of $i(t) = 0$, so

$$i(t) = 0 + (300 - 0)e^{-5t} = 300e^{-5t} \, \mu A, \quad t \geq 0^+.$$

We could have obtained this solution by differentiating the solution in (d) and multiplying by the capacitance. You may want to do so for yourself. Note that this alternative approach to finding $i(t)$ also predicts the discontinuity at $t = 0$.

(f) To find how long the switch must be in position b before the capacitor voltage becomes zero, we solve the equation derived in (d) for the time when $v_C(t) = 0$:

$$120e^{-5t} = 90 \quad \text{or} \quad e^{5t} = \frac{120}{90},$$

so

$$t = \frac{1}{5} \ln \left(\frac{4}{3} \right) = 57.54 \text{ ms}.$$

Note that when $v_C = 0$, $i = 225$ mA and the voltage drop across the 400 kΩ resistor is 90 V.

(g) Figure 5.44 shows the graphs of $v_C(t)$ and $i(t)$ versus t.

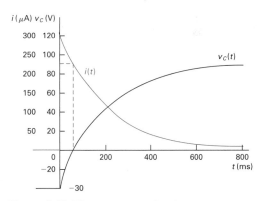

Figure 5.44 The current and voltage waveforms for Example 5.12.

EXAMPLE 5.13

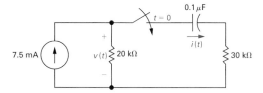

Figure 5.45 The circuit for Example 5.13.

The switch in the circuit shown in Fig. 5.45 has been open for a long time. The initial charge on the capacitor is zero. At $t = 0$, the switch is closed. Find the expression for

(a) $i(t)$ for $t \geq 0^+$

(b) $v(t)$ when $t \geq 0^+$

SOLUTION

(a) Because the initial voltage on the capacitor is zero, at the instant when the switch is closed the current in the 30 kΩ branch will be

$$i(0^+) = \frac{(7.5)(20)}{50} = 3 \text{ mA}.$$

The final value of the capacitor current will be zero because the capacitor eventually will appear as an open circuit in terms of dc current. Thus $i_f = 0$. The time constant of the circuit will equal the product of the Thévenin resistance (as seen from the capacitor) and the capacitance. Therefore $\tau = (20 + 30)10^3(0.1) \times 10^{-6} = 5$ ms. Substituting these values into Eq. (5.90) generates the expression

$$i(t) = 0 + (3 - 0)e^{-t/5 \times 10^{-3}} = 3e^{-200t} \text{ mA}, \quad t \geq 0^+.$$

(b) To find $v(t)$, we note from the circuit that it equals the sum of the voltage across the capacitor and the voltage across the 30 kΩ resistor. To find the capacitor voltage (which is a drop in the direction of the current), we note that its initial value is zero and its final value is (7.5)(20), or 150 V. The time constant is the same as before, or 5 ms. Therefore we use Eq. (5.90) to write

$$v_C(t) = 150 + (0 - 150)e^{-200t}$$

$$= (150 - 150e^{-200t}) \text{ V}, \quad t \geq 0.$$

Hence the expression for the voltage $v(t)$ is

$$v(t) = 150 - 150e^{-200t} + (30)(3)e^{-200t}$$

$$= (150 - 60e^{-200t}) \text{ V}, \quad t \geq 0^+.$$

As one check on this expression, note that it predicts the initial value of the voltage across the 20 Ω resistor as $150 - 60$, or 90 V. The instant the switch is closed, the current in the 20 kΩ resistor is (7.5)(30/50), or 4.5 mA. This current produces a 90 V drop across the 20 kΩ resistor, confirming the value predicted by the solution.

EXAMPLE 5.14

The switch in the circuit shown in Fig. 5.46 has been open for a long time. At $t = 0$ the switch is closed. Find the expression for

(a) $v(t)$ when $t \geq 0^+$

(b) $i(t)$ when $t \geq 0$

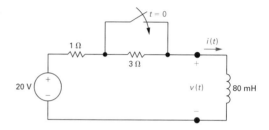

Figure 5.46 The circuit for Example 5.14.

SOLUTION

(a) The switch has been open for a long time, so the initial current in the inductor is 5 A, oriented from top to bottom. Immediately after the switch closes, the current still is 5 A, and therefore the initial voltage across the inductor becomes $20 - 5(1)$, or 15 V. The final value of the inductor voltage is 0 V. With the switch closed, the time constant is 80/1, or 80 ms. We use Eq. (5.90) to write the expression for $v(t)$:

$$v(t) = 0 + (15 - 0)e^{-t/80 \times 10^{-3}} = 15e^{-12.5t} \text{ V}, \quad t \geq 0^+.$$

(b) We have already noted that the initial value of the inductor current is 5 A. After the switch has been closed for a long time, the inductor current reaches 20/1, or 20 A. The circuit time constant is 80 ms, so the expression for $i(t)$ is

$$i(t) = 20 + (5 - 20)e^{-12.5t}$$

$$= (20 - 15e^{-12.5t}) \text{ A}, \quad t \geq 0.$$

We determine that the solutions for $v(t)$ and $i(t)$ agree by noting that

$$v(t) = L\frac{di}{dt} = 80 \times 10^{-3}[15(12.5)e^{-12.5t}]$$

$$= 15e^{-12.5t} \text{ V}, \quad t \geq 0^+.$$

DRILL EXERCISES

5.14 Assume that the switch in the circuit shown has
been in position a for a long time and that at
$t = 0$ it is moved to position b. Find (a) $v_C(0^+)$;
(b) $v_C(\infty)$; (c) τ for $t > 0$; (d) $i(0^+)$; (e) $v_C, t \geq 0$;
and (f) $i, t \geq 0^+$.

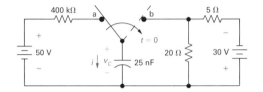

ANSWER: (a) 50 V; (b) −24 V; (c) 0.1 μs;
(d) −18.5 A; (e) $(-24 + 74e^{-10^7 t})$ V, $t \geq 0$;
(f) $-18.5e^{-10^7 t}$ A, $t \geq 0^+$.

5.15 The switch in the circuit shown has been in position
a for a long time. At $t = 0$ the switch is moved
to position b. Calculate (a) the initial voltage on
the capacitor; (b) the final voltage on the capacitor;
(c) the time constant (in microseconds) for $t > 0$;
and (d) the length of time (in microseconds) required for the capacitor voltage to reach zero after
the switch is moved to position b.

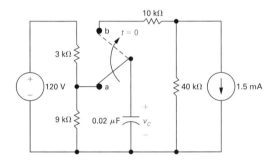

ANSWER: (a) 90 V; (b) −60 V; (c) 1000 μs;
(d) 916.3 μs.

5.16 After the switch in the circuit shown has been open
for a long time, it is closed at $t = 0$. Calculate
(a) the initial value of i; (b) the final value of i;
(c) the time constant for $t \geq 0$; and (d) the numerical expression for $i(t)$ when $t \geq 0$.

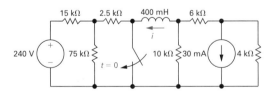

ANSWER: (a) −13 mA; (b) −12 mA; (c) 80 μs;
(d) $i = -(12 + e^{-12,500t})$ mA, $t \geq 0$.

5.6 THE INTEGRATING AMPLIFIER

Recall from the introduction to Chapter 4 that one reason for our interest in the operational amplifier is its use as an integrating amplifier. We are now ready to analyze an integrating-amplifier circuit, which is shown in Fig. 5.47. The purpose of such a circuit is to generate an output voltage proportional to the integral of the input voltage. In Fig. 5.47, we added the branch currents i_f and i_s, along with the node voltages v_n and v_p, to aid our analysis.

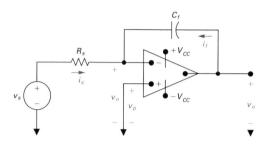

Figure 5.47 An integrating amplifier.

We assume that the operational amplifier is ideal. Thus we take advantage of the constraints

$$i_f + i_s = 0, \tag{5.91}$$

$$v_n = v_p. \tag{5.92}$$

Because $v_p = 0$,

$$i_s = \frac{v_s}{R_s}, \tag{5.93}$$

$$i_f = C_f \frac{dv_o}{dt}. \tag{5.94}$$

Hence, from Eqs. (5.91), (5.93), and (5.94),

$$\frac{dv_o}{dt} = -\frac{1}{R_s C_f} v_s. \tag{5.95}$$

Multiplying both sides of Eq. (5.95) by a differential time dt and then integrating from t_0 to t generates the equation

$$v_o(t) = -\frac{1}{R_s C_f} \int_{t_0}^{t} v_s \, dy + v_o(t_0). \tag{5.96}$$

In Eq. (5.96), t_0 represents the instant in time when we begin the integration. Thus $v_o(t_0)$ is the value of the output voltage at that time. Also, because $v_n = v_p = 0$, $v_o(t_0)$ is identical to the initial voltage on the feedback capacitor C_f.

Equation (5.96) states that the output voltage of an integrating amplifier equals the initial value of the voltage on the capacitor plus an inverted (minus sign), scaled $(1/R_s C_f)$ replica of the integral of the input voltage. If no energy is stored in the capacitor when integration commences, Eq. (5.96) reduces to

$$v_o(t) = -\frac{1}{R_s C_f} \int_{t_0}^{t} v_s \, dy. \tag{5.97}$$

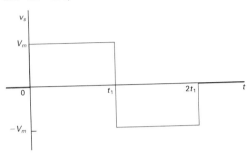

Figure 5.48 An input voltage signal.

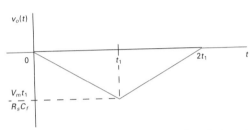

Figure 5.49 The output voltage of an integrating amplifier.

If v_s is a step change in a dc voltage level, the output voltage will vary linearly with time. For example, assume that the input voltage is the rectangular voltage pulse shown in Fig. 5.48. Assume also that the initial value of $v_o(t)$ is zero at the instant v_s steps from 0 to V_m. A direct application of Eq. (5.96) yields

$$v_o = -\frac{1}{R_s C_f} V_m t + 0, \quad 0 \le t \le t_1. \tag{5.98}$$

When t lies between t_1 and $2t_1$,

$$v_o = -\frac{1}{R_s C_f} \int_{t_1}^{t} (-V_m) \, dy - \frac{1}{R_s C_f} V_m t_1$$

$$= \frac{V_m}{R_s C_f} t - \frac{2V_m}{R_s C_f} t_1, \quad t_1 \le t \le 2t_1. \tag{5.99}$$

Figure 5.49 shows a sketch of $v_o(t)$ versus t. Clearly, the output voltage is an inverted, scaled replica of the integral of the input voltage.

The output voltage is proportional to the integral of the input voltage only if the op amp operates within its linear range, that is, if it doesn't saturate. Examples 5.15 and 5.16 further illustrate the analysis of the integrating amplifier.

EXAMPLE 5.15

Assume that the numerical values for the signal voltage shown in Fig. 5.48 are $V_m = 50$ mV and $t_1 = 1$ s. This signal voltage is applied to the integrating-amplifier circuit shown in Fig. 5.47. The circuit parameters of the amplifier are $R_s = 100 \, k\Omega$, $C_f = 0.1 \, \mu F$, and $V_{CC} = 6$ V. The initial voltage on the capacitor is zero.

(a) Calculate $v_o(t)$.

(b) Plot $v_o(t)$ versus t.

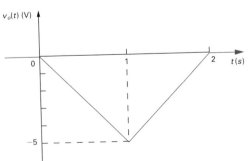

Figure 5.50 The output voltage for Example 5.15.

SOLUTION

(a) For $0 \le t \le 1$ s,

$$v_o = \frac{-1}{(100 \times 10^3)(0.1 \times 10^{-6})} 50 \times 10^{-3} t + 0$$

$$= -5t \text{ V}, \quad 0 \le t \le 1 \text{ s}.$$

For $1 \le t \le 2$ s,

$$v_o = (5t - 10) \text{ V}.$$

(b) Figure 5.50 shows a plot of $v_o(t)$ versus t.

EXAMPLE 5.16

At the instant the switch makes contact with terminal a in the circuit shown in Fig. 5.51, the voltage on the 0.1 μF capacitor is 5 V. The switch remains at terminal a for 9 ms and then moves instantaneously to terminal b. How many milliseconds after making contact with terminal a does the operational amplifier saturate?

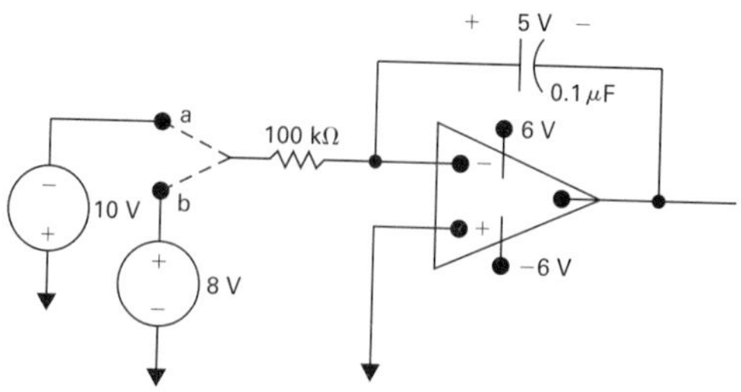

Figure 5.51 The circuit for Example 5.16.

SOLUTION

The expression for the output voltage during the time the switch is at terminal a is

$$v_o = -5 - \frac{1}{10^{-2}} \int_0^t (-10)\,dy = (-5 + 1000t)\text{ V.}$$

Thus, 9 ms after the switch makes contact with terminal a, the output voltage is $-5 + 9$, or 4 V.

The expression for the output voltage after the switch moves to terminal b is

$$v_o = 4 - \frac{1}{10^{-2}} \int_{9 \times 10^{-3}}^t 8\,dy$$

$$= 4 - 800(t - 9 \times 10^{-3}) = (11.2 - 800t)\text{ V.}$$

During this time interval, the voltage is decreasing, and the operational amplifier eventually saturates at -6 V. Therefore we set the expression for v_o equal to -6 V to obtain the saturation time t_s:

$$11.2 - 800t_s = -6, \quad \text{or} \quad t_s = 21.5 \text{ ms.}$$

Thus the integrating amplifier saturates 21.5 ms after making contact with terminal a.

From the examples, we see that the integrating amplifier can perform the integration function very well, but only within specified limits that avoid saturating the op amp.

Practical Perspective

Dual Slope Analog-to-Digital Converter

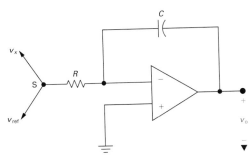

Figure 5.52 A dual-slope integrating amplifier.

We are now ready to analyze the dual-slope analog-to-digital converter introduced in the Practical Perspective at the beginning of this chapter. The dual-slope integrating-amplifier circuit shown in Fig. 5.52 is frequently used in digital multimeters to convert an unknown dc voltage v_x to a digital readout. Before we discuss the operation of this circuit, a few preliminary remarks are in order.

Digital multimeters are designed to measure dc voltage and current, ac voltage and current,[3] and resistance. This means that dc currents, ac voltages and currents, and resistance are all converted to an unknown dc voltage v_x by circuits inside the multimeter. The basic operation of the multimeter can be described with the aid of the simplified block diagram shown in Fig. 5.53.

The first two blocks in Fig. 5.53 indicate that the input variable and its range are guided to proper internal circuits by manually set switches. (Automatic ranging is available on some multimeters.) The third block emphasizes that internal circuits of the multimeter convert, when necessary, a selected input variable to a dc voltage v_x. (An input dc voltage is already in proper form and is guided directly through range selection to the analog-to-digital converter.) The fourth block, which includes the circuit shown in Fig. 5.52, indicates that the dc voltage v_x is converted to a digital form. Finally, the fifth block conveys the fact that this digital signal is converted to a digital display. We are now ready to discuss the operation of the dual-slope integrating amplifier.

The basic idea is to convert v_x to an interval of time. The time interval is then used to start and stop an oscillator. The output pulses from the oscillator are fed to a digital counter, and the resulting display is the value of the unknown voltage v_x. To see how v_x is converted to an interval of time, we first

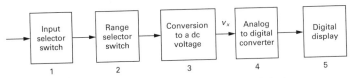

Figure 5.53 A simplified block diagram of the digital multimeter.

---

[3] AC voltage and current are introduced in Chapter 7.

note that the switch S in the circuit in Fig. 5.52 is controlled by a timing circuit internal to the multimeter, and the polarity of the reference voltage v_{ref} is always set opposite the polarity of v_x.

For purposes of showing how v_x is converted to an interval of time, assume that v_x is a negative dc voltage of V_a volts; the switch S is connected to v_x for t_1 seconds; and the initial value of v_o is zero. Then

$$v_o = -\frac{1}{RC} \int_0^{t_1} (-V_a)\, dx + 0$$

$$= \frac{V_a}{RC} t_1 \qquad\qquad (5.100)$$

After t_1 seconds switch S connects the integrating amplifier to v_{ref}. By hypothesis v_{ref} is positive, since we assumed v_x was negative. Hence for $t > t_1$ the output voltage is

$$v_o = -\frac{1}{RC} \int_{t_1}^{t} v_{\text{ref}}\, dx + \frac{V_a t_1}{RC}$$

$$= -\frac{v_{\text{ref}}(t - t_1)}{RC} + \frac{V_a t_1}{RC} \qquad\qquad (5.101)$$

The switch S remains at v_{ref} until v_o equal zero. It follows from Eq. (5.101) that

$$0 = -\frac{v_{\text{ref}}}{RC}(t_2 - t_1) + \frac{V_a t_1}{RC} \qquad\qquad (5.102)$$

or

$$t_2 = t_1 + \frac{V_a}{v_{\text{ref}}} t_1 \qquad\qquad (5.103)$$

A plot of v_o vs. t is shown in Fig. 5.54.

The important characteristic to observe about the plot in Fig. 5.54 is that the time required for v_o to reach zero, i.e., $t_2 - t_1$ is proportional to the unknown value of v_x, i.e., V_a. From Eq. (5.103) we note

$$t_2 - t_1 = \frac{V_a}{v_{\text{ref}}} t_1 \qquad\qquad (5.104)$$

In Eq. (5.104) t_1 and v_{ref} are fixed by the multimeter designer, thus the time it takes to discharge the capacitor in the circuit in Fig. 5.52 is proportional to V_a. Therefore the unknown analog voltage V_a is converted to a time interval $t_2 - t_1$, which in turn starts and stops the pulse oscillator.

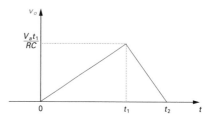

Figure 5.54 Output voltage vs. time for the dual-slope integrating amplifier.

EXAMPLE 5.17

Assume a pulse oscillator in a multimeter operates at 50,000 pulses per second.

(a) If t_1 equals 200 ms, how many pulses occur in the time interval t_1?

(b) Assume V_a is 2 V and v_{ref} in 5 V. How many pulses occur in the time interval $t_2 - t_1$?

(c) Assume the number of pulses in the time interval $t_2 - t_1$ is 5000. What is the value of V_a?

SOLUTION

(a) 200 ms is equal to 0.2 s; therefore the number of pulses in the time interval t_1 is $(0.2)(50,000)$ or 10,000 pulses.

(b) $t_2 - t_1 = \frac{V_a}{v_{ref}} t_1 = \frac{2}{5}(0.2) = 0.08$ s. Therefore the number of pulses in the time interval $t_2 - t_1$ is $(0.08)(50,000)$ or 4000 pulses.

(c) Since 4000 pulses corresponds to a voltage of 2 V, we have a scale factor of 2000 pulses per volt. Thus 5000 pulses corresponds to 2.5 V; i.e., $V_a = 5000/2000 = 2.5$ V.

SUMMARY

- **Inductance** is a linear circuit parameter that relates the voltage induced by a time-varying magnetic field to the current producing the field.

- **Capacitance** is a linear circuit parameter that relates the current induced by a time-varying electric field to the voltage producing the field.

- Inductors and capacitors are passive elements; they can store and release energy, but they cannot generate or dissipate energy.

- The instantaneous power at the terminals of an inductor or capacitor can be positive or negative, depending on whether energy is being delivered to or extracted from the element.

- An inductor:

  - does not permit an instantaneous change in its terminal current

  - does permit an instantaneous change in its terminal voltage

  - behaves as a short circuit in the presence of a constant terminal current

TABLE 5.2 Terminal equations for ideal inductors and capacitors.

INDUCTORS		CAPACITORS	
$v = L\frac{di}{dt}$	(V)	$v = \frac{1}{C}\int_{t_0}^{t} i\, d\tau + v(t_0)$	(V)
$i = \frac{1}{L}\int_{t_0}^{t} v\, d\tau + i(t_0)$	(A)	$i = C\frac{dv}{dt}$	(A)
$p = vi = Li\frac{di}{dt}$	(W)	$p = vi = Cv\frac{dv}{dt}$	(W)
$w = \frac{1}{2}Li^2$	(J)	$w = \frac{1}{2}Cv^2$	(J)

- A capacitor:

  - does not permit an instantaneous change in its terminal voltage
  - does permit an instantaneous change in its terminal current
  - behaves as an open circuit in the presence of a constant terminal voltage

- Equations for voltage, current, power, and energy in ideal inductors and capacitors are given in Table 5.2.

- Inductors in series or in parallel can be replaced by an equivalent inductor. Capacitors in series or in parallel can be replaced by an equivalent capacitor. The equations are summarized in Table 5.3. See Section 5.3 for a discussion on how to handle the initial conditions for series and parallel equivalent circuits involving inductors and capacitors.

- A first-order circuit may be reduced to a Thévenin (or Norton) equivalent connected to either a single equivalent inductor or capacitor.

- In the presence of a dc current, an inductor behaves like a short circuit.

TABLE 5.3 Equations for series- and parallel-connected inductors and capacitors.

SERIES-CONNECTED	PARALLEL-CONNECTED
$L_{eq} = L_1 + L_2 + \cdots + L_n$	$\frac{1}{L_{eq}} = \frac{1}{L_1} + \frac{1}{L_2} + \cdots + \frac{1}{L_n}$
$\frac{1}{C_{eq}} = \frac{1}{C_1} + \frac{1}{C_2} + \cdots + \frac{1}{C_n}$	$C_{eq} = C_1 + C_2 + \cdots + C_n$

- In the presence of a dc voltage, a capacitor behaves like an open circuit.

- The **time constant** of an RL circuit equals the equivalent inductance divided by the Thévenin resistance as viewed from the terminals of the equivalent inductor.

- The **time constant** of an RC circuit equals the equivalent capacitance times the Thévenin resistance as viewed from the terminals of the equivalent capacitor.

- The **natural response** is the currents and voltages that exist when stored energy is released to a circuit that contains no independent sources.

- The **step response** is the currents and voltages that result from abrupt changes in dc sources connected to a circuit. Stored energy may or may not be present at the time the abrupt changes take place.

- Capacitive voltages and inductive currents are continuous; that is, they have the same value at $t = t_0^-$ and $t = t_0^+$. Capacitive currents and inductive voltages may be discontinuous; that is, they may have different values at $t = t_0^-$ and $t = t_0^+$.

- The solution for either the natural or step response of both RL and RC circuits involves finding the initial and final value of the current or voltage of interest and the time constant of the circuit. Equations (5.89) and (5.90) summarize this approach.

- An integrating amplifier consists of an ideal op amp, a capacitor in the negative feedback branch, and a resistor in series with the signal source. It outputs the integral of the signal source, within specified limits that avoid saturating the op amp.

PROBLEMS

5.1. Evaluate the integral

$$\int_0^\infty p \, dt$$

for Example 5.2. Comment on the significance of the result.

P **5.2.** The current in the 4 mH inductor in Fig. P5.2 is known to be 2.5 A for $t \leq 0$. The inductor voltage for $t \geq 0^+$ is given by the expression

$$v_L(t) = 30e^{-3t} \text{ mV}, \quad 0^+ \leq t < \infty.$$

Sketch $v_L(t)$ and $i_L(t)$ for $0 \leq t \leq \infty$.

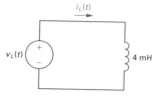

Figure P5.2

P **5.3.** The current in a 2 mH inductor is known to be

$$i_L = 50te^{-10t} \text{ A} \quad \text{for } t \geq 0.$$

(a) Find the voltage across the inductor for $t > 0$. (Assume the passive sign convention.)

(b) Find the power (in milliwatts) at the terminals of the inductor when $t = 200$ ms.

(c) Is the inductor absorbing or delivering power at 200 ms?

(d) Find the energy (in millijoules) stored in the inductor at 200 ms.

(e) Find the maximum energy (in millijoules) stored in the inductor and the time (in milliseconds) when it occurs.

M
P **5.4.** The voltage at the terminals of the 300 μH inductor in Fig. P5.4(a) is shown in Fig. P5.4(b). The inductor current i is known to be zero for $t \leq 0$.

(a) Derive the expressions for i for $t \geq 0$.

(b) Sketch i versus t for $0 \leq t \leq \infty$.

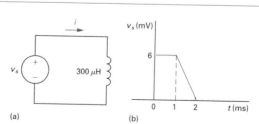

(a)   (b)

Figure P5.4

5.5. The triangular current pulse shown in Fig. P5.5 is applied to a 375 mH inductor.

(a) Write the expressions that describe $i(t)$ in the four intervals $t < 0, 0 \le t \le 25$ ms, 25 ms $\le t \le 50$ ms, and $t > 50$ ms.

(b) Derive the expressions for the inductor voltage, power, and energy. Use the passive sign convention.

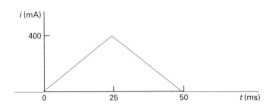

Figure P5.5

5.6. The current in and the voltage across a 5 H inductor are known to be zero for $t \le 0$. The voltage across the inductor is given by the graph in Fig. P5.6 for $t \ge 0$.

(a) Derive the expression for the current as a function of time in the intervals $0 \le t \le 1$ s, 1 s $\le t \le 3$ s, 3 s $\le t \le 5$ s, 5 s $\le t \le 6$ s, and 6 s $\le t \le \infty$.

(b) For $t > 0$, what is the current in the inductor when the voltage is zero?

(c) Sketch i versus t for $0 \le t \le \infty$.

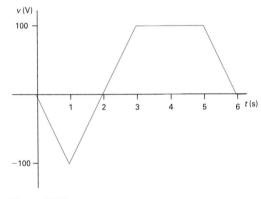

Figure P5.6

5.7. The current in a 20 mH inductor is known to be

$$i = 50 \text{ mA}, \quad t \le 0,$$

$$i = A_1 e^{-2500t} + A_2 e^{-7500t} \text{ A}, \quad t \ge 0.$$

The voltage across the inductor (passive sign convention) is 10 V at $t = 0$.

(a) Find the expression for the voltage across the inductor for $t > 0$.

(b) Find the time, greater than zero, when the power at the terminals of the inductor is zero.

5.8. Assume in Problem 5.7 that the value of the voltage across the inductor at $t = 0$ is -100 V instead of 10 V.

    (a) Find the numerical expressions for i and v for $t \geq 0$.

    (b) Specify the time intervals when the inductor is storing energy and the time intervals when the inductor is delivering energy.

    (c) Show that the total energy extracted from the inductor is equal to the total energy stored.

 5.9. The current in a 25 mH inductor is known to be -10 A for $t \leq 0$ and $-[10 \cos 400t + 5 \sin 400t]e^{-200t}$ A for $t \geq 0$. Assume the passive sign convention.

    (a) At what instant of time is the voltage across the inductor maximum?

    (b) What is the maximum voltage?

 5.10. **(a)** Find the inductor current in the circuit in Fig. P5.10 if $v = 250 \sin 1000t$ V, $L = 50$ mH, and $i(0) = -5$ A.

    (b) Sketch $v, i, p,$ and w versus t. In making these sketches, use the format used in Fig. 5.8. Plot over one complete cycle of the voltage waveform.

    (c) Describe the subintervals in the time interval between 0 and 2π ms when power is being absorbed by the inductor. Repeat for the subintervals when power is being delivered by the inductor.

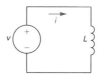

Figure P5.10

 5.11. The current in a 2 H inductor is

$$i = 25 \text{ A}, \quad t \leq 0,$$

$$i = (B_1 \cos 5t + B_2 \sin 5t)e^{-1t} \text{ A}, \quad t \geq 0.$$

The voltage across the inductor (passive sign convention) is 100 V at $t = 0$. Calculate the power at the terminals of the inductor at $t = 0.5$ s. State whether the inductor is absorbing or delivering power.

5.12. The expressions for voltage, power, and energy derived in Example 5.5 involved both integration and manipulation of algebraic expressions. As an engineer, you cannot accept such results on faith alone. That is, you should develop the habit of asking yourself, "Do these results make sense in terms of the known behavior of the circuit they purport to describe?" With these thoughts in mind, test the expressions of Example 5.5 by performing the following checks:

(a) Check the expressions to see whether the voltage is continuous in passing from one time interval to the next.

(b) Check the power expression in each interval by selecting a time within the interval and seeing whether it gives the same result as the corresponding product of v and i. For example, test at 10 and 30 μs.

(c) Check the energy expression within each interval by selecting a time within the interval and seeing whether the energy equation gives the same result as $\frac{1}{2}Cv^2$. Use 10 and 30 μs as test points.

5.13. The rectangular-shaped current pulse shown in Fig. P5.13 is applied to a 0.5 μF capacitor. The initial voltage on the capacitor is a 20 V drop in the reference direction of the current. Assume the passive sign convention. Derive the expression for the capacitor voltage for the time intervals in (a)–(c).

(a) $0 \leq t \leq 50 \, \mu$s

(b) $50 \, \mu$s $\leq t \leq 200 \, \mu$s

(c) $200 \, \mu$s $\leq t \leq \infty$

(d) Sketch $v(t)$ over the interval $-50 \, \mu$s $\leq t \leq 300 \, \mu$s

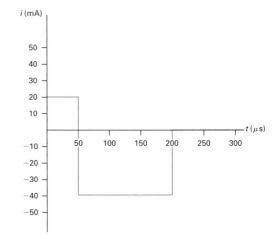

Figure P5.13

5.14. A 0.8 μF capacitor is subjected to a voltage pulse having a duration of 3 s. The pulse is described by the following equations:

$$v_c(t) = \begin{cases} 20t^3 \text{ V}, & 0 \leq t \leq 1 \text{ s}; \\ 2.5(3-t)^3 \text{ V}, & 1 \text{ s} \leq t \leq 3 \text{ s}; \\ 0 & \text{elsewhere.} \end{cases}$$

Sketch the current pulse that exists in the capacitor during the 3 s interval.

5.15. The initial voltage on the 0.2 μF capacitor shown in Fig. P5.15(a) is -60.6 V. The capacitor current has the waveform shown in Fig. P5.15(b).

(a) How much energy, in microjoules, is stored in the capacitor at $t = 250\,\mu s$?

(b) Repeat (a) for $t = \infty$.

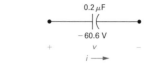

(a)

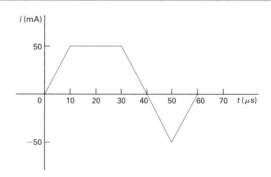

(b)

Figure P5.15

5.16. The voltage across the terminals of a 0.40 μF capacitor is

$$v = \begin{cases} 25 \text{ V}, & t \leq 0; \\ A_1 t e^{-1500t} + A_2 e^{-1500t} \text{ V}, & t \geq 0. \end{cases}$$

The initial current in the capacitor is 90 mA. Assume the passive sign convention.

(a) What is the initial energy stored in the capacitor?

(b) Evaluate the coefficients A_1 and A_2.

(c) What is the expression for the capacitor current?

5.17. The current pulse shown in Fig. P5.17 is applied to a 0.25 μF capacitor. The initial voltage on the capacitor is zero.

(a) Find the charge on the capacitor at $t = 30\,\mu s$.

(b) Find the voltage on the capacitor at $t = 50\,\mu s$.

(c) How much energy is stored in the capacitor by the current pulse?

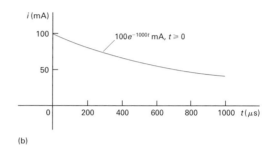

Figure P5.17

5.18. The voltage at the terminals of the capacitor in Fig. 5.10 is known to be

$$v = \begin{cases} -60 \text{ V}, & t \leq 0; \\ 15 - 15e^{-500t}(5\cos 2000t + \sin 2000t) \text{ V} & t \geq 0. \end{cases}$$

Assume $C = 0.4\,\mu\text{F}$.

   (a) Find the current in the capacitor for $t < 0$.

   (b) Find the current in the capacitor for $t > 0$.

   (c) Is there an instantaneous change in the voltage across the capacitor at $t = 0$?

   (d) Is there an instantaneous change in the current in the capacitor at $t = 0$?

   (e) How much energy (in microjoules) is stored in the capacitor at $t = \infty$?

5.19. Assume that the initial energy stored in the inductors of Fig. P5.19 is zero. Find the equivalent inductance with respect to the terminals a,b.

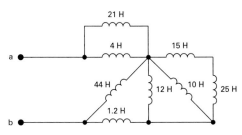

Figure P5.19

5.20. Assume that the initial energy stored in the inductors of Fig. P5.20 is zero. Find the equivalent inductance with respect to the terminals a,b.

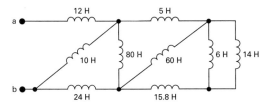

Figure P5.20

5.21. Derive the equivalent circuit for a series connection of ideal capacitors. Assume that each capacitor has its own initial voltage. Denote these initial voltages as $v_1(t_0)$, $v_2(t_0)$, and so on. (*Hint:* Sum the voltages across the string of capacitors, recognizing that the series connection forces the current in each capacitor to be the same.)

5.22. Derive the equivalent circuit for a parallel connection of ideal capacitors. Assume that the initial voltage across the paralleled capacitors is $v(t_0)$. (*Hint:* Sum the currents into the string of capacitors, recognizing that the parallel connection forces the voltage across each capacitor to be the same.)

5.23. Find the equivalent capacitance with respect to the terminals a,b for the circuit shown in Fig. P5.23.

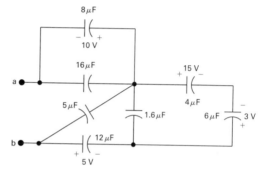

Figure P5.23

5.24. Find the equivalent capacitance with respect to the terminals a,b for the circuit shown in Fig. P5.24.

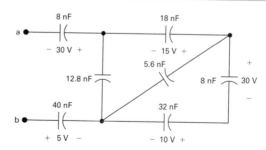

Figure P5.24

5.25. The three inductors in the circuit in Fig. P5.25 are connected across the terminals of a black box at $t = 0$. The resulting voltage for $t \geq 0$ is known to be

$$v_b = 1250e^{-25t} \text{ V}.$$

If $i_1(0) = 10$ A and $i_2(0) = -5$ A, find

(a) $i_o(0)$

(b) $i_o(t), t \geq 0$

(c) $i_1(t), t \geq 0$

(d) $i_2(t), t \geq 0$

(e) the initial energy stored in the three inductors

(f) the total energy delivered to the black box; and

(g) the energy trapped in the ideal inductors

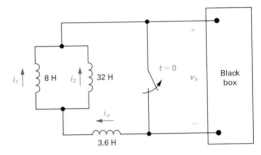

Figure P5.25

5.26. For the circuit shown in Fig. P5.25, how many milliseconds after the switch is opened is the energy delivered to the black box 80% of the total amount delivered?

5.27. The two parallel inductors in Fig. P5.27 are connected across the terminals of a black box at $t = 0$. The resulting voltage v for $t \geq 0$ is known to be $-1800e^{-20t}$ V. It is also known that $i_1(0) = 4$ A and $i_2(0) = -16$ A.

(a) Replace the original inductors with an equivalent inductor and find $i(t)$ for $t \geq 0$.

(b) Find $i_1(t)$ for $t \geq 0$.

(c) Find $i_2(t)$ for $t \geq 0$.

(d) How much energy is delivered to the black box in the time interval $0 \leq t \leq \infty$?

(e) How much energy was initially stored in the parallel inductors?

(f) How much energy is trapped in the ideal inductors?

(g) Show the solutions for i_1 and i_2 agree with the answer obtained in (f).

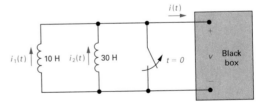

Figure P5.27

5.28. The two series-connected capacitors in Fig. P5.28 are connected to the terminals of a black box at $t = 0$. The resulting current $i(t)$ for $t \geq 0$ is known to be $900e^{-2500t}\ \mu\text{A}$.

(a) Replace the original capacitors with an equivalent capacitor and find $v_o(t)$ for $t \geq 0$.

(b) Find $v_1(t)$ for $t \geq 0$.

(c) Find $v_2(t)$ for $t \geq 0$.

(d) How much energy is delivered to the black box in the time interval $0 \leq t \leq \infty$?

(e) How much energy was initially stored in the series capacitors?

(f) How much energy is trapped in the ideal capacitors?

(g) Show the solutions for v_1 and v_2 agree with the answer obtained in (f).

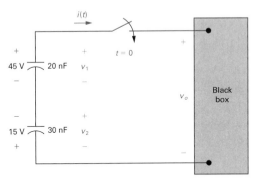

Figure P5.28

5.29. The four capacitors in the circuit in Fig. P5.29 are connected across the terminals of a black box at $t = 0$. The resulting current i_b for $t \geq 0$ is known to be

$$i_b = 50e^{-250t}\ \mu\text{A}.$$

If $v_a(0) = 15$ V, $v_c(0) = -45$ V, and $v_d(0) = 40$ V, find the following for $t \geq 0$: (a) $v_b(t)$, (b) $v_a(t)$, (c) $v_c(t)$, (d) $v_d(t)$, (e) $i_1(t)$, and (f) $i_2(t)$.

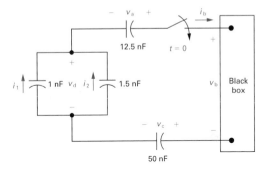

Figure P5.29

5.30. For the circuit in Fig. P5.29, calculate

(a) the initial energy stored in the capacitors

(b) the final energy stored in the capacitors

(c) the total energy delivered to the black box

(d) the percentage of the initial energy stored that is delivered to the black box

(e) the time, in milliseconds, it takes to deliver 5 μJ to the black box.

5.31. In the circuit in Fig. P5.31, the voltage and current expressions are

$$v = 100e^{-80t} \text{ V}, \quad t \geq 0^+;$$

$$i = 4e^{-80t} \text{ A}, \quad t \geq 0.$$

Figure P5.31

Find

(a) R

(b) τ (in milliseconds)

(c) L

(d) the initial energy stored in the inductor

(e) the time (in milliseconds) it takes to dissipate 80% of the initial stored energy

5.32. The switch in Fig. P5.32 has been closed for a long time before opening at $t = 0$. Find

(a) $i_L(t)$, $t \geq 0$

(b) $v_L(t)$, $t \geq 0^+$

(c) $i_\Delta(t)$, $t \geq 0^+$

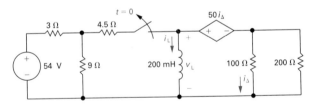

Figure P5.32

5.33. What percentage of the initial energy stored in the inductor in the circuit in Fig. P5.32 is dissipated by the current-controlled voltage source?

5.34. In the circuit shown in Fig. P5.34, the switch makes contact with position b just before breaking contact with position a. As already mentioned, this is known as a make-before-break switch and is designed so that the switch does not interrupt the current in an inductive circuit. The interval of time between "making" and "breaking" is assumed to be negligible. The switch has been in the a position for a long time. At $t = 0$ the switch is thrown from position a to position b.

(a) Determine the initial current in the inductor.

(b) Determine the time constant of the circuit for $t > 0$.

(c) Find i, v_1, and v_2 for $t \geq 0$.

(d) What percentage of the initial energy stored in the inductor is dissipated in the 20 Ω resistor 12 ms after the switch is thrown from position a to position b?

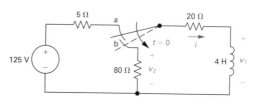

Figure P5.34

5.35. The switch in the circuit in Fig. P5.35 has been closed for a long time before opening at $t = 0$.

(a) Find $i_1(0^-)$ and $i_2(0^-)$.

(b) Find $i_1(0^+)$ and $i_2(0^+)$.

(c) Find $i_1(t)$ for $t \geq 0$.

(d) Find $i_2(t)$ for $t \geq 0^+$.

(e) Explain why $i_2(0^-) \neq i_2(0^+)$.

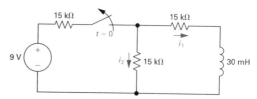

Figure P5.35

5.36. The switch in the circuit in Fig. P5.36 has been closed a long time. At $t = 0$ it is opened. Find $v_o(t)$ for $t \geq 0^+$.

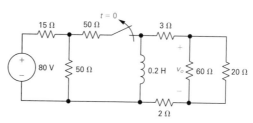

Figure P5.36

5.37. Assume that the switch in the circuit in Fig. P5.36 has been open for one time constant. At this instant, what percentage of the total energy stored in the 0.2 H inductor has been dissipated in the 20 Ω resistor?

5.38. The switch in the circuit in Fig. P5.38 has been closed for a long time before opening at $t = 0$. Find $v_o(t)$ for $t \geq 0^+$.

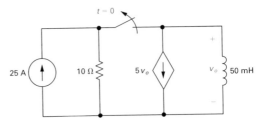

Figure P5.38

5.39. The 220 V, 1 Ω source in the circuit in Fig. P5.39 is inadvertently short-circuited at its terminals a,b. At the time the fault occurs, the circuit has been in operation for a long time.

(a) What is the initial value of the current i_{ab} in the short-circuit connection between terminals a,b?

(b) What is the final value of the current i_{ab}?

(c) How many microseconds after the short circuit has occurred is the current in the short equal to 210 A?

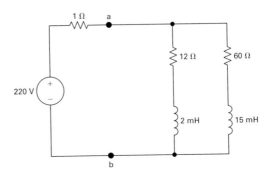

Figure P5.39

5.40. The switch shown in Fig. P5.40 has been open a long time before closing at $t = 0$.

(a) Find $i_o(0^-)$.

(b) Find $i_L(0^-)$.

(c) Find $i_o(0^+)$.

(d) Find $i_L(0^+)$.

(e) Find $i_o(\infty)$.

(f) Find $i_L(\infty)$.

(g) Write the expression for $i_L(t)$ for $t \geq 0$.

(h) Find $v_L(0^-)$.

(i) Find $v_L(0^+)$.

(j) Find $v_L(\infty)$.

(k) Write the expression for $v_L(t)$ for $t \geq 0^+$.

(l) Write the expression for $i_o(t)$ for $t \geq 0^+$.

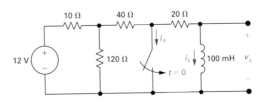

Figure P5.40

5.41. The two switches shown in the circuit in Fig. P5.41 operate simultaneously. Prior to $t = 0$ each switch has been in its indicated position for a long time. At $t = 0$ the two switches move instantaneously to their new positions. Find

(a) $v_o(t), t \geq 0^+$

(b) $i_o(t), t \geq 0$

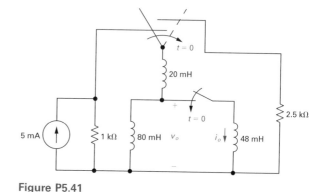

Figure P5.41

5.42. For the circuit seen in Fig. P5.41, find

(a) the total energy dissipated in the 2.5 kΩ resistor

(b) the energy trapped in the ideal inductors

5.43. The switch in the circuit in Fig. P5.43 has been open for a long time. At $t = 0$ the switch is closed.

(a) Determine $i_o(0^+)$ and $i_o(\infty)$.

(b) Determine $i_o(t)$ for $t \geq 0^+$.

(c) How many microseconds after the switch has been closed will the current in the switch equal 3.8 A?

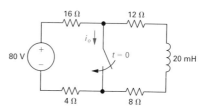

Figure P5.43

5.44. In the circuit shown in Fig. P5.44, the switch has been in position a for a long time. At $t = 0$, it moves instantaneously from a to b.

(a) Find $v_o(t)$ for $t \geq 0^+$.

(b) What is the total energy delivered to the 1 kΩ resistor?

(c) How many time constants does it take to deliver 95% of the energy found in (b)?

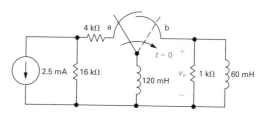

Figure P5.44

5.45. The switch in the circuit in Fig. P5.45 has been in position 1 for a long time. At $t = 0$, the switch moves instantaneously to position 2. Find $v_o(t)$ for $t \geq 0^+$.

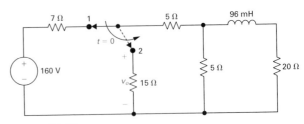

Figure P5.45

5.46. For the circuit of Fig. P5.45, what percentage of the initial energy stored in the inductor is eventually dissipated in the 20 Ω resistor?

5.47. The switch in the circuit seen in Fig. P5.47 has been in position 1 for a long time. At $t = 0$, the switch moves instantaneously to position 2. Find the value of R so that 50% of the initial energy stored in the 20 mH inductor is dissipated in R in 10 μs.

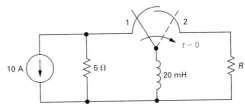

Figure P5.47

5.48. In the circuit in Fig. P5.47, let I_g represent the dc current source, σ represent the fraction of initial energy stored in the inductor that is dissipated in t_o seconds, and L represent the inductance.

(a) Show that

$$R = \frac{L \ln[1/(1 - \sigma)]}{2t_o}.$$

(b) Test the expression derived in (a) by using it to find the value of R in Problem 5.47.

5.49. In the circuit in Fig. P5.49, the switch has been closed for a long time before opening at $t = 0$.

(a) Find the value of L so that $v_o(t)$ equals $0.25 \, v_o(0^+)$ when $t = 5$ ms.

(b) Find the percentage of the stored energy that has been dissipated in the 50 Ω resistor when $t = 5$ ms.

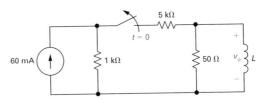

Figure P5.49

5.50. The two switches in the circuit seen in Fig. P5.50 are synchronized. The switches have been closed for a long time before opening at $t = 0$.

(a) How many microseconds after the switches are open is the energy dissipated in the 4 kΩ resistor 10% of the initial energy stored in the 6 H inductor?

(b) At the time calculated in (a), what percentage of the total energy stored in the inductor has been dissipated?

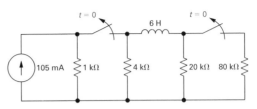

Figure P5.50

5.51. In the circuit in Fig. P5.51 the voltage and current expressions are

$$v = 100e^{-1000t} \text{ V}, \quad t \geq 0,$$

$$i = 5e^{-1000t} \text{ mA}, \quad t \geq 0^+.$$

Figure P5.51

Find

(a) R

(b) C

(c) τ (in milliseconds)

(d) the initial energy stored in the capacitor

(e) how many microseconds it takes to dissipate 80% of the initial energy stored in the capacitor

 5.52. The switch in the circuit in Fig. P5.52 has been in position a for a long time. At $t = 0$, the switch is thrown to position b. Calculate

(a) i, v_1, and v_2 for $t \geq 0^+$ if $v_2(0^-) = 0$ V

(b) the energy stored in the capacitor at $t = 0$

(c) the energy trapped in the circuit and the total energy dissipated in the 5 kΩ resistor if the switch remains in position b indefinitely

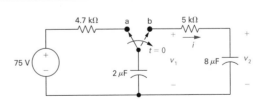

Figure P5.52

5.53. At the time the switch is closed in the circuit in Fig. P5.53, the voltage across the paralleled capacitors is 30 V and the voltage on the 200 nF capacitor is 10 V.

(a) What percentage of the initial energy stored in the three capacitors is dissipated in the 25 kΩ resistor?

(b) Repeat (a) for the 625 Ω and 15 kΩ resistors.

(c) What percentage of the initial energy is trapped in the capacitors?

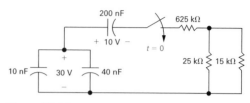

Figure P5.53

5.54. At the time the switch is closed in the circuit shown in Fig. P5.54, the capacitors are charged as shown.

(a) Find $v_o(t)$ for $t \geq 0^+$.

(b) What percentage of the total energy initially stored in the three capacitors is dissipated in the 25 kΩ resistor?

(c) Find $v_1(t)$ for $t \geq 0$.

(d) Find $v_2(t)$ for $t \geq 0$.

(e) Find the energy (in millijoules) trapped in the ideal capacitors.

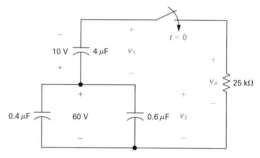

Figure P5.54

5.55. The switch in the circuit seen in Fig. P5.55 has been in position x for a long time. At $t = 0$, the switch moves instantaneously to position y.

(a) Find α so that the time constant for $t > 0$ is 1 ms.

(b) For the α found in (a), find v_Δ.

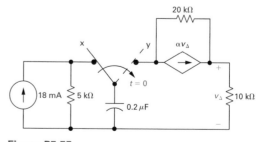

Figure P5.55

5.56. **(a)** In Problem 5.55, how many microjoules of energy are generated by the dependent current source during the time the capacitor discharges to 0 V?

(b) Show that for $t \geq 0$ the total energy stored and generated in the capacitive circuit equals the total energy dissipated.

5.57. In the circuit shown in Fig. P5.57, both switches operate together; that is, they either open or close at the same time. The switches are closed a long time before opening at $t = 0$.

(a) How many microjoules of energy have been dissipated in the 12 kΩ resistor 2 ms after the switches open?

(b) How long does it take to dissipate 95% of the initially stored energy?

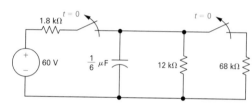

Figure P5.57

5.58. Both switches in the circuit in Fig. P5.58 have been closed for a long time. At $t = 0$, both switches open simultaneously.

(a) Find $i_o(t)$ for $t \geq 0^+$.

(b) Find $v_o(t)$ for $t \geq 0$.

(c) Calculate the energy (in microjoules) trapped in the circuit.

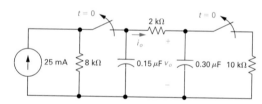

Figure P5.58

5.59. The switch in the circuit in Fig. P5.59 has been in position a for a long time. At $t = 0$, the switch is thrown to position b.

(a) Find $i_o(t)$ for $t \geq 0^+$.

(b) What percentage of the initial energy stored in the capacitor is dissipated in the 4 kΩ resistor 250 μs after the switch has been thrown?

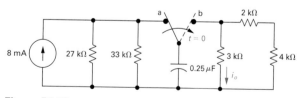

Figure P5.59

P **5.60.** After the circuit in Fig. P5.60 has been in operation for a long time, a screwdriver is inadvertently
connected across the terminals a,b. Assume the
resistance of the screwdriver is negligible.

(a) Find the current in the screwdriver at $t = 0^+$
and $t = \infty$.

(b) Derive the expression for the current in the
screwdriver for $t \geq 0^+$.

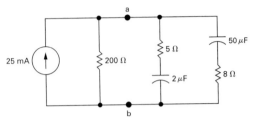

Figure P5.60

P **5.61.** The switch in the circuit in Fig. P5.61 has been in
position 1 for a long time before moving to posi-
tion 2 at $t = 0$. Find $i_o(t)$ for $t \geq 0^+$.

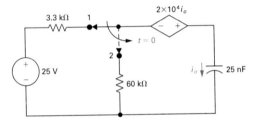

Figure P5.61

M
P **5.62.** The switch in the circuit seen in Fig. P5.62 has
been closed for a long time. The switch opens at
$t = 0$. Find the numerical expressions for $i_o(t)$
and $v_o(t)$ when $t \geq 0^+$.

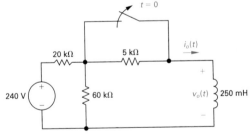

Figure P5.62

5.63. The switch in the circuit shown in Fig. P5.63 has been closed for a long time. The switch opens at $t = 0$. For $t \geq 0^+$:

 (a) Find $v_o(t)$ as a function of I_g, R_1, R_2, and L.

 (b) Verify your expression by using it to find $v_o(t)$ in the circuit of Fig. P5.63.

 (c) Explain what happens to $v_o(t)$ as R_2 gets larger and larger.

 (d) Find v_{SW} as a function of I_g, R_1, R_2, and L.

 (e) Explain what happens to v_{SW} as R_2 gets larger and larger.

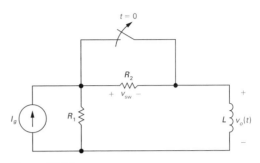

Figure P5.63

 5.64. The switch in the circuit in Fig. P5.64 has been in position 1 for a long time. At $t = 0$ it moves instantaneously to position 2. How many milliseconds after the switch operates does v_o equal -80 V?

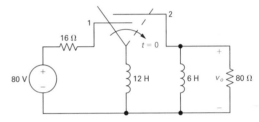

Figure P5.64

5.65. For the circuit in Fig. P5.64, find (in joules):

 (a) the total energy dissipated in the 80 Ω resistor

 (b) the energy trapped in the inductors

 (c) the initial energy stored in the inductors

 5.66. The switch in the circuit in Fig. P5.66 has been in position x for a long time. The initial charge on the 10 nF capacitor is zero. At $t = 0$, the switch moves instantaneously to position y.

 (a) Find $v_o(t)$ for $t \geq 0^+$.

 (b) Find $v_1(t)$ for $t \geq 0$.

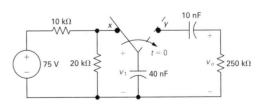

Figure P5.66

5.67. For the circuit in Fig. P5.66, find (in microjoules)

    (a) the energy delivered to the 250 kΩ resistor

    (b) the energy trapped in the capacitors

    (c) the initial energy stored in the capacitors

5.68. The circuit in Fig. P5.68 has been in operation for a long time. At $t = 0$, the voltage source reverses polarity and the current source drops from 3 mA to 2 mA. Find $v_o(t)$ for $t \geq 0$.

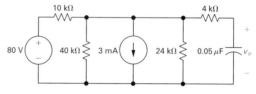

Figure P5.68

5.69. The switch in the circuit of Fig. P5.69 has been in position a for a long time. At $t = 0$, it moves instantaneously to position b. For $t \geq 0^+$, find

    (a) $v_o(t)$

    (b) $i_o(t)$

    (c) $v_1(t)$

    (d) $v_2(t)$

    (e) the energy trapped in the capacitors as $t \rightarrow \infty$

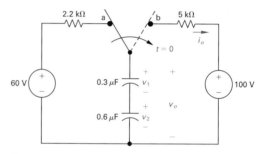

Figure P5.69

5.70. The switch in the circuit seen in Fig. P5.70 has been in position a for a long time. At $t = 0$, the switch moves instantaneously to position b. For $t \geq 0^+$, find

    (a) $v_o(t)$

    (b) $i_o(t)$

    (c) $v_g(t)$

    (d) $v_g(0^+)$

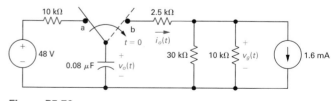

Figure P5.70

5.71. The voltage waveform shown in Fig. P5.71(a) is applied to the circuit of Fig. P5.71(b). The initial voltage on the capacitor is zero.

   (a) Calculate $v_o(t)$.

   (b) Make a sketch of $v_o(t)$ versus t.

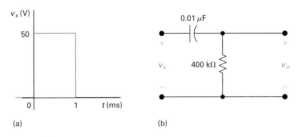

Figure P5.71

5.72. The voltage signal source in the circuit in Fig. P5.72(a) is generating the signal shown in Fig. P5.72(b). There is no stored energy at $t = 0$.

   (a) Derive the expressions for $v_o(t)$ that apply in the intervals $t < 0$; $0 \le t \le 10$ ms; 10 ms $\le t \le 20$ ms; and 20 ms $\le t \le \infty$.

   (b) Sketch v_o and v_s on the same coordinate axes.

   (c) Repeat (a) and (b) with R reduced to 10 kΩ.

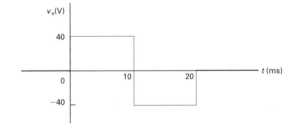

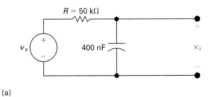

Figure P5.72

5.73. The input signal to the integrating amplifier circuit shown in Fig. 5.47 is $75 \cos 5000t$ V. Write the expression for the output voltage v_o if $R_s = 50$ kΩ; $C_f = 0.05 \mu$F; $V_{CC} = 10$ V; and the initial value of v_o is 0 V.

5.74. The input signal to the integrating amplifier circuit in Fig. 5.47 is the triangular voltage shown in Fig. P5.74. The integrating-amplifier parameters are: $R_s = 7.5\,k\Omega$; $C_f = 0.16\mu F$; and $V_{CC} = 15$ V. Assume v_o is zero when t is zero.

(a) Derive the expression for v_o when $0 \le t \le 25$ ms.

(b) Derive the expression for v_o when $25\,ms \le t \le 75$ ms.

(c) Derive the expression for v_o when $75\,ms \le t \le 100$ ms.

(d) Plot v_o vs. t for $0 \le t \le 100$ ms.

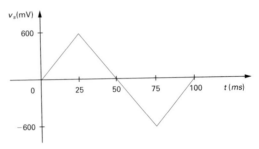

Figure P5.74

5.75. The switch S in the circuit in Fig. 5.52 connects to v_x for 250 ms. Assume v_x is -4 V and the op amp power supply is ± 10 V.

(a) If C equals 0.8 μF, what is the smallest value of R that keeps the op amp operating within its linear range?

(b) If $v_{ref} = 10$ V, how long does it take to discharge the capacitor when $v_x = -4$ V?

◆ **5.76.** Assume a digital multimeter containing the dual-slope integrating amplifier of Problem 5.75 has a pulse generator operating at 100,000 pulses/second.

(a) If $v_x = -3.6$ V, how many pulses are counted during the time the capacitor discharges?

(b) If 7000 pulses are counted during the time the capacitor discharges, what is the value of V_a?

◆ **5.77.** Show that the circuit in Fig. P5.77 can be used to convert an unknown resistance R_x to an unknown dc voltage v_x when v_{ref} is a known dc voltage.

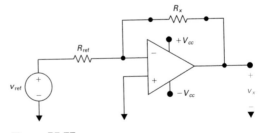

Figure P5.77

Parasitic Inductance

It is not possible to fabricate an integrated electronic circuit without introducing parasitic inductance. This inductance (associated with chip packaging and external wiring) is large enough to adversely affect digital switching times. At the end of this chapter we will analyze the effect inductance has on the step response of a series RLC circuit.

6 Natural and Step Responses of RLC Circuits

Chapter Contents

In this chapter, discussion of the natural response and step response of circuits containing both inductors and capacitors is limited to two simple structures: the parallel RLC circuit and the series RLC circuit. Finding the natural response of a parallel RLC circuit consists of finding the voltage created across the parallel branches by the release of energy stored in the inductor or capacitor or both. The task is defined in terms of the circuit shown in Fig. 6.1. The initial voltage on the capacitor, V_0, represents the initial energy stored in the capacitor. The initial current through the inductor, I_0, represents the initial energy stored in the inductor. If the individual branch currents are of interest, you can find them after determining the terminal voltage.

We derive the step response of a parallel RLC circuit by using Fig. 6.2. We are interested in the voltage that appears across the parallel branches as a result of the sudden application of a dc current source. Energy may or may not be stored in the circuit when the current source is applied.

Finding the natural response of a series RLC circuit consists of finding the current generated in the series-connected elements by the release of initially stored energy in the inductor, capacitor, or both. The task is defined by the circuit shown in Fig. 6.3. As before, the initial inductor current, I_0, and the initial capacitor voltage, V_0, represent the initially stored energy. If any of the

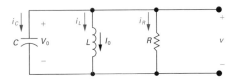

Figure 6.1 A circuit used to illustrate the natural response of a parallel RLC circuit.

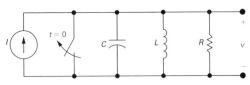

Figure 6.2 A circuit used to illustrate the step response of a parallel RLC circuit.

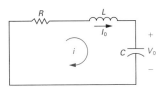

Figure 6.3 A circuit used to illustrate the natural response of a series RLC circuit.

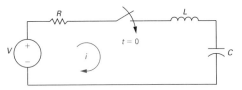

Figure 6.4 A circuit used to illustrate the step response of a series RLC circuit.

individual element voltages are of interest, you can find them after determining the current.

We describe the step response of a series RLC circuit in terms of the circuit shown in Fig. 6.4. We are interested in the current resulting from the sudden application of the dc voltage source. Energy may or may not be stored in the circuit when the switch is closed.

If you have not studied ordinary differential equations, derivation of the natural and step responses of parallel and series RLC circuits may be a bit difficult to follow. However, the results are important enough to warrant presentation at this time. We begin with the natural response of a parallel RLC circuit and cover this material over two sections: one to discuss the solution of the differential equation that describes the circuit and one to present the three distinct forms that the solution can take. After introducing these three forms, we show that the same forms apply to the step response of a parallel RLC circuit as well as to the natural and step responses of series RLC circuits.

6.1 INTRODUCTION TO THE NATURAL RESPONSE OF A PARALLEL RLC CIRCUIT

The first step in finding the natural response of the circuit shown in Fig. 6.1 is to derive the differential equation that the voltage v must satisfy. We choose to find the voltage first, because it is the same for each component. After that, a branch current can be found by using the current-voltage relationship for the branch component. We easily obtain the differential equation for the voltage by summing the currents away from the top node, where each current is expressed as a function of the unknown voltage v:

$$\frac{v}{R} + \frac{1}{L}\int_0^t v\,d\tau + I_0 + C\frac{dv}{dt} = 0. \tag{6.1}$$

We eliminate the integral in Eq. (6.1) by differentiating once with respect to t, and, because I_0 is a constant, we get

$$\frac{1}{R}\frac{dv}{dt} + \frac{v}{L} + C\frac{d^2v}{dt^2} = 0. \tag{6.2}$$

We now divide through Eq. (6.2) by the capacitance C and arrange the derivatives in descending order:

$$\frac{d^2v}{dt^2} + \frac{1}{RC}\frac{dv}{dt} + \frac{v}{LC} = 0. \tag{6.3}$$

Comparing Eq. (6.3) with the differential equations derived in Chapter 5 reveals that they differ by the presence of the term involving the second derivative. Equation (6.3) is an ordinary, second-order differential equation with constant coefficients. Circuits in this chapter contain both inductors and capacitors, so the differential equation describing these circuits is of the second order. Therefore, we sometimes call such circuits **second-order circuits**.

The General Solution of the Second-Order Differential Equation

We can't solve Eq. (6.3) by separating the variables and integrating as we were able to do with the first-order equations in Chapter 5. The classical approach to solving Eq. (6.3) is to assume that the solution is of exponential form, that is, to assume that the voltage is of the form

$$v = Ae^{st},$$
(6.4)

where A and s are unknown constants.

Before showing how this assumption leads to the solution of Eq. (6.3), we need to show that it is rational. The strongest argument we can make in favor of Eq. (6.4) is to note from Eq. (6.3) that the second derivative of the solution, plus a constant times the first derivative, plus a constant times the solution itself, must sum to zero for all values of t. This can occur only if higher-order derivatives of the solution have the same form as the solution. The exponential function satisfies this criterion. A second argument in favor of Eq. (6.4) is that the solutions of all the first-order equations we derived in Chapter 5 were exponential. It seems reasonable to assume that the solution of the second-order equation also involves the exponential function.

If Eq. (6.4) is a solution of Eq. (6.3), it must satisfy Eq. (6.3) for all values of t. Substituting Eq. (6.4) into Eq. (6.3) generates the expression

$$As^2e^{st} + \frac{As}{RC}e^{st} + \frac{Ae^{st}}{LC} = 0,$$

or

$$Ae^{st}\left(s^2 + \frac{s}{RC} + \frac{1}{LC}\right) = 0,$$
(6.5)

which can be satisfied for all values of t only if A is zero or the parenthetical term is zero, because $e^{st} \neq 0$ for any finite values of st. We cannot use $A = 0$ as a general solution because to do so implies that the voltage is zero for all time—a physical impossibility if energy is stored in either the inductor or capacitor. Therefore, in order for Eq. (6.4) to be a solution of Eq. (6.3), the parenthetical term in Eq. (6.5) must be zero, or

$$s^2 + \frac{s}{RC} + \frac{1}{LC} = 0. \tag{6.6}$$

Equation (6.6) is called the **characteristic equation** of the differential equation because the roots of this quadratic equation determine the mathematical character of $v(t)$.

The two roots of Eq. (6.6) are

$$s_1 = -\frac{1}{2RC} + \sqrt{\left(\frac{1}{2RC}\right)^2 - \frac{1}{LC}}, \tag{6.7}$$

$$s_2 = -\frac{1}{2RC} - \sqrt{\left(\frac{1}{2RC}\right)^2 - \frac{1}{LC}}. \tag{6.8}$$

If either root is substituted into Eq. (6.4), the assumed solution satisfies the given differential equation, that is, Eq. (6.3). Note from Eq. (6.5) that this result holds regardless of the value of A. Therefore, both

$$v = A_1 e^{s_1 t},$$

$$v = A_2 e^{s_2 t}$$

satisfy Eq. (6.3). Denoting these two solutions v_1 and v_2, respectively, we can show that their sum also is a solution. Specifically, if we let

$$v = v_1 + v_2 = A_1 e^{s_1 t} + A_2 e^{s_2 t}, \tag{6.9}$$

then

$$\frac{dv}{dt} = A_1 s_1 e^{s_1 t} + A_2 s_2 e^{s_2 t} \tag{6.10}$$

$$\frac{d^2 v}{dt^2} = A_1 s_1^2 e^{s_1 t} + A_2 s_2^2 e^{s_2 t}. \tag{6.11}$$

Substituting Eqs. (6.9)–(6.11) into Eq. (6.3) gives

$$A_1 e^{s_1 t}\left(s_1^2 + \frac{1}{RC}s_1 + \frac{1}{LC}\right) + A_2 e^{s_2 t}\left(s_2^2 + \frac{1}{RC}s_2 + \frac{1}{LC}\right) = 0. \tag{6.12}$$

But each parenthetical term is zero because by definition s_1 and s_2 are roots of the characteristic equation. Hence the natural response of the parallel RLC circuit shown in Fig. 6.1 is of the form

$$v = A_1 e^{s_1 t} + A_2 e^{s_2 t}. \tag{6.13}$$

Equation (6.13) is a repeat of the assumption made in Eq. (6.9). We have shown that v_1 is a solution, v_2 is a solution, and $v_1 + v_2$ is a solution. Therefore, the general solution of Eq. (6.3) has the form given in Eq. (6.13). The roots of the characteristic equation (s_1 and s_2) are determined by the circuit parameters R, L, and C. The initial conditions determine the values of the constants A_1 and A_2. Note that the form of Eq. (6.13) must be modified if the two roots s_1 and s_2 are equal. We discuss this modification when we turn to the critically damped voltage response in Section 6.2.

The behavior of $v(t)$ depends on the values of s_1 and s_2. Therefore the first step in finding the natural response is to determine the roots of the characteristic equation. We return to Eqs. (6.7) and (6.8) and rewrite them using a notation widely used in the literature:

$$s_1 = -\alpha + \sqrt{\alpha^2 - \omega_0^2}, \tag{6.14}$$

$$s_2 = -\alpha - \sqrt{\alpha^2 - \omega_0^2}, \tag{6.15}$$

where

$$\alpha = \frac{1}{2RC}, \tag{6.16}$$

$$\omega_0 = \frac{1}{\sqrt{LC}}. \tag{6.17}$$

These results are summarized in Table 6.1.

The exponent of e must be dimensionless, so both s_1 and s_2 (and hence α and ω_0) must have the dimension of the reciprocal of time, or frequency. To distinguish among the frequencies s_1, s_2, α, and ω_0, we use the following terminology: s_1 and s_2 are referred to as *complex frequencies*, α is called the *neper frequency*,

TABLE 6.1   **Natural Response Parameters of the Parallel RLC Circuit.**

PARAMETER	TERMINOLOGY	VALUE IN NATURAL RESPONSE
s_1, s_2	Characteristic roots	$s_1 = -\alpha + \sqrt{\alpha^2 - \omega_0^2}$
		$s_2 = -\alpha - \sqrt{\alpha^2 - \omega_0^2}$
α	Neper frequency	$\alpha = \frac{1}{2RC}$
ω_0	Resonant radian frequency	$\omega_0 = \frac{1}{\sqrt{LC}}$

and ω_0 is the *resonant radian frequency.* The full significance of this terminology unfolds as we move through the remaining chapters of this book. All these frequencies have the dimension of angular frequency per time. For complex frequencies, the neper frequency, and the resonant radian frequency, we specify values using the unit *radians per second* (rad/s). The nature of the roots s_1 and s_2 depends on the values of α and ω_0. There are three possible outcomes. First, if $\omega_0^2 < \alpha^2$, both roots will be real and distinct. For reasons to be discussed later, the voltage response is said to be **overdamped** in this case. Second, if $\omega_0^2 > \alpha^2$, both s_1 and s_2 will be complex and, in addition, will be conjugates of each other. In this situation, the voltage response is said to be **underdamped**. The third possible outcome is that $\omega_0^2 = \alpha^2$. In this case, s_1 and s_2 will be real and equal. Here the voltage response is said to be **critically damped**. As we shall see, damping affects the way the voltage response reaches its final (or steady-state) value. We discuss each case separately in Section 6.2.

Example 6.1 illustrates how the numerical values of s_1 and s_2 are determined by the values of R, L, and C.

EXAMPLE 6.1

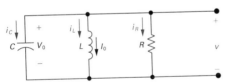

Figure 6.5 A circuit used to illustrate the natural response of a parallel *RLC* circuit.

(a) Find the roots of the characteristic equation that governs the transient behavior of the voltage shown in Fig. 6.5 if $R = 200\ \Omega$, $L = 50$ mH, and $C = 0.2\ \mu$F.

(b) Will the response be overdamped, underdamped, or critically damped?

(c) Repeat (a) and (b) for $R = 312.5\ \Omega$.

(d) What value of R causes the response to be critically damped?

SOLUTION

(a) For the given values of R, L, and C,

$$\alpha = \frac{1}{2RC} = \frac{10^6}{(400)(0.2)} = 1.25 \times 10^4 \text{ rad/s},$$

$$\omega_0^2 = \frac{1}{LC} = \frac{(10^3)(10^6)}{(50)(0.2)} = 10^8 \text{ rad}^2/\text{s}^2.$$

From Eqs. (6.14) and (6.15),

$$s_1 = -1.25 \times 10^4 + \sqrt{1.5625 \times 10^8 - 10^8}$$

$$= -12{,}500 + 7500 = -5000 \text{ rad/s},$$

$$s_2 = -1.25 \times 10^4 - \sqrt{1.5625 \times 10^8 - 10^8}$$

$$= -12{,}500 - 7500 = -20{,}000 \text{ rad/s}.$$

(b) The voltage response is overdamped because $\omega_0^2 < \alpha^2$.

(c) For $R = 312.5 \ \Omega$,

$$\alpha = \frac{10^6}{(625)(0.2)} = 8000 \text{ rad/s},$$

$$\alpha^2 = 64 \times 10^6 = 0.64 \times 10^8 \text{ rad}^2/\text{s}^2.$$

As ω_0^2 remains at $10^8 \text{ rad}^2/\text{s}^2$,

$$s_1 = -8000 + j6000 \text{ rad/s},$$

$$s_2 = -8000 - j6000 \text{ rad/s}.$$

(In electrical and computer engineering, the imaginary number $\sqrt{-1}$ is represented by the letter j, because the letter i represents current.)

In this case the voltage response is underdamped, since $\omega_0^2 > \alpha^2$.

(d) For critical damping, $\alpha^2 = \omega_0^2$, so

$$\left(\frac{1}{2RC}\right)^2 = \frac{1}{LC} = 10^8,$$

or

$$\frac{1}{2RC} = 10^4,$$

and

$$R = \frac{10^6}{(2 \times 10^4)(0.2)} = 250 \ \Omega.$$

DRILL EXERCISE

6.1 The resistance and inductance of the circuit in Fig. 6.5 are 100 Ω and 20 mH, respectively.

(a) Find the value of C that makes the voltage response critically damped.

(b) If C is adjusted to give a neper frequency of 5 krad/s, find the value of C and the roots of the characteristic equation.

(c) If C is adjusted to give a resonant frequency of 20 krad/s, find the value of C and the roots of the characteristic equation.

ANSWER: (a) $C = 500$ nF; (b) $C = 1 \mu$F, $s_1 = -5000 + j5000$ rad/s, $s_2 = -5000 - j5000$ rad/s; (c) $C = 125$ nF, $s_1 = -5359$ rad/s, $s_2 = -74{,}641$ rad/s.

6.2 THE FORMS OF THE NATURAL RESPONSE OF A PARALLEL RLC CIRCUIT

So far we have seen that the behavior of a second-order RLC circuit depends on the values of s_1 and s_2, which in turn depend on the circuit parameters R, L, and C. Therefore, the first step in finding the natural response is to calculate these values and, relatedly, determine whether the response is over-, under-, or critically damped.

Completing the description of the natural response requires finding two unknown coefficients, such as A_1 and A_2 in Eq. (6.13). The method used to do this is based on matching the solution for the natural response to the initial conditions imposed by the circuit, which are the initial value of the current (or voltage) and the initial value of the first derivative of the current (or voltage). Note that these same initial conditions, plus the final value of the variable, will also be needed when finding the step response of a second-order circuit.

In this section, we analyze the natural response form for each of the three types of damping, beginning with the overdamped response. As we will see, the response equations, as well as the equations for evaluating the unknown coefficients, are slightly different for each of the three damping configurations. This is why we want to determine at the outset of the problem whether the response is over-, under-, or critically damped.

The Overdamped Voltage Response

When the roots of the characteristic equation are real and distinct, the voltage response of a parallel RLC circuit is said to be overdamped. The solution for the voltage is of the form

$$v = A_1 e^{s_1 t} + A_2 e^{s_2 t}, \tag{6.18}$$

where s_1 and s_2 are the roots of the characteristic equation. The constants A_1 and A_2 are determined by the initial conditions, specifically from the values of $v(0^+)$ and $dv(0^+)/dt$, which in turn are determined from the initial voltage on the capacitor, V_0, and the initial current in the inductor, I_0.

Next, we show how to use the initial voltage on the capacitor and the initial current in the inductor to find A_1 and A_2. First we note from Eq. (6.18) that

$$v(0^+) = A_1 + A_2, \tag{6.19}$$

$$\frac{dv(0^+)}{dt} = s_1 A_1 + s_2 A_2. \tag{6.20}$$

With s_1 and s_2 known, the task of finding A_1 and A_2 reduces to finding $v(0^+)$ and $dv(0^+)/dt$. The value of $v(0^+)$ is the initial voltage on the capacitor V_0. We get the initial value of dv/dt by first finding the current in the capacitor branch at $t = 0^+$. Then,

$$\frac{dv(0^+)}{dt} = \frac{i_C(0^+)}{C}. \tag{6.21}$$

We use Kirchhoff's current law to find the initial current in the capacitor branch. We know that the sum of the three branch currents at $t = 0^+$ must be zero. The current in the resistive branch at $t = 0^+$ is the initial voltage V_0 divided by the resistance, and the current in the inductive branch is I_0. Using the reference system depicted in Fig. 6.5, we obtain

$$i_C(0^+) = \frac{-V_0}{R} - I_0. \tag{6.22}$$

After finding the numerical value of $i_C(0^+)$, we use Eq. (6.21) to find the initial value of dv/dt.

We can summarize the process for finding the overdamped response, $v(t)$, as follows:

1. Find the roots of the characteristic equation, s_1 and s_2, using the values of R, L, and C.
2. Find $v(0^+)$ and $dv(0^+)/dt$ using circuit analysis.

3. Find the values of A_1 and A_2 by solving Eqs. (6.23) and (6.24) simultaneously:

$$v(0^+) = A_1 + A_2, \qquad (6.23)$$

$$\frac{dv(0^+)}{dt} = \frac{i_C(0^+)}{C} = s_1 A_1 + s_2 A_2. \qquad (6.24)$$

4. Substitute the values for s_1, s_2, A_1, and A_2 into Eq. (6.18) to determine the expression for $v(t)$ for $t \geq 0$.

Examples 6.2 and 6.3 illustrate how to find the overdamped response of a parallel RLC circuit.

EXAMPLE 6.2

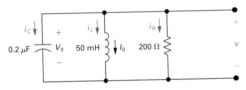

Figure 6.6 The circuit for Example 6.2.

For the circuit in Fig. 6.6, $v(0^+) = 12$ V, and $i_L(0^+) = 30$ mA.

(a) Find the initial current in each branch of the circuit.

(b) Find the initial value of dv/dt.

(c) Find the expression for $v(t)$.

(d) Sketch $v(t)$ in the interval $0 \leq t \leq 250\ \mu$s.

SOLUTION

(a) The inductor prevents an instantaneous change in its current, so the initial value of the inductor current is 30 mA:

$$i_L(0^-) = i_L(0) = i_L(0^+) = 30\text{ mA}.$$

The capacitor holds the initial voltage across the parallel elements to 12 V. Thus the initial current in the resistive branch, $i_R(0^+)$, is 12/200, or 60 mA. Kirchhoff's current law requires the sum of the currents leaving the top node to equal zero at every instant. Hence

$$i_C(0^+) = -i_L(0^+) - i_R(0^+)$$

$$= -90\text{ mA}.$$

Note that if we assumed the inductor current and capacitor voltage had reached their dc values at the instant that energy begins to be released, $i_C(0^-) = 0$. In other words, there is an instantaneous change in the capacitor current at $t = 0$.

(b) Because $i_C = C(dv/dt)$,

$$\frac{dv(0^+)}{dt} = \frac{-90 \times 10^{-3}}{0.2 \times 10^{-6}} = -450 \text{ kV/s}.$$

(c) The roots of the characteristic equation come from the values of R, L, and C. For the values specified and from Eqs. (6.14) and (6.15) along with (6.16) and (6.17),

$$s_1 = -1.25 \times 10^4 + \sqrt{1.5625 \times 10^8 - 10^8}$$

$$= -12{,}500 + 7500 = -5000 \text{ rad/s},$$

$$s_2 = -1.25 \times 10^4 - \sqrt{1.5625 \times 10^8 - 10^8}$$

$$= -12{,}500 - 7500 = -20{,}000 \text{ rad/s}.$$

Because the roots are real and distinct, we know that the response is overdamped and hence has the form of Eq. (6.18). We find the coefficients A_1 and A_2 from Eqs. (6.23) and (6.24). We've already determined s_1, s_2, $v(0^+)$, and $dv(0^+)/dt$, so

$$12 = A_1 + A_2,$$

$$-450 \times 10^3 = -5000A_1 - 20{,}000A_2.$$

We solve two equations for A_1 and A_2 to obtain $A_1 = -14$ V and $A_2 = 26$ V. Substituting these values into Eq. (6.18) yields the overdamped voltage response:

$$v(t) = (-14e^{-5000t} + 26e^{-20{,}000t}) \text{ V}, \quad t \geq 0.$$

As a check on these calculations, we note that the solution yields $v(0) = 12$ V and $dv(0^+)/dt = -450{,}000$ V/s.

(d) Figure 6.7 shows a plot of $v(t)$ versus t over the interval $0 \leq t \leq 250 \ \mu\text{s}$.

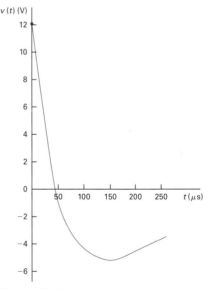

Figure 6.7 The voltage response for Example 6.2.

EXAMPLE 6.3

Derive the expressions that describe the three branch currents i_R, i_L, and i_C in Example 6.2 (Fig. 6.6) during the time the stored energy is being released.

SOLUTION

We know the voltage across the three branches from the solution in Example 6.2, namely,

$$v(t) = (-14e^{-5000t} + 26e^{-20,000t}) \text{ V}, \quad t \geq 0.$$

The current in the resistive branch is then

$$i_R(t) = \frac{v(t)}{200} = (-70e^{-5000t} + 130e^{-20,000t}) \text{ mA}, \quad t \geq 0.$$

There are two ways to find the current in the inductive branch. One way is to use the integral relationship that exists between the current and the voltage at the terminals of an inductor:

$$i_L(t) = \frac{1}{L} \int_0^t v_L(x)\, dx + I_0.$$

A second approach is to find the current in the capacitive branch first and then use the fact that $i_R + i_L + i_C = 0$. Let's use this approach. The current in the capacitive branch is

$$i_C(t) = C\frac{dv}{dt}$$

$$= 0.2 \times 10^{-6}(70,000e^{-5000t} - 520,000e^{-20,000t})$$

$$= (14e^{-5000t} - 104e^{-20,000t}) \text{ mA}, \quad t \geq 0^+.$$

Note that $i_C(0^+) = -90$ mA, which agrees with the result in Example 6.2.

Now we obtain the inductive branch current from the relationship

$$i_L(t) = -i_R(t) - i_C(t)$$

$$= (56e^{-5000t} - 26e^{-20,000t}) \text{ mA}, \quad t \geq 0.$$

We leave it to you, in Drill Exercise 6.2, to show that the integral relation alluded to leads to the same result. Note that the expression for i_L agrees with the initial inductor current, as it must.

DRILL EXERCISES

6.2   Use the integral relationship between i_L and v to find the expression for i_L in Fig. 6.6.

6.3   The element values in the circuit shown are $R = 2\ k\Omega$, $L = 250\ mH$, and $C = 10\ nF$. The initial current I_0 in the inductor is -4 A, and the initial voltage on the capacitor is 0 V. The output signal is the voltage v. Find (a) $i_R(0^+)$; (b) $i_C(0^+)$; (c) $dv(0^+)/dt$; (d) A_1; (e) A_2; and (f) $v(t)$ when $t \geq 0$.

ANSWER: $i_L(t) = (56e^{-5000t} - 26e^{-20,000t})$ mA, $t \geq 0$.

ANSWER: (a) 0; (b) 4 A; (c) 4×10^8 V/s; (d) 13,333 V; (e) $-13,333$ V; (f) $13,333\ (e^{-10,000t} - e^{-40,000t})$ V, $t \geq 0$.

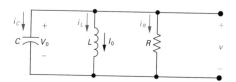

The Underdamped Voltage Response

When $\omega_0^2 > \alpha^2$, the roots of the characteristic equation are complex, and the response is underdamped. For convenience, we express the roots s_1 and s_2 as

$$s_1 = -\alpha + \sqrt{-(w_0^2 - \alpha^2)}$$

$$= \alpha + j\sqrt{\omega_0^2 - \alpha^2}$$

$$= -\alpha + j\omega_d \qquad (6.25)$$

$$s_2 = -\alpha - j\omega_d, \qquad (6.26)$$

where

$$\omega_d = \sqrt{\omega_0^2 - \alpha^2}. \qquad (6.27)$$

The term ω_d is called the **damped radian frequency**. We explain later the reason for this terminology.

The underdamped voltage response of a parallel RLC circuit is

$$v(t) = B_1 e^{-\alpha t} \cos \omega_d t + B_2 e^{-\alpha t} \sin \omega_d t, \qquad (6.28)$$

which follows from Eq. (6.18). In making the transition from Eq. (6.18) to Eq. (6.28), we use the Euler identity:

$$e^{\pm j\theta} = \cos \theta \pm j \sin \theta. \qquad (6.29)$$

Thus,

$$v(t) = A_1 e^{(-\alpha + j\omega_d)t} + A_2 e^{-(\alpha + j\omega_d)t}$$

$$= A_1 e^{-\alpha t} e^{j\omega_d t} + A_2 e^{-\alpha t} e^{-j\omega_d t}$$

$$= e^{-\alpha t}(A_1 \cos \omega_d t + j A_1 \sin \omega_d t + A_2 \cos \omega_d t$$

$$- j A_2 \sin \omega_d t)$$

$$= e^{-\alpha t}[(A_1 + A_2) \cos \omega_d t + j(A_1 - A_2) \sin \omega_d t].$$

At this point in the transition from Eq. (6.18) to (6.28), replace the arbitrary constants $A_1 + A_2$ and $j(A_1 - A_2)$ with new arbitrary constants denoted B_1 and B_2 to get

$$v = e^{-\alpha t}(B_1 \cos \omega_d t + B_2 \sin \omega_d t)$$

$$= B_1 e^{-\alpha t} \cos \omega_d t + B_2 e^{-\alpha t} \sin \omega_d t.$$

The constants B_1 and B_2 are real, not complex, because the voltage is a real function. Don't be misled by the fact that $B_2 = j(A_1 - A_2)$. In this underdamped case, A_1 and A_2 are complex conjugates, and thus B_1 and B_2 are real. (See Problems 6.13 and 6.14.) The reason for defining the underdamped response in terms of the coefficients B_1 and B_2 is that it yields a simpler expression for the voltage, v. We determine B_1 and B_2 by the initial energy stored in the circuit, in the same way that we found A_1 and A_2 for the overdamped response: by evaluating v at $t = 0^+$ and its derivative at $t = 0^+$. As with s_1 and s_2, α and ω_d are fixed by the circuit parameters R, L, and C.

For the underdamped response, the two simultaneous equations that determine B_1 and B_2 are

$$v(0^+) = V_0 = B_1, \tag{6.30}$$

$$\frac{dv(0^+)}{dt} = \frac{i_c(0^+)}{C} = -\alpha B_1 + \omega_d B_2. \tag{6.31}$$

Let's look at the general nature of the underdamped response. First, the trigonometric functions indicate that this response is oscillatory; that is, the voltage alternates between positive and negative values. The rate at which the voltage oscillates is fixed by ω_d. Second, the amplitude of the oscillation decreases exponentially. The rate at which the amplitude falls off is determined by α. Because α determines how quickly the oscillations subside, it is also referred to as the **damping factor** or **damping coefficient**. That explains why ω_d is called the damped radian frequency. If there is no damping, $\alpha = 0$ and the frequency of oscillation is ω_0. Whenever there is a dissipative element, R, in the circuit, α is not zero and the frequency of oscillation, ω_d, is less than ω_0. Thus when α is not zero, the frequency of oscillation is said to be damped.

The oscillatory behavior is possible because of the two types of energy-storage elements in the circuit: the inductor and the capacitor. (A mechanical analogy of this electric circuit is that of a mass suspended on a spring, where oscillation is possible because energy can be stored in both the spring and the moving mass.) We say more about the characteristics of the underdamped response following Example 6.4, which examines a circuit whose response is underdamped. In summary, note that the overall process for finding the underdamped response is the same as that for the overdamped response, although the response equations and the simultaneous equations used to find the constants are slightly different.

EXAMPLE 6.4

In the circuit shown in Fig. 6.8, $V_0 = 0$, and $I_0 = -12.25$ mA.

(a) Calculate the roots of the characteristic equation.

(b) Calculate v and dv/dt at $t = 0^+$.

(c) Calculate the voltage response for $t \geq 0$.

(d) Plot $v(t)$ versus t for the time interval $0 \leq t \leq 11$ ms.

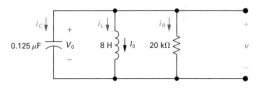

Figure 6.8 The circuit for Example 6.4.

SOLUTION

(a) Because

$$\alpha = \frac{1}{2RC} = \frac{10^6}{2(20)10^3(0.125)} = 200 \text{ rad/s},$$

$$\omega_0 = \frac{1}{\sqrt{LC}} = \sqrt{\frac{10^6}{(8)(0.125)}} = 10^3 \text{ rad/s},$$

we have

$$\omega_0^2 > \alpha^2.$$

Therefore, the response is underdamped. Now,

$$\omega_d = \sqrt{\omega_0^2 - \alpha^2} = \sqrt{10^6 - 4 \times 10^4} = 100\sqrt{96}$$

$$= 979.80 \text{ rad/s},$$

$$s_1 = -\alpha + j\omega_d = -200 + j979.80 \text{ rad/s},$$

$$s_2 = -\alpha - j\omega_d = -200 - j979.80 \text{ rad/s}.$$

For the underdamped case, we do not ordinarily solve for s_1 and s_2 because we do not use them explicitly. However, this example emphasizes why s_1 and s_2 are known as complex frequencies.

(b) Because v is the voltage across the terminals of a capacitor, we have

$$v(0) = v(0^+) = V_0 = 0.$$

Because $v(0^+) = 0$, the current in the resistive branch is zero at $t = 0^+$. Hence the current in the capacitor at $t = 0^+$ is the negative of the inductor current:

$$i_C(0^+) = -(-12.25) = 12.25 \text{ mA}.$$

Therefore the initial value of the derivative is

$$\frac{dv(0^+)}{dt} = \frac{(12.25)(10^{-3})}{(0.125)(10^{-6})} = 98,000 \text{ V/s}.$$

(c) From Eqs. (6.30) and (6.31), $B_1 = 0$ and

$$B_2 = \frac{98,000}{\omega_d} \approx 100 \text{ V}.$$

Substituting the numerical values of α, ω_d, B_1, and B_2 into the expression for $v(t)$ gives

$$v(t) = 100e^{-200t} \sin 979.80t \text{ V}, \quad t \geq 0.$$

(d) Figure 6.9 shows the plot of $v(t)$ versus t for the first 11 ms after the stored energy is released. It clearly indicates the damped oscillatory nature of the underdamped response. The voltage $v(t)$ approaches its final value, alternating between values that are greater than and less than the final value. Furthermore, these swings about the final value decrease exponentially with time.

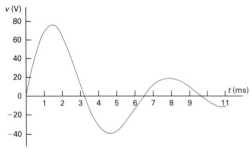

Figure 6.9 The voltage response for Example 6.4.

Characteristics of the Underdamped Response

The underdamped response has several important characteristics. First, as the dissipative losses in the circuit decrease, the persistence of the oscillations increases, and the frequency of the oscillations approaches ω_0. In other words, as $R \to \infty$, the dissipation in the circuit in Fig. 6.8 approaches zero because $p = v^2/R$. As $R \to \infty, \alpha \to 0$, which tells us that $\omega_d \to \omega_0$. When $\alpha = 0$, the maximum amplitude of the voltage remains constant; thus the oscillation at ω_0 is sustained. In Example 6.4, if R were increased to infinity, the solution for $v(t)$ would become

$$v(t) = 98 \sin 1000t \text{ V}, \quad t \geq 0.$$

Thus, in this case the oscillation is sustained, the maximum amplitude of the voltage is 98 V, and the frequency of oscillation is 1000 rad/s.

We may now describe qualitatively the difference between an underdamped and an overdamped response. In an underdamped system, the response oscillates, or "bounces," about its final value. This oscillation is also referred to as *ringing*. In an overdamped system, the response approaches its final value without ringing or in what is sometimes described as a "sluggish" manner. When specifying the desired response of a second-order system, you may want to reach the final value in the shortest time possible, and you may not be concerned with small oscillations about that final value. If so, you would design the system components to achieve an underdamped response. On the other hand, you may be concerned that the response not exceed its final value, perhaps to ensure that components are not damaged. In such a case, you would design the system components to achieve an overdamped response, and you would have to accept a relatively slow rise to the final value.

DRILL EXERCISE

6.4   A 10 mH inductor, a 1 μF capacitor, and a variable resistor are connected in parallel in the circuit shown. The resistor is adjusted so that the roots of the characteristic equation are $-8000 \pm j6000$ rad/s. The initial voltage on the capacitor is 10 V, and the initial current in the inductor is 80 mA. Find (a) R; (b) $dv(0^+)/dt$; (c) B_1 and B_2 in the solution for v; and (d) $i_L(t)$.

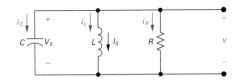

ANSWER: (a) 62.5 Ω; (b) $-240{,}000$ V/s;
(c) $B_1 = 10$ V, $B_2 = -80/3$ V;
(d) $i_L(t) = 10e^{-8000t}[8 \cos 6000t + (82/3) \sin 6000t]$ mA when $t \geq 0$.

The Critically Damped Voltage Response

The second-order circuit in Fig. 6.8 is critically damped when $\omega_0^2 = \alpha^2$, or $\omega_0 = \alpha$. When a circuit is critically damped, the response is on the verge of oscillating. In addition, the two roots of the characteristic equation are real and equal; that is,

$$s_1 = s_2 = -\alpha = -\frac{1}{2RC}. \qquad (6.32)$$

When this occurs, the solution for the voltage no longer takes the form of Eq. (6.18). This equation breaks down because if $s_1 = s_2 = -\alpha$, it predicts that

$$v = (A_1 + A_2)e^{-\alpha t} = A_0 e^{-\alpha t}, \qquad (6.33)$$

where A_0 is an arbitrary constant. Equation (6.33) cannot satisfy two independent initial conditions (V_0, I_0) with only one arbitrary constant, A_0. Recall that the circuit parameters R and C fix α.

We can trace this dilemma back to the assumption that the solution takes the form of Eq. (6.18). When the roots of the characteristic equation are equal, the solution for the differential equation takes a different form, namely

$$v(t) = D_1 t e^{-\alpha t} + D_2 e^{-\alpha t}. \qquad (6.34)$$

Thus in the case of a repeated root, the solution involves a simple exponential term plus the product of a linear and an exponential term. The justification of Eq. (6.34) is left for an introductory course in differential equations. Finding the solution involves obtaining D_1 and D_2 by following the same pattern set in the overdamped and underdamped cases: We use the initial values of the voltage and the derivative of the voltage with respect to time to write two equations containing D_1 and/or D_2.

From Eq. (6.34), the two simultaneous equations needed to determine D_1 and D_2 are

$$v(0^+) = V_0 = D_2, \qquad (6.35)$$

$$\frac{dv(0^+)}{dt} = \frac{i_C(0^+)}{C} = D_1 - \alpha D_2. \qquad (6.36)$$

As we can see, in the case of a critically damped response, both the equation for $v(t)$ and the simultaneous equations for the constants D_1 and D_2 differ from those for over- and underdamped responses, but the general approach is the same. You will

rarely encounter critically damped systems in practice, largely because ω_0 must equal α exactly. Both of these quantities depend
on circuit parameters, and in a real circuit it is very difficult to
choose component values that satisfy an exact equality relationship.

Example 6.5 illustrates the approach for finding the critically
damped response of a parallel RLC circuit.

EXAMPLE 6.5

(a) For the circuit in Example 6.4 (Fig. 6.8), find the value of R
that results in a critically damped voltage response.

(b) Calculate $v(t)$ for $t \geq 0$.

(c) Plot $v(t)$ versus t for $0 \leq t \leq 7$ ms.

SOLUTION

(a) From Example 6.4, we know that $\omega_0^2 = 10^6$. Therefore for
critical damping,

$$\alpha = 10^3 = \frac{1}{2RC},$$

or

$$R = \frac{10^6}{(2000)(0.125)} = 4000 \ \Omega.$$

(b) From the solution of Example 6.4, we know that $v(0^+) = 0$
and $dv(0^+)/dt = 98{,}000$ V/s. From Eqs. (6.35) and (6.36),
$D_2 = 0$ and $D_1 = 98{,}000$ V/s. Substituting these values for
α, D_1, and D_2 into Eq. (6.34) gives

$$v(t) = 98{,}000te^{-1000t} \ \text{V}, \quad t \geq 0.$$

(c) Figure 6.10 shows a plot of $v(t)$ versus t in the interval $0 \leq$
$t \leq 7$ ms.

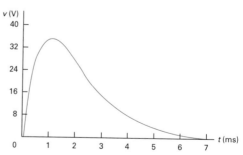

Figure 6.10 The voltage response for Example 6.5.

DRILL EXERCISE

6.5 The resistor in the circuit in Figure 6.1 is adjusted for critical damping. The inductance and capacitance values are 0.4 H and 10 μF, respectively. The initial energy stored in the circuit is 25 mJ and is distributed equally between the inductor and capacitor. Assume V_0 and I_0 are positive. Find (a) R; (b) V_0; (c) I_0; (d) D_1 and D_2 in the solution for v; and (e) i_R, $t \geq 0^+$.

ANSWER: (a) 100 Ω; (b) 50 V; (c) 250 mA; (d) $-50{,}000$ V/s, 50 V; (e) $i_R(t) = (-500te^{-500t} + 0.50e^{-500t})$ A, $t \geq 0^+$.

A Summary of the Results

We conclude our discussion of the parallel RLC circuit's natural response with a brief summary of the results. The first step in finding the natural response is to calculate the roots of the characteristic equation. You then know immediately whether the response is overdamped, underdamped, or critically damped.

If the roots are real and distinct ($\omega_0^2 < \alpha^2$), the response is overdamped and the voltage is

$$v(t) = A_1 e^{s_1 t} + A_2 e^{s_2 t},$$

where

$$s_1 = -\alpha + \sqrt{\alpha^2 - \omega_0^2},$$

$$s_2 = -\alpha - \sqrt{\alpha^2 - \omega_0^2},$$

$$\alpha = \frac{1}{2RC},$$

$$\omega_0^2 = \frac{1}{LC}.$$

The values of A_1 and A_2 are determined by solving the following simultaneous equations:

$$v(0^+) = A_1 + A_2,$$

$$\frac{dv(0^+)}{dt} = \frac{i_C(0^+)}{C} = s_1 A_1 + s_2 A_2.$$

If the roots are complex ($\omega_0^2 > \alpha^2$), the response is under-damped and the voltage is

$$v(t) = B_1 e^{-\alpha t} \cos \omega_d t + B_2 e^{-\alpha t} \sin \omega_d t,$$

where

$$\omega_d = \sqrt{\omega_0^2 - \alpha^2}.$$

The values of B_1 and B_2 are found by solving the following simultaneous equations:

$$v(0^+) = V_0 = B_1,$$

$$\frac{dv(0^+)}{dt} = \frac{i_C(0^+)}{C} = -\alpha B_1 + \omega_d B_2.$$

If the roots of the characteristic equation are real and equal ($\omega_0^2 = \alpha^2$), the voltage response is

$$v(t) = D_1 t e^{-\alpha t} + D_2 e^{-\alpha t},$$

where α is as in the other solution forms. To determine values for the constants D_1 and D_2, solve the following simultaneous equations:

$$v(0^+) = V_0 = D_2,$$

$$\frac{dv(0^+)}{dt} = \frac{i_C(0^+)}{C} = D_1 - \alpha D_2.$$

6.3 THE STEP RESPONSE OF A PARALLEL RLC CIRCUIT

Finding the step response of a parallel RLC circuit involves finding the voltage across the parallel branches or the current in the individual branches as a result of the sudden application of a dc current source. There may or may not be energy stored in the circuit when the current source is applied. The task is represented by the circuit shown in Fig. 6.11. To develop a general approach to finding the step response of a second-order circuit, we focus on finding the current in the inductive branch (i_L). This current is of particular interest because it does not approach zero as t increases. Rather, after the switch has been open for a long time, the inductor current equals the dc source current I. Because we want to focus on the technique for finding the step response, we assume that the initial energy stored in the circuit is zero. This assumption simplifies the calculations and doesn't alter the basic process involved. In Example 6.10 we will see how the presence of initially stored energy enters into the general procedure.

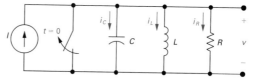

Figure 6.11 A circuit used to describe the step response of a parallel RLC circuit.

To find the inductor current i_L, we must solve a second-order differential equation equated to the forcing function I, which we derive as follows. From Kirchhoff's current law, we have

$$i_L + i_R + i_C = I,$$

or

$$i_L + \frac{v}{R} + C\frac{dv}{dt} = I. \tag{6.37}$$

Because

$$v = L\frac{di_L}{dt}, \tag{6.38}$$

we get

$$\frac{dv}{dt} = L\frac{d^2i_L}{dt^2}. \tag{6.39}$$

Substituting Eqs. (6.38) and (6.39) into Eq. (6.37) gives

$$i_L + \frac{L}{R}\frac{di_L}{dt} + LC\frac{d^2i_L}{dt^2} = I. \tag{6.40}$$

For convenience, we divide through by LC and rearrange terms:

$$\frac{d^2i_L}{dt^2} + \frac{1}{RC}\frac{di_L}{dt} + \frac{i_L}{LC} = \frac{I}{LC}. \tag{6.41}$$

Comparing Eq. (6.41) with Eq. (6.3) reveals that the presence of a nonzero term on the right-hand side of the equation alters the task. Before showing how to solve Eq. (6.41) directly, we obtain the solution indirectly. When we know the solution of Eq. (6.41), explaining the direct approach will be easier.

The Indirect Approach

We can solve for i_L indirectly by first finding the voltage v. We do this with the techniques introduced in Section 6.2, because the differential equation that v must satisfy is identical to Eq. (6.3). To see this, we simply return to Eq. (6.37) and express i_L as a function of v; thus

$$\frac{1}{L}\int_0^t v d\tau + \frac{v}{R} + C\frac{dv}{dt} = I. \tag{6.42}$$

Differentiating Eq. (6.42) once with respect to t reduces the right-hand side to zero because I is a constant. Thus

$$\frac{v}{L} + \frac{1}{R}\frac{dv}{dt} + C\frac{d^2v}{dt^2} = 0,$$

or

$$\frac{d^2v}{dt^2} + \frac{1}{RC}\frac{dv}{dt} + \frac{v}{LC} = 0. \qquad (6.43)$$

As discussed in Section 6.2, the solution for v depends on the roots of the characteristic equation. Thus the three possible solutions are

$$v = A_1 e^{s_1 t} + A_2 e^{s_2 t}, \qquad (6.44)$$

$$v = B_1 e^{-\alpha t}\cos\omega_d t + B_2 e^{-\alpha t}\sin\omega_d t, \qquad (6.45)$$

$$v = D_1 t e^{-\alpha t} + D_2 e^{-\alpha t}. \qquad (6.46)$$

A word of caution: Because there is a source in the circuit for $t > 0$, you must take into account the value of the source current at $t = 0^+$ when you evaluate the coefficients in Eqs. (6.44)–(6.46).

To find the three possible solutions for i_L, we substitute Eqs. (6.44)–(6.46) into Eq. (6.37). You should be able to verify, when this has been done, that the three solutions for i_L will be

$$i_L = I + A_1' e^{s_1 t} + A_2' e^{s_2 t}, \qquad (6.47)$$

$$i_L = I + B_1' e^{-\alpha t}\cos\omega_d t + B_2' e^{-\alpha t}\sin\omega_d t, \qquad (6.48)$$

$$i_L = I + D_1' t e^{-\alpha t} + D_2' e^{-\alpha t}, \qquad (6.49)$$

where A_1', A_2', B_1', B_2', D_1', and D_2' are arbitrary constants.

In each case, the primed constants can be found indirectly in terms of the arbitrary constants associated with the voltage solution. However, this approach is cumbersome.

The Direct Approach

It is much easier to find the primed constants directly in terms of the initial values of the response function. For the circuit being discussed, we would find the primed constants from $i_L(0)$ and $di_L(0)/dt$.

The solution for a second-order differential equation with a constant forcing function equals the forced response plus a response function identical in form to the natural response. Thus we can always write the solution for the step response in the form

$$i = I_f + \left\{ \begin{array}{c} \text{function of the same form} \\ \text{as the natural response} \end{array} \right\}, \qquad (6.50)$$

or

$$v = V_f + \left\{ \begin{array}{c} \text{function of the same form} \\ \text{as the natural response} \end{array} \right\}, \qquad (6.51)$$

where I_f and V_f represent the final value of the response function. The final value may be zero, as was, for example, the case with the voltage v in the circuit in Fig. 6.8.

Examples 6.6–6.10 illustrate the technique of finding the step response of a parallel RLC circuit using the direct approach.

EXAMPLE 6.6

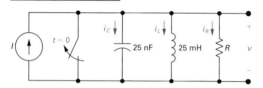

Figure 6.12 The circuit for Example 6.6.

The initial energy stored in the circuit in Fig. 6.12 is zero. At $t = 0$, a dc current source of 24 mA is applied to the circuit. The value of the resistor is 400 Ω.

(a) What is the initial value of i_L?

(b) What is the initial value of di_L/dt?

(c) What are the roots of the characteristic equation?

(d) What is the numerical expression for $i_L(t)$ when $t \geq 0$?

SOLUTION

(a) No energy is stored in the circuit prior to the application of the dc current source, so the initial current in the inductor is zero. The inductor prohibits an instantaneous change in inductor current; therefore $i_L(0) = 0$ immediately after the switch has been opened.

(b) The initial voltage on the capacitor is zero before the switch has been opened; therefore it will be zero immediately after. Now, because $v = L di_L/dt$,

$$\frac{di_L}{dt}(0^+) = 0.$$

(c) From the circuit elements, we obtain

$$\omega_0^2 = \frac{1}{LC} = \frac{10^{12}}{(25)(25)} = 16 \times 10^8,$$

$$\alpha = \frac{1}{2RC} = \frac{10^9}{(2)(400)(25)} = 5 \times 10^4 \text{ rad/s},$$

or

$$\alpha^2 = 25 \times 10^8.$$

Because $\omega_0^2 < \alpha^2$, the roots of the characteristic equation are real and distinct. Thus

$$s_1 = -5 \times 10^4 + 3 \times 10^4 = -20{,}000 \text{ rad/s},$$

$$s_2 = -5 \times 10^4 - 3 \times 10^4 = -80{,}000 \text{ rad/s}.$$

(d) Because the roots of the characteristic equation are real and distinct, the inductor current response will be overdamped. Thus $i_L(t)$ takes the form of Eq. (6.47), namely,

$$i_L = I_f + A_1' e^{s_1 t} + A_2' e^{s_2 t}.$$

Hence, from this solution, the two simultaneous equations that determine A_1' and A_2' are

$$i_L(0) = I_f + A_1' + A_2' = 0$$

$$\frac{di_L}{dt}(0) = s_1 A_1' + s_2 A_2' = 0.$$

Solving for A_1' and A_2' gives

$$A_1' = -32 \text{ mA} \quad \text{and} \quad A_2' = 8 \text{ mA}.$$

The numerical solution for $i_L(t)$ is

$$i_L(t) = (24 - 32e^{-20,000t} + 8e^{-80,000t}) \text{ mA}, \quad t \geq 0.$$

EXAMPLE 6.7

The resistor in the circuit in Example 6.6 (Fig. 6.12) is increased to 625 Ω. Find $i_L(t)$ for $t \geq 0$.

SOLUTION

Because L and C remain fixed, ω_0^2 has the same value as in Example 6.6; that is, $\omega_0^2 = 16 \times 10^8$. Increasing R to 625 Ω decreases α to 3.2×10^4 rad/s. With $\omega_0^2 > \alpha^2$, the roots of the characteristic equation are complex. Hence

$$s_1 = -32,000 + j24,000 \text{ rad/s},$$

$$s_2 = -32,000 - j24,000 \text{ rad/s}.$$

The current response is now underdamped and given by Eq. (6.48):

$$i_L(t) = I_f + B_1' e^{-\alpha t} \cos \omega_d t + B_2' e^{-\alpha t} \sin \omega_d t.$$

Here, α is 32,000 rad/s, ω_d is 24,000 rad/s, and I_f is 24 mA.

As in Example 6.6, B'_1 and B'_2 are determined from the initial conditions. Thus the two simultaneous equations are

$$i_L(0) = I_f + B'_1 = 0,$$

$$\frac{di_L}{dt}(0) = \omega_d B'_2 - \alpha B'_1 = 0.$$

Then,

$$B'_1 = -24 \text{ mA} \quad \text{and} \quad B'_2 = -32 \text{ mA}.$$

The numerical solution for $i_L(t)$ is

$$i_L(t) = (24 - 24e^{-32,000t} \cos 24,000t - 32e^{-32,000t} \sin 24,000t) \text{ mA},$$

$$t \geq 0.$$

EXAMPLE 6.8

The resistor in the circuit in Example 6.6 (Fig. 6.12) is set at 500 Ω. Find i_L for $t \geq 0$.

SOLUTION

We know that ω_0^2 remains at 16×10^8. With R set at 500 Ω, α becomes 4×10^4 s^{-1}, which corresponds to critical damping. Therefore the solution for $i_L(t)$ takes the form of Eq. (6.49):

$$i_L(t) = I_f + D'_1 t e^{-\alpha t} + D'_2 e^{-\alpha t}.$$

Again, D'_1 and D'_2 are computed from initial conditions, or

$$i_L(0) = I_f + D'_2 = 0,$$

$$\frac{di_L}{dt}(0) = D'_1 - \alpha D'_2 = 0.$$

Thus

$$D'_1 = -960,000 \text{ mA/s} \quad \text{and} \quad D'_2 = -24 \text{ mA}.$$

The numerical expression for $i_L(t)$ is

$$i_L(t) = (24 - 960,000t e^{-40,000t} - 24e^{-40,000t}) \text{ mA}, \ t \geq 0.$$

EXAMPLE 6.9

(a) Plot on a single graph, over a range from 0 to 200 μs, the over-damped, underdamped, and critically damped responses derived in Examples 6.6–6.8.

(b) Use the plots of (a) to find the time required for i_L to reach 90% of its final value.

(c) On the basis of the results obtained in (b), which response would you specify in a design that puts a premium on reaching 90% of the final value of the output in the shortest time?

(d) Which response would you specify in a design that must ensure that the final value of the current is never exceeded?

SOLUTION

(a) See Fig. 6.13.

(b) The final value of i_L is 24 mA, so we can read the times off the plots corresponding to $i_L = 21.6$ mA. Thus $t_{od} = 130 \, \mu$s, $t_{cd} = 97 \mu$s, and $t_{ud} = 74 \, \mu$s.

(c) The underdamped response reaches 90% of the final value in the fastest time, so it is the desired response type when speed is the most important design specification.

(d) From the plot, you can see that the underdamped response overshoots the final value of current, whereas neither the critically damped nor the overdamped response produces currents in excess of 24 mA. Although specifying either of the latter two responses would meet the design specification, it is best to use the overdamped response. It would be impractical to require a design to achieve the exact component values that ensure a critically damped response.

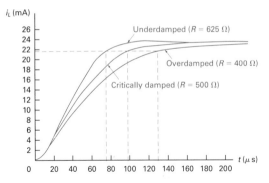

Figure 6.13 The current plots for Example 6.9.

EXAMPLE 6.10

Energy is stored in the circuit in Example 6.8 (Fig. 6.12, with $R = 500 \, \Omega$) at the instant the dc current source is applied. The initial current in the inductor is 29 mA, and the initial voltage across the capacitor is 50 V. Find (a) $i_L(0)$; (b) $di_L(0)/dt$; (c) $i_L(t)$ for $t \geq 0$; (d) $v(t)$ for $t \geq 0^+$.

SOLUTION

(a) There cannot be an instantaneous change of current in an inductor, so the initial value of i_L in the first instant after the dc current source has been applied must be 29 mA.

(b) The capacitor holds the initial voltage across the inductor to 50 V. Therefore

$$L\frac{di_L}{dt}(0^+) = 50,$$

$$\frac{di_L}{dt}(0^+) = \frac{50}{25} \times 10^3 = 2000 \text{ A/s.}$$

(c) From the solution of Example 6.8, we know that the current response is critically damped. Thus

$$i_L(t) = I_f + D_1' t e^{-\alpha t} + D_2' e^{-\alpha t},$$

where

$$\alpha = \frac{1}{2RC} = 40,000 \text{ rad/s} \quad \text{and} \quad I_f = 24 \text{ mA.}$$

Notice that the effect of the nonzero initial stored energy is on the calculations for the constants D_1' and D_2', which we obtain from the initial conditions. First we use the initial value of the inductor current:

$$i_L(0) = I_f + D_2' = 29 \text{ mA,}$$

from which we get

$$D_2' = 29 - 24 = 5 \text{ mA.}$$

The solution for D_1' is

$$\frac{di_L}{dt}(0^+) = D_1' - \alpha D_2' = 2000,$$

or

$$D_1' = 2000 + \alpha D_2'$$

$$= 2000 + (40,000)(5 \times 10^{-3})$$

$$= 2200 \text{ A/s} = 2.2 \times 10^6 \text{ mA/s.}$$

Thus the numerical expression for $i_L(t)$ is

$$i_L(t) = (24 + 2.2 \times 10^6 t e^{-40,000t} + 5e^{-40,000t}) \text{ mA,} \quad t \geq 0.$$

(d) We can get the expression for $v(t)$, $t \geq 0^+$ by using the relationship between the voltage and current in an inductor:

$$v(t) = L\frac{di_L}{dt}$$

$$= (25 \times 10^{-3})[(2.2 \times 10^6)(-40,000)te^{-40,000t}$$

$$+ 2.2 \times 10^6 e^{-40,000t} + (5)(-40,000)e^{-40,000t}] \times 10^{-3}$$

$$= -2.2 \times 10^6 te^{-40,000t} + 50e^{-40,000t} \text{ V}.$$

To check this result, let's verify that the initial voltage across the inductor is 50 V:

$$v(0^+) = -2.2 \times 10^6(0)(1) + 50(1) = 50 \text{ V}.$$

DRILL EXERCISE

6.6   In the circuit shown, $R = 500 \ \Omega$, $L = 0.64$ H, $C = 1 \ \mu$F, $I_0 = 0.5$ A, $V_0 = 40$ V, and $I = -1$ A. Find (a) $i_R(0^+)$; (b) $i_C(0^+)$; (c) $di_L(0^+)/dt$; (d) s_1, s_2; (e) $i_L(t)$ for $t \geq 0$; and (f) $v(t)$ for $t \geq 0^+$.

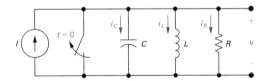

ANSWER: (a) 80 mA; (b) -1.58 A; (c) 62.5 A/s; (d) $(-1000 + j750)$ rad/s, $(-1000 - j750)$ rad/s; (e) $i_L(t) = [-1 + e^{-1000t}[1.5 \cos 750t + 2.0833 \sin 750t]$ A, for $t \geq 0$; (f) $v(t) = e^{-1000t}(40 \cos 750t - 2053.33 \sin 750t)$ V, for $t \geq 0^+$.

6.4   THE NATURAL AND STEP RESPONSE OF A SERIES RLC CIRCUIT

The procedures for finding the natural or step responses of a series RLC circuit are the same as those used to find the natural or step responses of a parallel RLC circuit, because both circuits are described by differential equations that have the same form.

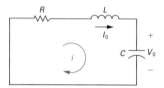

Figure 6.14 A circuit used to illustrate the natural response of a series RLC circuit.

We begin by summing the voltages around the closed path in the circuit shown in Fig. 6.14. Thus

$$Ri + L\frac{di}{dt} + \frac{1}{C}\int_0^t i\,d\tau + V_o = 0. \tag{6.52}$$

We now differentiate Eq. (6.52) once with respect to t to get

$$R\frac{di}{dt} + L\frac{d^2i}{dt^2} + \frac{i}{C} = 0, \tag{6.53}$$

which we can rearrange as

$$\frac{d^2i}{dt^2} + \frac{R}{L}\frac{di}{dt} + \frac{i}{LC} = 0. \tag{6.54}$$

Comparing Eq. (6.54) with Eq. (6.3) reveals that they have the same form. Therefore, to find the solution of Eq. (6.54), we follow the same process that led us to the solution of Eq. (6.3).

From Eq. (6.54), the characteristic equation for the series RLC circuit is

$$s^2 + \frac{R}{L}s + \frac{1}{LC} = 0. \tag{6.55}$$

The roots of the characteristic equation are

$$s_{1,2} = -\frac{R}{2L} \pm \sqrt{\left(\frac{R}{2L}\right)^2 - \frac{1}{LC}}, \tag{6.56}$$

or

$$s_{1,2} = -\alpha \pm \sqrt{\alpha^2 - \omega_0^2}. \tag{6.57}$$

The neper frequency (α) for the series RLC circuit is

$$\alpha = \frac{R}{2L} \text{ rad/s}, \tag{6.58}$$

and the expression for the resonant radian frequency is

$$\omega_0 = \frac{1}{\sqrt{LC}} \text{ rad/s}. \tag{6.59}$$

Note that the neper frequency of the series RLC circuit differs from that of the parallel RLC circuit, but the resonant radian frequencies are the same.

The current response will be overdamped, underdamped, or critically damped according to whether $\omega_0^2 < \alpha^2$, $\omega_0^2 > \alpha^2$, or

$\omega_0^2 = \alpha^2$, respectively. Thus the three possible solutions for the current are as follows:

$$i(t) = A_1 e^{s_1 t} + A_2 e^{s_2 t} \text{ (overdamped)}, \qquad (6.60)$$

$$i(t) = B_1 e^{-\alpha t} \cos \omega_d t + B_2 e^{-\alpha t} \sin \omega_d t \text{ (underdamped)}, \qquad (6.61)$$

and

$$i(t) = D_1 t e^{-\alpha t} + D_2 e^{-\alpha t} \text{ (critically damped)}. \qquad (6.62)$$

When you have obtained the natural current response, you can find the natural voltage response across any circuit element.

To verify that the procedure for finding the step response of a series RLC circuit is the same as that for a parallel RLC circuit, we show that the differential equation that describes the capacitor voltage in Fig. 6.15 has the same form as the differential equation that describes the inductor current in Fig. 6.11. For convenience, we assume that zero energy is stored in the circuit at the instant the switch is closed.

Applying Kirchhoff's voltage law to the circuit shown in Fig. 6.15 gives

$$V = Ri + L\frac{di}{dt} + v_C. \qquad (6.63)$$

The current (i) is related to the capacitor voltage (v_C) by the expression

$$i = C\frac{dv_C}{dt}, \qquad (6.64)$$

from which

$$\frac{di}{dt} = C\frac{d^2 v_C}{dt^2}. \qquad (6.65)$$

Substitute Eqs. (6.64) and (6.65) into Eq. (6.63) and write the resulting expression as

$$\frac{d^2 v_C}{dt^2} + \frac{R}{L}\frac{dv_C}{dt} + \frac{v_C}{LC} = \frac{V}{LC}. \qquad (6.66)$$

Equation (6.66) has the same form as Eq. (6.41); therefore the procedure for finding v_C parallels that for finding i_L. The three possible solutions for v_C are as follows:

$$v_C = V_f + A_1' e^{s_1 t} + A_2' e^{s_2 t} \text{ (overdamped)}, \qquad (6.67)$$

$$v_C = V_f + B_1' e^{-\alpha t} \cos \omega_d t + B_2' e^{-\alpha t} \sin \omega_d t \text{ (underdamped)}, \qquad (6.68)$$

$$v_C = V_f + D_1' t e^{-\alpha t} + D_2' e^{-\alpha t} \text{ (critically damped)}, \qquad (6.69)$$

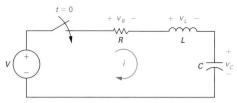

Figure 6.15 A circuit used to illustrate the step response of a series RLC circuit.

where V_f is the final value of v_C. Hence, from the circuit shown in Fig. 6.15, the final value of v_C is the dc source voltage V.

Examples 6.11 and 6.12 illustrate the mechanics of finding the natural and step responses of a series RLC circuit.

EXAMPLE 6.11

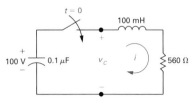

Figure 6.16 The circuit for Example 6.11.

The 0.1 μF capacitor in the circuit shown in Fig. 6.16 is charged to 100 V. At $t = 0$ the capacitor is discharged through a series combination of a 100 mH inductor and a 560 Ω resistor.

(a) Find $i(t)$ for $t \geq 0$.

(b) Find $v_C(t)$ for $t \geq 0$.

SOLUTION

(a) The first step to finding $i(t)$ is to calculate the roots of the characteristic equation. For the given element values,

$$\omega_0^2 = \frac{1}{LC} = \frac{(10^3)(10^6)}{(100)(0.1)} = 10^8,$$

$$\alpha = \frac{R}{2L} = \frac{560}{2(100)} \times 10^3 = 2800 \text{ rad/s.}$$

Next, we compare ω_0^2 to α^2 and note that $\omega_0^2 > \alpha^2$, because

$$\alpha^2 = 7.84 \times 10^6 = 0.0784 \times 10^8.$$

At this point, we know that the response is underdamped and that the solution for $i(t)$ is of the form

$$i(t) = B_1 e^{-\alpha t} \cos \omega_d t + B_2 e^{-\alpha t} \sin \omega_d t,$$

where $\alpha = 2800$ rad/s and $\omega_d = 9600$ rad/s. The numerical values of B_1 and B_2 come from the initial conditions. The inductor current is zero before the switch has been closed, and hence it is zero immediately after. Therefore

$$i(0) = 0 = B_1.$$

To find B_2, we evaluate $di(0^+)/dt$. From the circuit, we note that, because $i(0) = 0$ immediately after the switch has been closed, there will be no voltage drop across the resistor.

Thus the initial voltage on the capacitor appears across the terminals of the inductor, which leads to the expression,

$$L\frac{di(0^+)}{dt} = V_0,$$

or

$$\frac{di(0^+)}{dt} = \frac{V_0}{L} = \frac{100}{100} \times 10^3 = 1000 \text{ A/s}.$$

Because $B_1 = 0$,

$$\frac{di}{dt} = 400 B_2 e^{-2800t} (24 \cos 9600t - 7 \sin 9600t).$$

Thus

$$\frac{di(0^+)}{dt} = 9600 B_2,$$

$$B_2 = \frac{1000}{9600} \approx 0.1042 \text{ A}.$$

The solution for $i(t)$ is

$$i(t) = 0.1042 e^{-2800t} \sin 9600t \text{ A}, \quad t \geq 0.$$

(b) To find $v_C(t)$, we can use either of the following relationships:

$$v_C = -\frac{1}{C} \int_0^t i \, d\tau + 100$$

$$v_C = iR + L\frac{di}{dt}.$$

Whichever expression is used (the second is recommended), the result is

$$v_C(t) = (100 \cos 9600t + 29.17 \sin 9600t) e^{-2800t} \text{ V}, \quad t \geq 0.$$

EXAMPLE 6.12

No energy is stored in the 100 mH inductor or the 0.4 μF capacitor when the switch in the circuit shown in Fig. 6.17 is closed. Find $v_C(t)$ for $t \geq 0$.

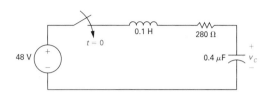

Figure 6.17 The circuit for Example 6.12.

SOLUTION

The roots of the characteristic equation are

$$s_1 = -\frac{280}{0.2} + \sqrt{\left(\frac{280}{0.2}\right)^2 - \frac{10^6}{(0.1)(0.4)}}$$

$$= (-1400 + j4800) \text{ rad/s},$$

$$s_2 = (-1400 - j4800) \text{ rad/s}.$$

The roots are complex, so the voltage response is underdamped. Thus

$$v_C(t) = 48 + B_1'e^{-1400t} \cos 4800t + B_2'e^{-1400t} \sin 4800t, \quad t \geq 0.$$

No energy is stored in the circuit initially, so both $v_C(0)$ and $dv_C(0^+)/dt$ are zero. Then,

$$v_C(0) = 0 = 48 + B_1'$$

$$\frac{dv_C(0^+)}{dt} = 0 = 4800B_2' - 1400B_1'.$$

Solving for B_1' and B_2' yields

$$B_1' = -48 \text{ V},$$

$$B_2' = -14 \text{ V}.$$

Therefore, the solution for $v_C(t)$ is

$$v_C(t) = (48 - 48e^{-1400t} \cos 4800t - 14e^{-1400t} \sin 4800t) \text{ V}, \quad t \geq 0.$$

DRILL EXERCISES

6.7 The switch in the circuit shown has been in position a for a long time. At $t = 0$, it moves to position b. Find (a) $i(0^+)$; (b) $v_C(0^+)$; (c) $di(0^+)/dt$; (d) s_1, s_2; and (e) $i(t)$ for $t \geq 0$.

ANSWER: (a) 0; (b) 50 V; (c) 10,000 A/s; (d) $(-8000 + j6000)$ rad/s, $(-8000 - j6000)$ rad/s; (e) $i(t) = (1.67e^{-8000t} \sin 6000t)$ A for $t \geq 0$.

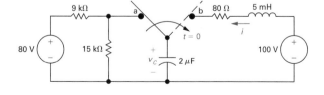

6.8 Find $v_C(t)$ for $t \geq 0$ for the circuit in Drill Exercise 6.7.

ANSWER: $v_C = [100 - e^{-8000t}(50 \cos 6000t + 66.67 \sin 6000t)]$ V for $t \geq 0$.

Practical Perspective

Parasitic Inductance

We are now ready to study the effect inductance can have on the step response of an RLC series circuit. Assume the switch in the circuit in Fig. 6.18 represents the transition from a high voltage (a binary 1) to low voltage (a binary 0) and vice versa. That is, when the switch is in position a, the signal v_o goes high and when the switch is in position b, v_o goes low.

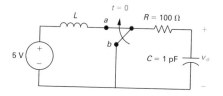

Figure 6.18 A series RLC circuit.

To see what effect L has on how fast v_o rises toward 5 V we begin by assuming that L is zero and the initial value of v_o is 1 V. Using the approach set forth in Eq. (5.90), we can write the expression for v_o directly; i.e., noting that 1/RC equals 10^{10}, we have

$$v_o = 5 + (1 - 5)e^{-10^{10}t}$$

$$= 5 - 4e^{-10^{10}t} \text{ V}.$$

Now let us assume v_o will be recognized as a binary 1 when it reaches 4 V. Let t_x represent the time it takes for v_o to reach 4 V; then

$$4 = 5 - 4e^{-10^{10}t_x}$$

or

$$e^{-10^{10}t_x} = 4$$

from which

$$t_x = 100 \ln 4 \text{ ps}$$

$$= 138.63 \text{ ps}$$

The inductance will have an adverse effect on the transition time when it is large enough to cause an undamped response. For our illustrative circuit the critical value of L is 2.5 nH. We will leave to you via Problem 6.49 to show that the solution for v_o for the values of L specified in Table 6.2 is as tabulated.

We also leave to you via Problem 6.50 to determine, for each value of L, the time it takes to rise from 1 V to 4 V (i.e., the transition time t_x) as given in Table 6.3. Note from Table 6.3 that the transition times are shorter when the response is overdamped or critically damped and longer when the response is underdamped.

TABLE 6.2 Voltage Response for Various Values of Inductance in the Circuit in Figure 6.18.

L (nH)	v_o(V)
1.6	$5 + \left(\dfrac{4}{3}\right)e^{-50 \times 10^9 t} - \left(\dfrac{16}{3}\right)e^{-12.5 \times 10^9 t}$
2.5	$5 - (8 \times 10^{10}t + 4)e^{-2 \times 10^{10}t}$
5	$5 - 4e^{-10^{10}t}\left[\cos(10^{10}t) + \sin(10^{10}t)\right]$
25	$5 - 4e^{-2 \times 10^9 t}\left[\cos(6 \times 10^9 t) + \left(\dfrac{1}{3}\right)\sin(6 \times 10^9 t)\right]$

TABLE 6.3 Transition Times.

L (nH)	t_x (ps)
0	138.63
1.6	133.79
2.5	134.64
5	147.41
25	248.64

SUMMARY

- The **characteristic equation** for both the parallel and series *RLC* circuits has the form

$$s^2 + 2\alpha s + \omega_0^2 = 0,$$

where $\alpha = 1/2RC$ for the parallel circuit, $\alpha = R/2L$ for the series circuit, and $\omega_0^2 = 1/LC$ for both the parallel and series circuits.

- The roots of the characteristic equation are

$$s_{1,2} = -\alpha \pm \sqrt{\alpha^2 - \omega_0^2}.$$

- The form of the natural and step responses of series and parallel *RLC* circuits depends on the values of α^2 and ω_0^2; such responses can be **overdamped**, **underdamped**, or **critically damped**. These terms describe the impact of the dissipative element (R) on the response. The **neper frequency**, α, reflects the effect of R.

- The response of a second-order circuit is overdamped, underdamped, or critically damped as follows:

THE CIRCUIT IS	WHEN	QUALITATIVE NATURE OF THE RESPONSE
Overdamped	$\alpha^2 > \omega_o^2$	The voltage or current approaches its final value without oscillation
Underdamped	$\alpha^2 < \omega_o^2$	The voltage or current oscillates about its final value
Critically damped	$\alpha^2 = \omega_o^2$	The voltage or current is on the verge of oscillating about its final value

- In determining the **natural response** of a second-order circuit, we first determine whether it is over-, under-, or critically damped, and then we solve the appropriate equations as follows:

DAMPING	NATURAL-RESPONSE EQUATIONS	COEFFICIENT EQUATIONS OVERDAMPED
Overdamped	$x(t) = A_1 e^{s_1 t} + A_2 e^{s_2 t}$	$x(0) = A_1 + A_2;$ $dx/dt(0) = A_1 s_1 + A_2 s_2$
Underdamped	$x(t) = (B_1 \cos \omega_d t + B_2 \sin \omega_d t)e^{-\alpha t}$	$x(0) = B_1;$ $dx/dt(0) = -\alpha B_1 + \omega_d B_2,$ where $\omega_d = \sqrt{\omega_0^2 - \alpha^2}$
Critically damped	$x(t) = (D_1 t + D_2)e^{-\alpha t}$	$x(0) = D_2,$ $dx/dt(0) = D_1 - \alpha D_2$

- In determining the **step response** of a second-order circuit, we apply the appropriate equations depending on the damping, as follows:

DAMPING	NATURAL-RESPONSE EQUATIONS[1]	COEFFICIENT EQUATIONS OVERDAMPED
Overdamped	$x(t) = X_f + A_1' e^{s_1 t} + A_2' e^{s_2 t}$	$x(0) = X_f + A_1' + A_2';$ $dx/dt(0) = A_1' s_1 + A_2' s_2$
Underdamped	$x(t) = X_f + (B_1' \cos \omega_d t + B_2' \sin \omega_d t)e^{-\alpha t}$	$x(0) = X_f + B_1';$ $dx/dt(0) = -\alpha B_1' + \omega_d B_2'$
Critically damped	$x(t) = X_f + D_1' t e^{-\alpha t} + D_2' e^{-\alpha t}$	$x(0) = X_f + D_2';$ $dx/dt(0) = D_1' - \alpha D_2'$

[1] where X_f is the final value of $x(t)$.

- For each of the three forms of response, the unknown coefficients (i.e., the As, Bs, and Ds) are obtained by evaluating the initial value of the response, $x(0)$, and the initial value of the first derivative of the response, $dx(0)/dt$.

PROBLEMS

 6.1. In the circuit in Fig. 6.1, $R = 2\ \Omega$, $L = 0.4$ H, $C = 0.25$ F, $V_0 = 0$ V, and $I_0 = -3$ A.

 (a) Find $v(t)$ for $t \geq 0$.

 (b) Find the first three values of t for which dv/dt is zero. Let these values of t be denoted t_1, t_2, and t_3.

 (c) Show that $t_3 - t_1 = T_d$.

 (d) Show that $t_2 = t_1 = T_d/2$.

 (e) Calculate $v(t_1)$, $v(t_2)$, and $v(t_3)$.

 (f) Sketch $v(t)$ versus t for $0 \leq t \leq t_2$.

 6.2.

 (a) Find $v(t)$ for $t \geq 0$ in the circuit in Problem 6.1 if the 2 Ω resistor is removed from the circuit.

 (b) Calculate the frequency of $v(t)$ in hertz.

 (c) Calculate the maximum amplitude of $v(t)$ in volts.

6.3. The natural voltage response of the circuit in Fig. 6.1 is

$$v = 125e^{-4000t}(\cos 3000t - 2\sin 3000t) \text{ V}, \quad t \geq 0,$$

when the capacitor is 50 nF. Find (a) L; (b) R; (c) V_0; (d) I_0; and (e) $i_L(t)$.

6.4. The voltage response for the circuit in Fig. 6.1 is known to be

$$v(t) = D_1 t\, e^{-4000t} + D_2 e^{-4000t}, \quad t \geq 0.$$

The initial current in the inductor (I_0) is 5 mA, and the initial voltage on the capacitor (V_0) is 25 V. The inductor has an inductance of 5 H.

 (a) Find the value of R, C, D_1, and D_2.

 (b) Find $i_C(t)$ for $t \geq 0^+$.

6.5. The initial voltage on the 0.05 μF capacitor in the circuit shown in Fig. 6.1 is 15 V. The initial current in the inductor is zero. The voltage response for $t \geq 0$ is

$$v(t) = -5e^{-5000t} + 20e^{-20,000t} \text{ V}.$$

 (a) Determine the numerical values of R, L, α, and ω_0.

 (b) Calculate $i_R(t)$, $i_L(t)$, and $i_C(t)$ for $t \geq 0^+$.

 6.6. In the circuit shown in Fig. 6.1, a 5 H inductor is shunted by a 8 nF capacitor, the resistor R is adjusted for critical damping, $V_0 = -25$ V, and $I_0 = -1$ mA.

 (a) Calculate the numerical value of R.

 (b) Calculate $v(t)$ for $t \geq 0$.

 (c) Find $v(t)$ when $i_C(t) = 0$.

 (d) What percentage of the initially stored energy remains stored in the circuit at the instant $i_C(t)$ is 0?

 6.7. The circuit elements in the circuit in Fig. 6.1 are $R = 2000\Omega$, $C = 10$ nF, and $L = 250$ mH. The initial inductor current is -30 mA, and the initial capacitor voltage is 90 V.

 (a) Calculate the initial current in each branch of the circuit.

 (b) Find $v(t)$ for $t \geq 0$.

 (c) Find $i_L(t)$ for $t \geq 0$.

 6.8. The resistance in Problem 6.7 is increased to 12,500/3 Ω. Find the expression for $v(t)$ for $t \geq 0$.

 6.9. The resistance in Problem 6.7 is increased to 2500 Ω. Find the expression for $v(t)$ for $t \geq 0$.

6.10. The resistance, inductance, and capacitance in a parallel RLC circuit are 5000 Ω, 1.25 H, and 8 nF, respectively.

  (a) Calculate the roots of the characteristic equation that describe the voltage response of the circuit.

  (b) Will the response be over-, under-, or critically damped?

  (c) What value of R will yield a damped frequency of 6 krad/s?

  (d) What are the roots of the characteristic equation for the value of R found in (c)?

  (e) What value of R will result in a critically damped response?

6.11. The natural response for the circuit shown in Fig. 6.1 is known to be

$$v = -12\left(e^{-200t} + e^{-1800t}\right) \text{ V}, \quad t \geq 0.$$

If $C = 18 \; \mu\text{F}$, find $i_L(0^+)$ in milliamperes.

6.12. The initial value of the voltage v in the circuit in Fig. 6.1 is zero, and the initial value of the capacitor current, $i_c(0^+)$ is 15 mA. The expression for the capacitor current is known to be

$$i_c(t) = A_1 e^{-40t} + A_2 e^{-160t}, \quad t \geq 0^+,$$

when R is 200 Ω. Find the numerical

  (a) value of α, ω_0, L, C, A_1, and A_2

  (b) expression for $v(t), t \geq 0$

  (c) expression for $i_R(t) \geq 0^+$

  (d) expression for $i_L(t) \geq 0$

6.13. Assume the underdamped voltage response of the circuit in Fig. 6.1 is written as

$$v(t) = (A_1 + A_2)e^{-\alpha t} \cos \omega_d t + j(A_1 - A_2)e^{-\alpha t} \sin \omega_d t$$

The initial value of the inductor current is I_0, and the initial value of the capacitor voltage is V_0. Show that A_2 is the conjugate of A_1. (**Hint:** Use the same process as outlined in the text to find A_1 and A_2.)

6.14. Show that the results obtained from Problem 6.13—that is, the expressions for A_1 and A_2—are consistent with Eqs. (6.30) and (6.31) in the text.

6.15. The switch in the circuit of Fig. P6.15 has been in position a for a long time. At $t = 0$ the switch moves instantaneously to position b. Find $v_o(t)$ for $t \geq 0$.

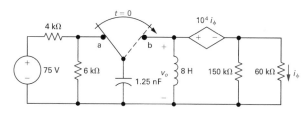

Figure P6.15

6.16. The resistor in the circuit in Example 6.4 is changed to 3200 Ω.

  (a) Find the numerical expression for $v(t)$ when $t \geq 0$.

  (b) Plot $v(t)$ versus t for the time interval $0 \leq t \leq 7$ ms. Compare this response with the one in Example 6.4 ($R = 20$ kΩ) and Example 6.5 ($R = 4$ kΩ). In particular, compare peak values of $v(t)$ and the times when these peak values occur.

6.17. For the circuit in Example 6.6, find, for $t \geq 0$, (a) $v(t)$; (b) $i_R(t)$; and (c) $i_C(t)$.

6.18. For the circuit in Example 6.7, find, for $t \geq 0$, (a) $v(t)$ and (b) $i_C(t)$.

6.19. For the circuit in Example 6.8, find $v(t)$ for $t \geq 0$.

6.20. The two switches in the circuit seen in Fig. P6.20 operate synchronously. When switch 1 is in position a, switch 2 is in position d. When switch 1 moves to position b, switch 2 moves to position c, and vice versa. Switch 1 has been in position a for a long time. At $t = 0$, the switches move to their alternate positions. Find $v_o(t)$ for $t \geq 0$.

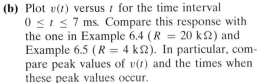

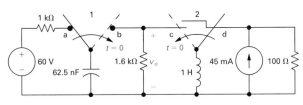

Figure P6.20

6.21. At the same time the resistor in the circuit in Fig. P6.20 is increased from 1.6 kΩ to 2.0 kΩ, the inductor is decreased from 1 H to 0.64 H. Find $v_o(t)$ for $t > 0$.

6.22. At the same time the resistor in the circuit in Fig. P6.20 is decreased from 1.6 kΩ to 800 Ω the inductor is decreased from 1 H to 160 mH. Find $v_o(t)$ for $t \geq 0$.

6.23. The initial energy stored in the 50 nF capacitor in the circuit in Fig. P6.23 is 90 μJ. Assume v_o is positive. The initial energy stored in the inductor is zero. The roots of the characteristic equation that describes the natural behavior of the current i are -1000 s^{-1} and -4000 s^{-1}.

   (a) Find the numerical values of R and L.

   (b) Find the numerical values of $i(0)$ and $di(0)/dt$ immediately after the switch has been closed.

   (c) Find $i(t)$ for $t \geq 0$.

   (d) How many microseconds after the switch closes does the current reach its maximum value?

   (e) What is the maximum value of i in milliamperes?

   (f) Find $v_L(t)$ for $t \geq 0$.

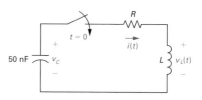

Figure P6.23

6.24. The switch in the circuit in Fig. P6.24 has been open a long time before closing at $t = 0$. The initial energy stored in the capacitor is zero. Find v_o for $t \geq 0$.

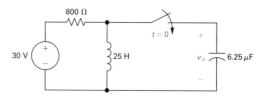

Figure P6.24

6.25. The circuit shown in Fig. P6.25 has been in operation for a long time. At $t = 0$, the voltage suddenly drops to 100 V. Find $v_o(t)$ for $t \geq 0$.

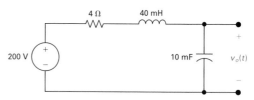

Figure P6.25

6.26. The current in the circuit in Fig. 6.3 is known to be

$$i = B_1 e^{-800t} \cos 600t + B_2 e^{-800t} \sin 600t, \ t \geq 0.$$

The capacitor has a value of 500 μF; the initial value of the current is zero; and the initial voltage on the capacitor is 12 V. Find the values of R, L, B_1, and B_2.

6.27. Find the voltage across the 500 μF capacitor for the circuit described in Problem 6.26. Assume the reference polarity for the capacitor voltage is positive at the upper terminal.

6.28. Switches 1 and 2 in the circuit in Fig. P6.28 are synchronized. When switch 1 is opened, switch 2 closes, and vice versa. Switch 1 has been open a long time before closing at $t = 0$. Find $i_L(t)$ for $t \geq 0$.

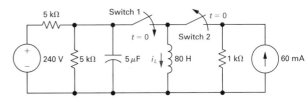

Figure P6.28

6.29. The initial energy stored in the circuit in Fig. P6.29 is zero. Find $v_o(t)$ for $t \geq 0$.

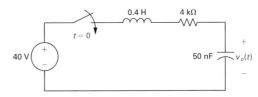

Figure P6.29

6.30. The switch in the circuit in Fig. P6.30 has been open for a long time before closing at $t = 0$. Find $v_o(t)$ for $t \geq 0$.

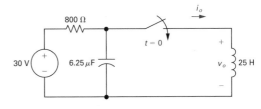

Figure P6.30

6.31. **(a)** For the circuit in Fig. P6.30, find i_o for $t \geq 0$.

(b) Show that your solution for i_o is consistent with the solution for v_o in Problem 6.30.

6.32. There is no energy stored in the circuit in Fig. P6.32 when the switch is closed at $t = 0$. Find $v_o(t)$ for $t \geq 0$.

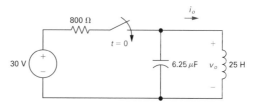

Figure P6.32

6.33. **(a)** For the circuit in Fig. P6.32, find i_o for $t \geq 0$.

(b) Show that your solution for i_o is consistent with the solution for v_o in Problem 6.32.

6.34. The switch in the circuit in Fig. P6.34 has been open a long time before closing at $t = 0$. Find $i_L(t)$ for $t \geq 0$.

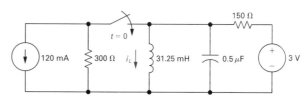

Figure P6.34

6.35. The make-before-break switch in the circuit shown in Fig. P6.35 has been in position a for a long time. At $t = 0$, the switch is moved instantaneously to position b. Find $i(t)$ for $t \geq 0$.

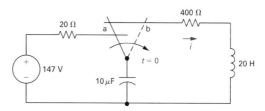

Figure P6.35

6.36. The switch in the circuit of Fig. P6.36 has been in position a for a long time. At $t = 0$ the switch moves instantaneously to position b. Find

(a) $v_o(0^+)$

(b) $dv_o(0^+)/dt$

(c) $v_o(t)$ for $t \geq 0$

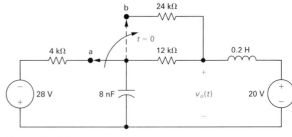

Figure P6.36

6.37. The switch in the circuit shown in Fig. P6.37 has been closed for a long time. The switch opens at $t = 0$. Find

(a) $i_o(t)$ for $t \geq 0$

(b) $v_o(t)$ for $t \geq 0$

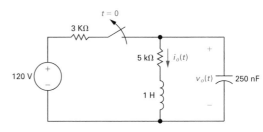

Figure P6.37

6.38. Assume that at the instant the 15 mA dc current source is applied to the circuit in Fig. P6.38, the initial current in the 20 H inductor is -30 mA, and the initial voltage on the capacitor is 60 V (positive at the upper terminal).
Find the expression for $i_L(t)$ for $t \geq 0$ if R equals 800 Ω.

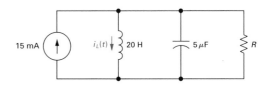

Figure P6.38

6.39. The resistance in the circuit in Fig. P6.38 is increased to 1250 Ω. Find $i_L(t)$ for $t \geq 0$.

6.40. The resistance in the circuit in Fig. P6.38 is changed to 1000 Ω. Find $i_L(t)$ for $t \geq 0$.

6.41. The switch in the circuit shown in Fig. P6.41 has been closed for a long time. The switch opens at $t = 0$. Find $v_o(t)$ for $t \geq 0$.

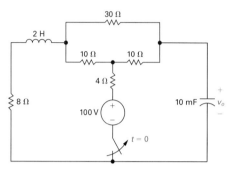

Figure P6.41

6.42. The switch in the circuit shown in Fig. P6.42 has been closed for a long time before it is opened at $t = 0$. Assume that the circuit parameters are such that the response is underdamped.

  (a) Derive the expression for $v_o(t)$ as a function of v_g, α, ω_d, C, and R for $t \geq 0$.

  (b) Derive the expression for the value of t when the magnitude of v_o is maximum.

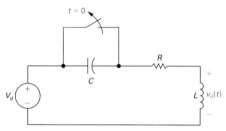

Figure P6.42

6.43. The circuit parameters in the circuit of Fig. P6.42 are $R = 120 \ \Omega$, $L = 5$ mH, $C = 500$ nF, and $v_g = -600$ V.

  (a) Express $v_o(t)$ numerically for $t \geq 0$.

  (b) How many microseconds after the switch opens is the inductor voltage maximum?

  (c) What is the maximum value of the inductor voltage?

  (d) Repeat (a)–(c) with R reduced to 12 Ω.

6.44. The switch in the circuit in Fig. P6.44 has been open a long time before closing at $t = 0$. Find

  (a) $v_o(t)$ for $t \geq 0^+$

  (b) $i_L(t)$ for $t \geq 0$

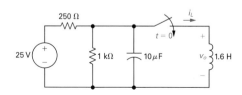

Figure P6.44

6.45. Use the circuit in Fig. P6.44

(a) Find the total energy delivered to the inductor.

(b) Find the total energy delivered to the equivalent resistor.

(c) Find the total energy delivered to the capacitor.

(d) Find the total energy delivered by the equivalent current source.

(e) Check the results of parts (a) through (d) against the conservation of energy principle.

6.46. In the circuit in Fig. P6.46, the resistor is adjusted for critical damping. The initial capacitor voltage is 90 V, and the initial inductor current is 24 mA.

(a) Find the numerical value of R

(b) Find the numerical values of i and di/dt immediately after the switch is closed.

(c) Find $v_C(t)$ for $t \geq 0$.

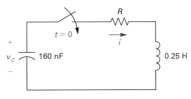

Figure P6.46

6.47. The two switches in the circuit seen in Fig. P6.47 operate synchronously. When switch 1 is in position a, switch 2 is closed. When switch 1 is in position b, switch 2 is open. Switch 1 has been in position a for a long time. At $t = 0$, it moves instantaneously to position b. Find $v_c(t)$ for $t \geq 0$.

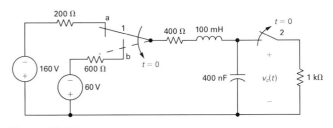

Figure P6.47

6.48. The switch in the circuit in Fig. P6.48 has been in position a for a long time. At $t = 0$, the switch moves instantaneously to position b.

(a) What is the initial value of v_a?

(b) What is the initial value of dv_a/dt?

(c) What is the numerical expression for $v_a(t)$ for $t \geq 0$?

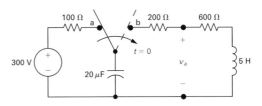

Figure P6.48

◆ **6.49.** **(a)** The value of L in the circuit in Fig. 6.18 is 1.6 nH. Assume $v_o(0) = 1$ V. Derive the expression for $v_o(t)$ for $t \geq 0$.

(b) Repeat (a) when $L = 2.5$ nH.

(c) Repeat (a) when $L = 5$ nH.

(d) Repeat (a) when $L = 25$ nH.

◆ **6.50.** Verify the transition times (t_x) given in Table 6.3 for $L = 1.6$ nH; $L = 2.5$ nH; $L = 5$ nH; and $L = 25$ nH. (*Hint:* All these solutions require a trial-and-error approach to find t_x.)

Household Distribution Circuit

Power systems that generate, transmit, and distribute electric power are designed to operate in the sinusoidal steady state. The standard household distribution circuit used in the United States is the three-wire 240/120 V circuit shown in the accompanying figure. The operating frequency of power systems in the United States is 60 Hz. Both 50 and 60 Hz systems are found outside the United States. The voltage ratings alluded to above are rms values. The reason for defining an rms value of a time-varying signal is explained in Section 7.1.

At the end of this chapter we will discuss the reason why an overcurrent device (either a circuit breaker or fuse, which is represented by the symbol ⌒ in the figure) is not placed in the neutral conductor of the distribution circuit.

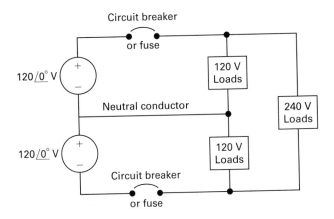

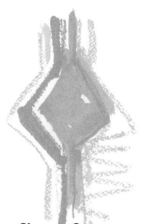

7 Sinusoidal Steady-State Analysis

Chapter Contents

Thus far, we have focused on circuits with constant sources; in this chapter we are now ready to consider circuits energized by time-varying voltage or current sources. In particular, we are interested in sources in which the value of the voltage or current varies sinusoidally. Sinusoidal sources and their effect on circuit behavior form an important area of study for several reasons. First, the generation, transmission, distribution, and consumption of electric energy occur under essentially sinusoidal steady-state conditions. Second, an understanding of sinusoidal behavior makes it possible to predict the behavior of circuits with nonsinusoidal sources. Third, steady-state sinusoidal behavior often simplifies the design of electrical systems. Thus a designer can spell out specifications in terms of a desired steady-state sinusoidal response and design the circuit or system to meet those characteristics. If the device satisfies the specifications, the designer knows that the circuit will respond satisfactorily to nonsinusoidal inputs.

The subsequent chapters of this book are largely based on a thorough understanding of the techniques needed to analyze circuits driven by sinusoidal sources. Fortunately, the circuit analysis and simplification techniques first introduced in Chapters 1–3 work for circuits with sinusoidal as well as dc sources, so some of the material in this chapter will be very familiar to you. The challenges in first approaching sinusoidal analysis include developing the appropriate modeling equations and working in the mathematical realm of complex numbers.

7.1 THE SINUSOIDAL SOURCE

A **sinusoidal voltage source** (independent or dependent) produces a voltage that varies sinusoidally with time. A **sinusoidal current source** (independent or dependent) produces a current that varies sinusoidally with time. In reviewing the sinusoidal function, we use a voltage source, but our observations also apply to current sources.

We can express a sinusoidally varying function with either the sine function or the cosine function. Although either works equally well, we cannot use both functional forms simultaneously. We will use the cosine function throughout our discussion. Hence, we write a sinusoidally varying voltage as

$$v = V_m \cos(\omega t + \phi). \tag{7.1}$$

To aid discussion of the parameters in Eq. (7.1), we show the voltage-versus-time plot in Fig. 7.1.

Note that the sinusoidal function repeats at regular intervals. Such a function is called periodic. One parameter of interest is the length of time required for the sinusoidal function to pass through all its possible values. This time is referred to as the **period** of the function and is denoted T. It is measured in seconds. The reciprocal of T gives the number of cycles per second, or the frequency, of the sine function and is denoted f, or

$$f = \frac{1}{T}. \tag{7.2}$$

A cycle per second is referred to as a hertz, abbreviated Hz. (The term *cycles per second* rarely is used in contemporary technical literature.) The coefficient of t in Eq. (7.1) contains the numerical value of T or f. Omega (ω) represents the angular frequency of the sinusoidal function, or

$$\omega = 2\pi f = 2\pi / T \text{ (radians/second)}. \tag{7.3}$$

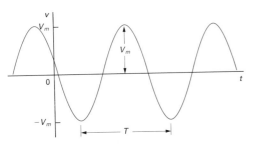

Figure 7.1 A sinusoidal voltage.

Equation 7.3 is based on the fact that the cosine (or sine) function passes through a complete set of values each time its argument, ωt, passes through 2π rad (360°). From Eq. (7.3), note that, whenever t is an integral multiple of T, the argument ωt increases by an integral multiple of 2π rad.

The coefficient V_m gives the maximum amplitude of the sinusoidal voltage. Because ± 1 bounds the cosine function, $\pm V_m$ bounds the amplitude. Figure 7.1 shows these characteristics.

The angle ϕ in Eq. (7.1) is known as the **phase angle** of the sinusoidal voltage. It determines the value of the sinusoidal function at $t = 0$; therefore, it fixes the point on the periodic wave at which we start measuring time. Changing the phase angle ϕ shifts the sinusoidal function along the time axis but has no effect on either the amplitude (V_m) or the angular frequency (ω). Note, for example, that reducing ϕ to zero shifts the sinusoidal function shown in Fig. 7.1 ϕ/ω time units to the right, as shown in Fig. 7.2. Note also that if ϕ is positive, the sinusoidal function shifts to the left, whereas if ϕ is negative, the function shifts to the right. (See Problem 7.2.)

A comment with regard to the phase angle is in order: ωt and ϕ must carry the same units, because they are added together in the argument of the sinusoidal function. With ωt expressed in radians, you would expect ϕ to be also. However, ϕ normally is given in degrees, and ωt is converted from radians to degrees before the two quantities are added. We continue this bias toward degrees by expressing the phase angle in degrees. Recall from your studies of trigonometry that the conversion from radians to degrees is given by

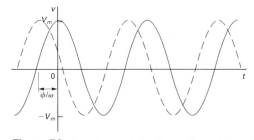

Figure 7.2 The sinusoidal voltage from Fig. 7.1 shifted to the right when $\phi = 0$.

$$(\text{number of degrees}) = \frac{180°}{\pi}(\text{number of radians}). \qquad (7.4)$$

Another important characteristic of the sinusoidal voltage (or current) is its **rms value.** The rms value of a periodic function is defined as the square *root* of the *mean* value of the *squared* function. Hence, if $v = V_m \cos(\omega t + \phi)$, the rms value of v is

$$V_{\text{rms}} = \sqrt{\frac{1}{T}\int_{t_0}^{t_0+T} V_m^2 \cos^2(\omega t + \phi)\, dt}. \qquad (7.5)$$

Note from Eq. (7.5) that we obtain the mean value of the squared voltage by integrating v^2 over one period (that is, from t_0 to t_0+T) and then dividing by the range of integration, T. Note further that the starting point for the integration t_0 is arbitrary.

The quantity under the radical sign in Eq. (7.5) reduces to $V_m^2/2$. (See Problem 7.7.) Hence the rms value of v is

$$V_{\text{rms}} = \frac{V_m}{\sqrt{2}}. \qquad (7.6)$$

The rms value of the sinusoidal voltage depends only on the maximum amplitude of v, namely, V_m. The rms value is not a function of either the frequency or the phase angle. We stress the importance of the rms value as it relates to power calculations in Section 7.10.

Thus, we can completely describe a specific sinusoidal signal if we know its frequency, phase angle, and amplitude (either the maximum or the rms value). Examples 7.1, 7.2, and 7.3 illustrate these basic properties of the sinusoidal function. In Example 7.4, we calculate the rms value of a periodic function, and in so doing we clarify the meaning of *root mean square*.

EXAMPLE 7.1

A sinusoidal current has a maximum amplitude of 20 A. The current passes through one complete cycle in 1 ms. The magnitude of the current at zero time is 10 A.

(a) What is the frequency of the current in hertz?

(b) What is the frequency in radians per second?

(c) Write the expression for $i(t)$ using the cosine function. Express ϕ in degrees.

(d) What is the rms value of the current?

SOLUTION

(a) From the statement of the problem, $T = 1$ ms; hence $f = 1/T = 1000$ Hz.

(b) $\omega = 2\pi f = 2000\pi$ rad/s.

(c) We have $i(t) = I_m \cos(\omega t + \phi) = 20 \cos(2000\pi t + \phi)$, but $i(0) = 10$ A. Therefore $10 = 20 \cos \phi$ and $\phi = 60°$. Thus the expression for $i(t)$ becomes

$$i(t) = 20 \cos(2000\pi t + 60°) \text{ A}.$$

(d) From the derivation of Eq. (7.6), the rms value of a sinusoidal current is $I_m/\sqrt{2}$. Therefore the rms value is $20/\sqrt{2}$, or 14.14 A.

EXAMPLE 7.2

A sinusoidal voltage is given by the expression
$v = 300 \cos(120\pi t + 30°)$ V.

(a) What is the period of the voltage in milliseconds?

(b) What is the frequency in hertz?

(c) What is the magnitude of v at $t = 2.778$ ms?

(d) What is the rms value of v?

SOLUTION

(a) From the expression for v, $\omega = 120\pi$ rad/s. Because $\omega = 2\pi/T$, $T = 2\pi/\omega = \frac{1}{60}$ s, or 16.667 ms.

(b) The frequency is $1/T$, or 60 Hz.

(c) From (a), $\omega = 2\pi/16.667$; thus, at $t = 2.778$ ms, ωt is nearly 1.047 rad, or 60°. Therefore, $v(2.778$ ms$) = 300 \cos(60° + 30°) = 0$ V.

(d) $V_{rms} = 300/\sqrt{2} = 212.13$ V.

EXAMPLE 7.3

We can translate the sine function to the cosine function by subtracting 90° ($\pi/2$ rad) from the argument of the sine function.

(a) Verify this translation by showing that

$$\sin(\omega t + \theta) = \cos(\omega t + \theta - 90°).$$

(b) Use the result in (a) to express $\sin(\omega t + 30°)$ as a cosine function.

SOLUTION

(a) Verification involves direct application of the trigonometric identity

$$\cos(\alpha - \beta) = \cos \alpha \cos \beta + \sin \alpha \sin \beta.$$

We let $\alpha = \omega t + \theta$ and $\beta = 90°$. As $\cos 90° = 0$ and $\sin 90° = 1$, we have

$$\cos(\alpha - \beta) = \sin \alpha = \sin(\omega t + \theta) = \cos(\omega t + \theta - 90°).$$

(b) From (a) we have

$$\sin(\omega t + 30°) = \cos(\omega t + 30° - 90°) = \cos(\omega t - 60°).$$

EXAMPLE 7.4

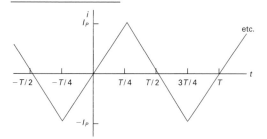

Figure 7.3 Periodic triangular current.

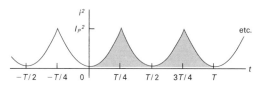

Figure 7.4 i^2 versus t.

Calculate the rms value of the periodic triangular current shown in Fig. 7.3. Express your answer in terms of the peak current I_p.

SOLUTION

From Eq. (7.5), the rms value of i is

$$I_{\text{rms}} = \sqrt{\frac{1}{T} \int_{t_0}^{t_0+T} i^2 \, dt}.$$

Interpreting the integral under the radical sign as the area under the squared function for an interval of one period is helpful in finding the rms value. The squared function with the area between 0 and T shaded is shown in Fig. 7.4, which also indicates that for this particular function, the area under the squared current for an interval of one period is equal to four times the area under the squared current for the interval 0 to $T/4$ seconds; that is,

$$\int_{t_0}^{t_0+T} i^2 \, dt = 4 \int_{0}^{T/4} i^2 \, dt.$$

The analytical expression for i in the interval 0 to $T/4$ is

$$i = \frac{4I_p}{T} t, \quad 0 < t < T/4.$$

The area under the squared function for one period is

$$\int_{t_0}^{t_0+T} i^2 \, dt = 4 \int_{0}^{T/4} \frac{16I_p^2}{T^2} t^2 \, dt = \frac{I_p^2 T}{3}.$$

The mean, or average, value of the function is simply the area for one period divided by the period. Thus

$$i_{\text{mean}} = \frac{1}{T} \frac{I_p^2 T}{3} = \frac{1}{3} I_p^2.$$

The rms value of the current is the square root of this mean value. Hence

$$I_{\text{rms}} = \frac{I_p}{\sqrt{3}}.$$

DRILL EXERCISES

7.1   A sinusoidal voltage is given by the expression

$$v = 10\cos(3769.91t - 53.13°) \text{ V.}$$

Find (a) f in hertz; (b) T in milliseconds; (c) V_m; (d) $v(0)$; (e) ϕ in degrees and radians; (f) the smallest positive value of t at which $v = 0$; and (g) the smallest positive value of t at which $dv/dt = 0$.

7.2   Find the rms value of the half-wave rectified sinusoidal voltage shown.

ANSWER: $V_{\text{rms}} = V_m/2$.

ANSWER: (a) 600 Hz; (b) 1.67 ms; (c) 10 V; (d) 6 V; (e) −53.13°, or −0.9273 rad; (f) 662.62 μs; (g) 245.97 μs.

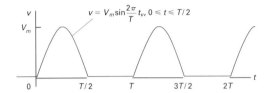

7.2 THE SINUSOIDAL RESPONSE

Before focusing on the steady-state response to sinusoidal sources, let's consider the problem in broader terms, that is, in terms of the total response. Such an overview will help you keep the steady-state solution in perspective. The circuit shown in Fig. 7.5 describes the general nature of the problem. There, v_s is a sinusoidal voltage, or

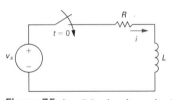

Figure 7.5 An RL circuit excited by a sinusoidal voltage source.

$$v_s = V_m \cos(\omega t + \phi). \qquad (7.7)$$

For convenience, we assume the initial current in the circuit to be zero and measure time from the moment the switch is closed. The task is to derive the expression for $i(t)$ when $t \geq 0$. It is similar to finding the step response of an RL circuit, as in Chapter 5. The only difference is that the voltage source is now a time-varying sinusoidal voltage rather than a constant, or dc, voltage. Direct application of Kirchhoff's voltage law to the circuit shown in Fig. 7.5 leads to the ordinary differential equation

$$L\frac{di}{dt} + Ri = V_m \cos(\omega t + \phi), \qquad (7.8)$$

the formal solution of which is discussed in an introductory course in differential equations. We ask those of you who have not yet studied differential equations to accept that the solution for i is

$$i = \frac{-V_m}{\sqrt{R^2 + \omega^2 L^2}} \cos(\phi - \theta) e^{-(R/L)t}$$

$$+ \frac{V_m}{\sqrt{R^2 + \omega^2 L^2}} \cos(\omega t + \phi - \theta), \qquad (7.9)$$

where θ is defined as the angle whose tangent is $\omega L / R$. Thus we can easily determine θ for a circuit driven by a sinusoidal source of known frequency.

We can check the validity of Eq. (7.9) by determining that it satisfies Eq. (7.8) for all values of $t \geq 0$; this exercise is left for your exploration in Problem 7.5.

The first term on the right-hand side of Eq. (7.9) is referred to as the **transient component** of the current because it becomes infinitesimal as time elapses. The second term on the right-hand side is known as the **steady-state component** of the solution. It exists as long as the switch remains closed and the source continues to supply the sinusoidal voltage. In this chapter, we develop a technique for calculating the steady-state response directly, thus avoiding the problem of solving the differential equation. However, in using this technique we forfeit obtaining either the transient component or the total response, which is the sum of the transient and steady-state components.

We now focus on the steady-state portion of Eq. (7.9). It is important to remember the following characteristics of the steady-state solution:

1. The steady-state solution is a sinusoidal function.

2. The frequency of the response signal is identical to the frequency of the source signal. This condition is always true in a linear circuit when the circuit parameters, R, L, and C, are constant. (If frequencies in the response signals are not present in the source signals, there is a nonlinear element in the circuit.)

3. The maximum amplitude of the steady-state response, in general, differs from the maximum amplitude of the source. For the circuit being discussed, the maximum amplitude of the response signal is $V_m/\sqrt{R^2 + \omega^2 L^2}$, and the maximum amplitude of the signal source is V_m.

4. The phase angle of the response signal, in general, differs from the phase angle of the source. For the circuit being discussed, the phase angle of the current is $\phi - \theta$ and that of the voltage source is ϕ.

These characteristics are worth remembering because they help you understand the motivation for the phasor method, which we introduce in Section 7.3. In particular, note that once the decision has been made to find only the steady-state response, the task is reduced to finding the maximum amplitude and phase angle of the response signal. The waveform and frequency of the response are already known.

DRILL EXERCISE

7.3   The voltage applied to the circuit shown in Fig. 7.5 at $t = 0$ is $20\cos(800t + 25°)$ V. The circuit resistance is 80 Ω, and the initial current in the 75 mH inductor is zero.

(a) Find $i(t)$ for $t \geq 0$.

(b) Write the expressions for the transient and steady-state components of $i(t)$.

(c) Find the numerical value of i after the switch has been closed for 1.875 ms.

(d) What are the maximum amplitude, frequency (in radians per second), and phase angle of the steady-state current?

(e) By how many degrees are the voltage and the steady-state current out of phase?

ANSWER:
(a) $-195.72e^{-1066.67t} + 200\cos(800t - 11.87°)$ mA;
(b) $-195.72e^{-1066.67t}$ mA, $200\cos(800t - 11.87°)$ mA;
(c) 28.39 mA; (d) 0.2 A, 800 rad/s, $-11.87°$;
(e) 36.87°.

7.3 THE PHASOR

The **phasor** is a complex number that carries the amplitude and phase angle information of a sinusoidal function.[1] The phasor concept is rooted in Euler's identity, which relates the exponential function to the trigonometric function:

$$e^{\pm j\theta} = \cos\theta \pm j\sin\theta. \qquad (7.10)$$

Equation (7.10) is important here because it gives us another way of expressing the cosine and sine functions. We can think of the cosine function as the real part of the exponential function and the sine function as the imaginary part of the exponential function; that is,

$$\cos\theta = \Re\{e^{j\theta}\}, \qquad (7.11)$$

---

[1] If you feel a bit uneasy about complex numbers, peruse Appendix B.

and

$$\sin \theta = \Im\{e^{j\theta}\}, \tag{7.12}$$

where $\Re$ means "the real part of" and $\Im$ means "the imaginary part of."

Because we have already chosen to use the cosine function in analyzing the sinusoidal steady state (see Section 7.1), we can apply Eq. (7.11) directly. In particular, we write the sinusoidal voltage function given by Eq. (7.1) in the form suggested by Eq. (7.11):

$$v = V_m \cos(\omega t + \phi)$$

$$= V_m \Re\{e^{j(\omega t + \phi)}\}$$

$$= V_m \Re\{e^{j\omega t} e^{j\phi}\}. \tag{7.13}$$

We can move the coefficient V_m inside the argument of the real part of the function without altering the result. We can also reverse the order of the two exponential functions inside the argument and write Eq. (7.13) as

$$v = \Re\{V_m e^{j\phi} e^{j\omega t}\}. \tag{7.14}$$

In Eq. (7.14), note that the quantity $V_m e^{j\phi}$ is a complex number that carries the amplitude and phase angle of the given sinusoidal function. This complex number is by definition the **phasor representation,** or **phasor transform,** of the given sinusoidal function. Thus

$$\mathbf{V} = V_m e^{j\phi} = \mathcal{P}\{V_m \cos(\omega t + \phi)\}, \tag{7.15}$$

where the notation $\mathcal{P}\{V_m \cos(\omega t + \phi)\}$ is read "the phasor transform of $V_m \cos(\omega t + \phi)$." Thus the phasor transform transfers the sinusoidal function from the time domain to the complex-number domain, which is also called the **frequency domain,** since the response depends, in general, on ω. As in Eq. (7.15), throughout this book we represent a phasor quantity by using a boldface letter.

Equation (7.15) is the polar form of a phasor, but we also can express a phasor in rectangular form. Thus we rewrite Eq. (7.15) as

$$\mathbf{V} = V_m \cos \phi + j V_m \sin \phi. \tag{7.16}$$

Both polar and rectangular forms are useful in circuit applications of the phasor concept.

One additional comment regarding Eq. (7.15) is in order. The frequent occurrence of the exponential function $e^{j\phi}$ has led to an abbreviation that lends itself to text material. This abbreviation is the angle notation

$$1 \underline{/\phi^\circ} \equiv 1e^{j\phi}.$$

We use this notation extensively in the material that follows.

Inverse Phasor Transform

So far we have emphasized moving from the sinusoidal function to its phasor transform. However, we may also reverse the process. That is, for a phasor we may write the expression for the sinusoidal function. Thus for $\mathbf{V} = 100 \underline{/-26^\circ}$, the expression for v is $100\cos(\omega t - 26^\circ)$ because we have decided to use the cosine function for all sinusoids. Observe that we cannot deduce the value of ω from the phasor. The phasor carries only amplitude and phase information. The step of going from the phasor transform to the time-domain expression is referred to as **finding the inverse phasor transform** and is formalized by the equation

$$\mathcal{P}^{-1}\{V_m e^{j\phi}\} = \Re\{V_m e^{j\phi} e^{j\omega t}\}, \qquad (7.17)$$

where the notation $\mathcal{P}^{-1}\{V_m e^{j\phi}\}$ is read as "the inverse phasor transform of $V_m e^{j\phi}$." Equation (7.17) indicates that to find the inverse phasor transform, we multiply the phasor by $e^{j\omega t}$ and then extract the real part of the product.

The phasor transform is useful in circuit analysis because it reduces the task of finding the maximum amplitude and phase angle of the steady-state sinusoidal response to the algebra of complex numbers. The following observations verify this conclusion:

1. The transient component vanishes as time elapses, so the steady-state component of the solution must also satisfy the differential equation. [See Problem 7.5[b].]

2. In a linear circuit driven by sinusoidal sources, the steady-state response also is sinusoidal, and the frequency of the sinusoidal response is the same as the frequency of the sinusoidal source.

3. Using the notation introduced in Eq. (7.11), we can postulate that the steady-state solution is of the form $\Re\{Ae^{j\beta} e^{j\omega t}\}$, where A is the maximum amplitude of the response and β is the phase angle of the response.

4. When we substitute the postulated steady-state solution into the differential equation, the exponential term $e^{j\omega t}$ cancels out, leaving the solution for A and β in the domain of complex numbers.

We illustrate these observations with the circuit shown in Fig. 7.5. We know that the steady-state solution for the current i is of the form

$$i_{ss}(t) = \Re\{I_m e^{j\beta} e^{j\omega t}\}, \tag{7.18}$$

where the subscript "ss" emphasizes that we are dealing with the steady-state solution. When we substitute Eq. (7.18) into Eq. (7.8), we generate the expression

$$\Re\{j\omega L I_m e^{j\beta} e^{j\omega t}\} + \Re\{R I_m e^{j\beta} e^{j\omega t}\} = \Re\{V_m e^{j\phi} e^{j\omega t}\}. \tag{7.19}$$

In deriving Eq. (7.19) we recognized that both differentiation and multiplication by a constant can be taken inside the real part of an operation. We also rewrote the right-hand side of Eq. (7.8), using the notation of Eq. (7.11). From the algebra of complex numbers, we know that the sum of the real parts is the same as the real part of the sum. Therefore we may reduce the left-hand side of Eq. (7.19) to a single term:

$$\Re\{(j\omega L + R) I_m e^{j\beta} e^{j\omega t}\} = \Re\{V_m e^{j\phi} e^{j\omega t}\}. \tag{7.20}$$

Recall that our decision to use the cosine function in analyzing the response of a circuit in the sinusoidal steady state results in the use of the $\Re$ operator in deriving Eq. (7.20). If instead we had chosen to use the sine function in our sinusoidal steady-state analysis, we would have applied Eq. (7.12) directly, in place of Eq. (7.11), and the result would be Eq. (7.21):

$$\Im\{(j\omega L + R) I_m e^{j\beta} e^{j\omega t}\} = \Im\{V_m e^{j\phi} e^{j\omega t}\}. \tag{7.21}$$

Note that the complex quantities on either side of Eq. (7.21) are identical to those on either side of Eq. (7.20). When both the real and imaginary parts of two complex quantities are equal, then the complex quantities are themselves equal. Therefore, from Eqs. (7.20) and (7.21),

$$(j\omega L + R) I_m e^{j\beta} = V_m e^{j\phi},$$

or

$$I_m e^{j\beta} = \frac{V_m e^{j\phi}}{R + j\omega L}. \tag{7.22}$$

Note that $e^{j\omega t}$ has been eliminated from the determination of the amplitude (I_m) and phase angle (β) of the response. Thus, for this circuit, the task of finding I_m and β involves the algebraic manipulation of the complex quantities $V_m e^{j\phi}$ and $R + j\omega L$. Note that we encountered both polar and rectangular forms.

An important warning is in order: The phasor transform, along with the inverse phasor transform, allows you to go back and forth between the time domain and the frequency domain. Therefore, when you obtain a solution, you are either in the time domain or the frequency domain. You cannot be in both domains simultaneously. Any solution that contains a mixture of time domain and phasor domain nomenclature is nonsensical.

The phasor transform is also useful in circuit analysis because it applies directly to the sum of sinusoidal functions. Circuit analysis involves summing currents and voltages, so the importance of this observation is obvious. We can formalize this property as follows: If

$$v = v_1 + v_2 + \cdots + v_n \qquad (7.23)$$

where all the voltages on the right-hand side are sinusoidal voltages of the same frequency, then

$$\mathbf{V} = \mathbf{V}_1 + \mathbf{V}_2 + \cdots + \mathbf{V}_n. \qquad (7.24)$$

Thus the phasor representation is the sum of the phasors of the individual terms. We discuss the development of Eq. (7.24) in Section 7.5.

Before applying the phasor transform to circuit analysis, we illustrate its usefulness in solving a problem with which you are already familiar: adding sinusoidal functions via trigonometric identities. Example 7.5 shows how the phasor transform greatly simplifies this type of problem.

EXAMPLE 7.5

If $y_1 = 20\cos(\omega t - 30°)$ and $y_2 = 40\cos(\omega t + 60°)$, express $y = y_1 + y_2$ as a single sinusoidal function.

(a) Solve by using trigonometric identities.

(b) Solve by using the phasor concept.

SOLUTION

(a) First we expand both y_1 and y_2, using the cosine of the sum of two angles, to get

$$y_1 = 20\cos \omega t \cos 30° + 20\sin \omega t \sin 30°;$$

$$y_2 = 40\cos \omega t \cos 60° - 40\sin \omega t \sin 60°.$$

Adding y_1 and y_2, we obtain

$$y = (20\cos 30° + 40\cos 60°)\cos \omega t + (20\sin 30° - 40\sin 60°)\sin \omega t$$

$$= 37.32\cos \omega t - 24.64\sin \omega t.$$

To combine these two terms we treat the coefficients of the cosine and sine as sides of a right triangle (Fig. 7.6) and then multiply and divide the right-hand side by the hypotenuse. Our expression for y becomes

$$y = 44.72 \left(\frac{37.32}{44.72}\cos \omega t - \frac{24.64}{44.72}\sin \omega t \right)$$

$$= 44.72(\cos 33.43°\cos \omega t - \sin 33.43°\sin \omega t).$$

Again, we invoke the identity involving the cosine of the sum of two angles and write

$$y = 44.72\cos(\omega t + 33.43°).$$

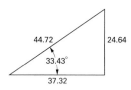

Figure 7.6 A right triangle used in the solution for y.

(b) We can solve the problem by using phasors as follows: Because

$$y = y_1 + y_2,$$

then, from Eq. (7.24),

$$\mathbf{Y} = \mathbf{Y}_1 + \mathbf{Y}_2$$

$$= 20\ \angle{-30°} + 40\ \angle{60°}$$

$$= (17.32 - j10) + (20 + j34.64)$$

$$= 37.32 + j24.64$$

$$= 44.72\ \angle{33.43°}.$$

Once we know the phasor $\mathbf{Y}$, we can write the corresponding trigonometric function for y by taking the inverse phasor transform:

$$y = \mathscr{P}^{-1}\{44.72e^{j33.43}\} = \Re\{44.72e^{j33.43}e^{j\omega t}\}$$

$$= 44.72\cos(\omega t + 33.43°).$$

The superiority of the phasor approach for adding sinusoidal functions should be apparent. Note that it requires the ability to move back and forth between the polar and rectangular forms of complex numbers.

DRILL EXERCISES

7.4   Find the phasor transform of each trigonometric function:

    (a) $v = 170\cos(377t - 40°)$ V.

    (b) $i = 10\sin(1000t + 20°)$ A.

    (c) $i = [5\cos(\omega t + 36.87°) + 10\cos(\omega t - 53.13°)]$ A.

    (d) $v = [300\cos(20,000\pi t + 45°) - 100\sin(20,000\pi t + 30°)]$ mV.

    ANSWER: (a) $\mathbf{V} = 170 \ \angle{-40°}$ V;
(b) $\mathbf{I} = 10 \ \angle{-70°}$ A; (c) $\mathbf{I} = 11.18 \ \angle{-26.57°}$ A;
(d) $\mathbf{V} = 339.90 \ \angle{61.51°}$ mV.

7.5   Find the time-domain expression corresponding to each phasor:

    (a) $\mathbf{V} = 18.6 \ \angle{-54°}$ V.

    (b) $\mathbf{I} = (20 \ \angle{45°} - 50 \ \angle{-30°})$ mA.

    (c) $\mathbf{V} = (20 + j80 - 30 \ \angle{15°})$ V.

    ANSWER: (a) $v = 18.6\cos(\omega t - 54°)$ V;
(b) $i = 48.81\cos(\omega t + 126.68°)$ mA;
(c) $v = 72.79\cos(\omega t + 97.08°)$ V.

7.4   THE PASSIVE CIRCUIT ELEMENTS IN THE FREQUENCY DOMAIN

The systematic application of the phasor transform in circuit analysis requires two steps. First, we must establish the relationship between the phasor current and the phasor voltage at the terminals of the passive circuit elements. Second, we must develop the phasor-domain version of Kirchhoff's laws, which we discuss in Section 7.5. In this section, we establish the relationship between the phasor current and voltage at the terminals of the resistor, inductor, and capacitor. We begin with the resistor and use the passive sign convention in all the derivations.

The V-I Relationship for a Resistor

From Ohm's law, if the current in a resistor varies sinusoidally with time—that is, if $i = I_m\cos(\omega t + \theta_i)$—the voltage at the terminals of the resistor, as shown in Fig. 7.7, is

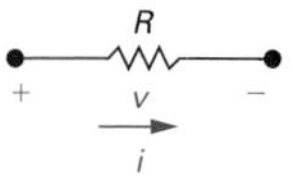

Figure 7.7 A resistive element carrying a sinusoidal current.

$$v = R[I_m\cos(\omega t + \theta_i)]$$

$$= RI_m[\cos(\omega t + \theta_i)], \qquad (7.25)$$

where I_m is the maximum amplitude of the current in amperes and θ_i is the phase angle of the current.

Figure 7.8 The frequency-domain equivalent circuit of a resistor.

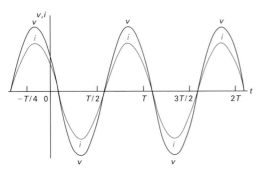

Figure 7.9 A plot showing that the voltage and current at the terminals of a resistor are in phase.

The phasor transform of this voltage is

$$\mathbf{V} = RI_m e^{j\theta_i} = RI_m \; \underline{/\theta_i} \;. \tag{7.26}$$

But $I_m \; \underline{/\theta_i}$ is the phasor representation of the sinusoidal current, so we can write Eq. (7.26) as

$$\mathbf{V} = R\mathbf{I}, \tag{7.27}$$

which states that the phasor voltage at the terminals of a resistor is simply the resistance times the phasor current. Figure 7.8 shows the circuit diagram for a resistor in the frequency domain.

Equations (7.25) and (7.27) both contain another important piece of information—namely, that at the terminals of a resistor, there is no phase shift between the current and voltage. Figure 7.9 depicts this phase relationship, where the phase angle of both the voltage and the current waveforms is $60°$. The signals are said to be **in phase** because they both reach corresponding values on their respective curves at the same time (for example, they are at their positive maxima at the same instant).

The V-I Relationship for an Inductor

We derive the relationship between the phasor current and phasor voltage at the terminals of an inductor by assuming a sinusoidal current and using $L\,di/dt$ to establish the corresponding voltage. Thus, for $i = I_m \cos(\omega t + \theta_i)$, the expression for the voltage is

$$v = L\frac{di}{dt} = -\omega L I_m \sin(\omega t + \theta_i). \tag{7.28}$$

We now rewrite Eq. (7.28) using the cosine function:

$$v = -\omega L I_m \cos(\omega t + \theta_i - 90°). \tag{7.29}$$

The phasor representation of the voltage given by Eq. (7.29) is

$$\mathbf{V} = -\omega L I_m e^{j(\theta_i - 90°)}$$

$$= -\omega L I_m e^{j\theta_i} e^{-j90°}$$

$$= j\omega L I_m e^{j\theta_i}$$

$$= j\omega L \mathbf{I}. \tag{7.30}$$

Note that in deriving Eq. (7.30) we used the identity

$$e^{-j90°} = \cos 90° - j \sin 90° = -j.$$

Equation (7.30) states that the phasor voltage at the terminals of an inductor equals $j\omega L$ times the phasor current. Figure 7.10 shows the frequency-domain equivalent circuit for the inductor.

Figure 7.10 The frequency-domain equivalent circuit for an inductor.

We can rewrite Eq. (7.30) as

$$\mathbf{V} = \omega L \ \underline{/90^\circ} \ I_m \ \underline{/\theta_i}$$
$$= \omega L I_m \ \underline{/(\theta_i + 90)^\circ}, \qquad (7.31)$$

which indicates that the voltage and current are out of phase by exactly 90°. In particular, the voltage leads the current by 90°, or, equivalently, the current lags behind the voltage by 90°. Figure 7.11 illustrates this concept of **voltage leading current** or **current lagging voltage**. For example, the voltage reaches its negative peak exactly 90° before the current reaches its negative peak. The same observation can be made with respect to the zero-going positive crossing or the positive peak.

We can also express the phase shift in seconds. A phase shift of 90° corresponds to one-fourth of a period; hence the voltage leads the current by $T/4$, or $\frac{1}{4f}$ second.

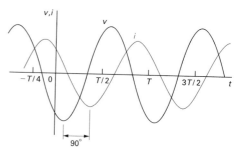

Figure 7.11 A plot showing the phase relationship between the current and voltage at the terminals of an inductor ($\theta_i = 60^\circ$).

The V-I Relationship for a Capacitor

We obtain the relationship between the phasor current and phasor voltage at the terminals of a capacitor from the derivation of Eq. (7.30). In other words, if we note that for a capacitor that

$$i = C\frac{dv}{dt},$$

and assume that

$$v = V_m \cos(\omega t + \theta_v),$$

then

$$\mathbf{I} = j\omega C\mathbf{V}. \qquad (7.32)$$

Now if we solve Eq. (7.32) for the voltage as a function of the current, we get

$$\mathbf{V} = \frac{1}{j\omega C}\mathbf{I}. \qquad (7.33)$$

Figure 7.12 The frequency-domain equivalent circuit of a capacitor.

Equation (7.33) demonstrates that the equivalent circuit for the capacitor in the phasor domain is as shown in Fig. 7.12.

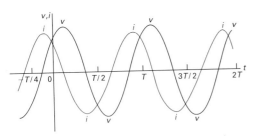

Figure 7.13 A plot showing the phase relationship between the current and voltage at the terminals of a capacitor ($\theta_i = 60°$).

The voltage across the terminals of a capacitor lags behind the current by exactly $90°$. We can easily show this relationship by rewriting Eq. (7.33) as

$$\mathbf{V} = \frac{1}{\omega C} \underline{/{-90°}} \; I_m \; \underline{/\theta_i^\circ}$$

$$= \frac{I_m}{\omega C} \underline{/(\theta_i - 90)^\circ}. \tag{7.34}$$

The alternative way to express the phase relationship contained in Eq. (7.34) is to say that the current leads the voltage by $90°$. Figure 7.13 shows the phase relationship between the current and voltage at the terminals of a capacitor.

Impedance and Reactance

We conclude this discussion of passive circuit elements in the frequency domain with an important observation. When we compare Eqs. (7.27), (7.30), and (7.33), we note that they are all of the form

$$\mathbf{V} = Z\mathbf{I}, \tag{7.35}$$

where Z represents the **impedance** of the circuit element. Solving for Z in Eq. (7.35), you can see that impedance is the ratio of a circuit element's voltage phasor to its current phasor. Thus the impedance of a resistor is R, the impedance of an inductor is $j\omega L$ and the impedance of a capacitor is $j(-1/\omega C)$. In all cases, impedance is measured in ohms. Note that, although impedance is a complex number, it is not a phasor. Remember, a phasor is a complex number that shows up as the coefficient of $e^{j\omega t}$. Thus, although all phasors are complex numbers, not all complex numbers are phasors.

Impedance in the frequency domain is the quantity analogous to resistance, inductance, and capacitance in the time domain. The imaginary part of the impedance is called **reactance**. The values of impedance and reactance for each of the component values are summarized in Table 7.1.

TABLE 7.1   Impedance and reactance values

CIRCUIT ELEMENT	IMPEDANCE	REACTANCE
Resistor	R	—
Inductor	$j\omega L$	ωL
Capacitor	$j(-1/\omega C)$	$-1/\omega C$

And finally, a reminder. If the reference direction for the current in a passive circuit element is in the direction of the voltage rise across the element, you must insert a minus sign into the equation that relates the voltage to the current.

DRILL EXERCISES

7.6 The current in the 20 mH inductor is $10\cos(10{,}000t + 30°)$ mA. Calculate (a) the inductive reactance; (b) the impedance of the inductor; (c) the phasor voltage $\mathbf{V}$; and (d) the steady-state expression for $v(t)$.

ANSWER: (a) 200 Ω; (b) $j200$ Ω; (c) $2 \ \angle 120°$ V; (d) $2\cos(10{,}000t + 120°)$ V.

7.7 The voltage across the terminals of the 5 μF capacitor is $30\cos(4000t + 25°)$ V. Calculate (a) the capacitive reactance; (b) the impedance of the capacitor; (c) the phasor current $\mathbf{I}$; and (d) the steady-state expression for $i(t)$.

ANSWER: (a) -50 Ω; (b) $-j50$ Ω; (c) $0.6 \ \angle 115°$ A; (d) $0.6\cos(4000t + 115°)$ A.

7.5   KIRCHHOFF'S LAWS IN THE FREQUENCY DOMAIN

We pointed out in Section 7.3, with reference to Eqs. (7.23) and (7.24), that the phasor transform is useful in circuit analysis because it applies to the sum of sinusoidal functions. We illustrated this usefulness in Example 7.5. We now formalize this observation by developing Kirchhoff's laws in the frequency domain.

Kirchhoff's Voltage Law in the Frequency Domain

We begin by assuming that $v_1 - v_n$ represent voltages around a closed path in a circuit. We also assume that the circuit is operating in a sinusoidal steady state. Thus Kirchhoff's voltage law requires that

$$v_1 + v_2 + \cdots + v_n = 0, \tag{7.36}$$

which in the sinusoidal steady state becomes

$$V_{m_1} \cos(\omega t + \theta_1) + V_{m_2} \cos(\omega t + \theta_2) + \cdots + V_{m_n} \cos(\omega t + \theta_n) = 0.$$
$$(7.37)$$

We now use Euler's identity to write Eq. (7.37) as

$$\Re\{V_{m_1} e^{j\theta_1} e^{j\omega t}\} + \Re\{V_{m_2} e^{j\theta_2} e^{j\omega t}\} + \cdots + \Re\{V_{m_n} e^{j\theta_n} e^{j\omega t}\} = 0,$$
$$(7.38)$$

which we rewrite as

$$\Re\{V_{m_1} e^{j\theta_1} e^{j\omega t} + V_{m_2} e^{j\theta_2} e^{j\omega t} + \cdots + V_{m_n} e^{j\theta_n} e^{j\omega t}\} = 0. \quad (7.39)$$

Factoring the term $e^{j\omega t}$ from each term yields

$$\Re\{(V_{m_1} e^{j\theta_1} + V_{m_2} e^{j\theta_2} + \cdots + V_{m_n} e^{j\theta_n})e^{j\omega t}\} = 0,$$

or

$$\Re\{(\mathbf{V}_1 + \mathbf{V}_2 + \cdots + \mathbf{V}_n)e^{j\omega t}\} = 0. \quad (7.40)$$

But $e^{j\omega t} \neq 0$, so

$$\mathbf{V}_1 + \mathbf{V}_2 + \cdots + \mathbf{V}_n = 0, \quad (7.41)$$

which is the statement of Kirchhoff's voltage law as it applies to phasor voltages. In other words, Eq. (7.36) applies to a set of sinusoidal voltages in the time domain, and Eq. (7.41) is the equivalent statement in the frequency domain.

Kirchhoff's Current Law in the Frequency Domain

A similar derivation applies to a set of sinusoidal currents. Thus if

$$i_1 + i_2 + \cdots + i_n = 0, \quad (7.42)$$

then

$$\mathbf{I}_1 + \mathbf{I}_2 + \cdots + \mathbf{I}_n = 0, \quad (7.43)$$

where $\mathbf{I}_1, \mathbf{I}_2, \ldots, \mathbf{I}_n$ are the phasor representations of the individual currents $i_1, i_2, \ldots, i_n$.

Equations (7.35), (7.41), and (7.43) form the basis for circuit analysis in the frequency domain. Note that Eq. (7.35) has the same algebraic form as Ohm's law, and that Eqs. (7.41) and (7.43) state Kirchhoff's laws for phasor quantities. Therefore you may use all the techniques developed for analyzing resistive circuits to find phasor currents and voltages. You need learn no new analytic techniques; the basic circuit analysis and simplification tools covered in Chapters 2 and 3 can all be used to analyze circuits in the frequency domain. Phasor circuit analysis consists of two fundamental tasks: (1) You must be able to construct the frequency-domain model of a circuit; and (2) you must be able to manipulate complex numbers and/or quantities algebraically. We illustrate these aspects of phasor analysis in the discussion that follows.

DRILL EXERCISE

7.8 Four branches terminate at a common node. The reference direction of each branch current (i_1, i_2, i_3, and i_4) is toward the node. If $i_1 = 100\cos(\omega t + 25°)$ A, $i_2 = 100\cos(\omega t + 145°)$ A, and $i_3 = 100\cos(\omega t - 95°)$ A, find i_4.

ANSWER: $i_4 = 0$.

7.6 CIRCUIT SIMPLIFICATIONS

The rules for combining impedances in series or parallel are the same as those for resistors. The only difference is that combining impedances involves the algebraic manipulation of complex numbers.

Combining Impedances in Series and Parallel

Impedances in series can be combined into a single impedance by simply adding the individual impedances. The circuit shown in Fig. 7.14 defines the problem in general terms. The impedances $Z_1, Z_2, \ldots, Z_n$ are connected in series between terminals a,b. When impedances are in series, they carry the same phasor current **I**. From Eq. (7.35), the voltage drop across each impedance is $Z_1\mathbf{I}, Z_2\mathbf{I}, \ldots, Z_n\mathbf{I}$, and from Kirchhoff's voltage law,

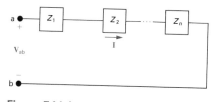

Figure 7.14 Impedances in series.

$$\mathbf{V}_{ab} = Z_1\mathbf{I} + Z_2\mathbf{I} + \cdots + Z_n\mathbf{I}$$

$$= (Z_1 + Z_2 + \cdots + Z_n)\mathbf{I}. \qquad (7.44)$$

The equivalent impedance between terminals a,b is

$$Z_{ab} = \frac{\mathbf{V}_{ab}}{\mathbf{I}} = Z_1 + Z_2 + \cdots + Z_n. \qquad (7.45)$$

Example 7.6 illustrates a numerical application of Eq. (7.45).

EXAMPLE 7.6

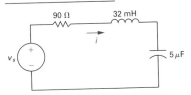

Figure 7.15 The circuit for Example 7.6.

A 90 Ω resistor, a 32 mH inductor, and a 5 μF capacitor are connected in series across the terminals of a sinusoidal voltage source, as shown in Fig. 7.15. The steady-state expression for the source voltage v_s is $750\cos(5000t + 30°)$ V.

(a) Construct the frequency-domain equivalent circuit.

(b) Calculate the steady-state current i by the phasor method.

SOLUTION

(a) From the expression for v_s, we have $\omega = 5000$ rad/s. Therefore the impedance of the 32 mH inductor is

$$Z_L = j\omega L = j(5000)(32 \times 10^{-3}) = j160 \ \Omega,$$

and the impedance of the capacitor is

$$Z_C = j\frac{-1}{\omega C} = -j\frac{10^6}{(5000)(5)} = -j40 \ \Omega.$$

The phasor transform of v_s is

$$\mathbf{V}_s = 750 \ \underline{/30°} \ \text{V}.$$

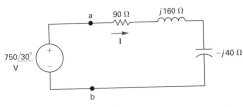

Figure 7.16 The frequency-domain equivalent circuit of the circuit shown in Fig. 7.15.

Figure 7.16 illustrates the frequency-domain equivalent circuit of the circuit shown in Fig. 7.15.

(b) We compute the phasor current simply by dividing the voltage of the voltage source by the equivalent impedance between the terminals a,b. From Eq. (7.45),

$$Z_{ab} = 90 + j160 - j40$$
$$= 90 + j120 = 150 \ \underline{/53.13°} \ \Omega.$$

Thus

$$\mathbf{I} = \frac{750 \ \underline{/30°}}{150 \ \underline{/53.13°}} = 5 \ \underline{/-23.13°} \ \text{A}.$$

We may now write the steady-state expression for i directly:

$$i = 5\cos(5000t - 23.13°) \ \text{A}.$$

DRILL EXERCISE

7.9   For the circuit in Fig. 7.15, with $\mathbf{V}_s = 125 \; \underline{/{-}60^\circ}$ V and $\omega = 5000$ rad/s, find

(a) the value of capacitance that yields a steady-state output current i with a phase angle of -105°

(b) the magnitude of the steady-state output current i

ANSWER: (a) 2.86 μF; (b) 0.982 A.

Impedances connected in parallel may be reduced to a single equivalent impedance by the reciprocal relationship

$$\frac{1}{Z_{ab}} = \frac{1}{Z_1} + \frac{1}{Z_2} + \cdots + \frac{1}{Z_n}. \tag{7.46}$$

Figure 7.17 depicts the parallel connection of impedances. Note that when impedances are in parallel, they have the same voltage across their terminals. We derive Eq. (7.46) directly from Fig. 7.17 by simply combining Kirchhoff's current law with the phasor-domain version of Ohm's law, that is, Eq. (7.35). From Fig. 7.17,

$$\mathbf{I} = \mathbf{I}_1 + \mathbf{I}_2 + \cdots + \mathbf{I}_n,$$

or

$$\frac{\mathbf{V}}{Z_{ab}} = \frac{\mathbf{V}}{Z_1} + \frac{\mathbf{V}}{Z_2} + \cdots + \frac{\mathbf{V}}{Z_n}. \tag{7.47}$$

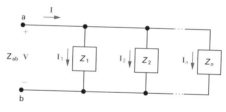

Figure 7.17 Impedances in parallel.

Canceling the common voltage term out of Eq. (7.47) reveals Eq. (7.46).

From Eq. (7.46), for the special case of just two impedances in parallel,

$$Z_{ab} = \frac{Z_1 Z_2}{Z_1 + Z_2}. \tag{7.48}$$

We can also express Eq. (7.46) in terms of **admittance,** defined as the reciprocal of impedance and denoted Y. Thus

$$Y = \frac{1}{Z} = G + jB. \tag{7.49}$$

Admittance is, of course, a complex number, whose real part, G, is called **conductance** and whose imaginary part, B, is called

TABLE 7.2 Admittance and Susceptance Values

CIRCUIT ELEMENT	ADMITTANCE	SUSCEPTANCE
Resistor	G	—
Inductor	$j(-1/\omega L)$	$-1/\omega L$
Capacitor	$j\omega C$	ωC

susceptance. Admittance, conductance and susceptance are measured in siemens (S). Using Eq. (7.49) in Eq. (7.46), we get

$$Y_{ab} = Y_1 + Y_2 + \cdots + Y_n. \tag{7.50}$$

The admittance of each of the ideal passive circuit elements also is worth noting and is summarized in Table 7.2.

Example 7.7 illustrates the application of Eqs. (7.49) and (7.50) to a specific circuit.

EXAMPLE 7.7

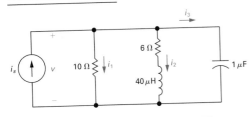

Figure 7.18 The circuit for Example 7.7.

The sinusoidal current source in the circuit shown in Fig. 7.18 produces the current $i_s = 8\cos 200{,}000t$ A.

(a) Construct the frequency-domain equivalent circuit.

(b) Find the steady-state expressions for v, i_1, i_2, and i_3.

SOLUTION

(a) The phasor transform of the current source is $8\ \angle 0°$ A; the resistors transform directly to the frequency domain as 10 and 6 Ω; the 40 μH inductor has an impedance of $j8$ Ω at the given frequency of 200,000 rad/s; and at this frequency the 1 μF capacitor has an impedance of $-j5$ Ω. Figure 7.19 shows the frequency-domain equivalent circuit and symbols representing the phasor transforms of the unknowns.

(b) The circuit shown in Fig. 7.19 indicates that we can easily obtain the voltage across the current source once we know the equivalent impedance of the three parallel branches. Moreover, once we know **V**, we can calculate the three phasor currents $\mathbf{I}_1$, $\mathbf{I}_2$, and $\mathbf{I}_3$ by using Eq. (7.35). To find the equivalent impedance of the three branches, we first find the equivalent admittance simply by adding the admittances

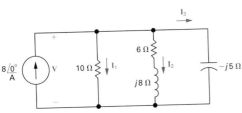

Figure 7.19 The frequency-domain equivalent circuit.

of each branch. The admittance of the first branch is

$$Y_1 = \frac{1}{10} = 0.1 \text{ S},$$

the admittance of the second branch is

$$Y_2 = \frac{1}{6+j8} = \frac{6-j8}{100} = 0.06 - j0.08 \text{ S},$$

and the admittance of the third branch is

$$Y_3 = \frac{1}{-j5} = j0.2 \text{ S}.$$

The admittance of the three branches is

$$Y = Y_1 + Y_2 + Y_3$$
$$= 0.16 + j0.12$$
$$= 0.2 \; \underline{/36.87°} \text{ S}.$$

The impedance at the current source is

$$Z = \frac{1}{Y} = 5 \; \underline{/-36.87°} \; \Omega.$$

The voltage **V** is

$$\mathbf{V} = Z\mathbf{I} = 40 \; \underline{/-36.87°} \text{ V}.$$

Hence

$$\mathbf{I}_1 = \frac{40 \; \underline{/-36.87°}}{10} = 4 \; \underline{/-36.87°} = 3.2 - j2.4 \text{ A},$$

$$\mathbf{I}_2 = \frac{40 \; \underline{/-36.87°}}{6+j8} = 4 \; \underline{/-90°} = -j4 \text{ A},$$

and

$$\mathbf{I}_3 = \frac{40 \; \underline{/-36.87°}}{5 \; \underline{/-90°}} = 8 \; \underline{/53.13°} = 4.8 + j6.4 \text{ A}.$$

We check the computations at this point by verifying that

$$\mathbf{I}_1 + \mathbf{I}_2 + \mathbf{I}_3 = \mathbf{I}.$$

Specifically,

$$3.2 - j2.4 - j4 + 4.8 + j6.4 = 8 + j0.$$

The corresponding steady-state time-domain expressions are

$$v = 40\cos(200{,}000t - 36.87°) \text{ V},$$
$$i_1 = 4\cos(200{,}000t - 36.87°) \text{ A},$$
$$i_2 = 4\cos(200{,}000t - 90°) \text{ A},$$
$$i_3 = 8\cos(200{,}000t + 53.13°) \text{ A}.$$

DRILL EXERCISES

7.10 A 20 Ω resistor is connected in parallel with a 5 mH inductor. This parallel combination is connected in series with a 5 Ω resistor and a 25 μF capacitor.

 (a) Calculate the impedance of this interconnection if the frequency is 2 krad/s.

 (b) Repeat (a) for a frequency of 8 krad/s.

7.11 The interconnection described in Drill Exercise 7.10 is connected across the terminals of a voltage source that is generating $v = 150 \cos 4000t$ V. What is the maximum amplitude of the current in the 5 mH inductor?

7.12 Three branches having impedances of $3 + j4$ Ω, $16 - j12$ Ω, and $-j4$ Ω, respectively, are connected in parallel. What are the equivalent (a) admittance, (b) conductance, and (c) susceptance of the parallel connection in millisiemens? (d) If the parallel branches are excited from a sinusoidal current source where $i = 8 \cos \omega t$ A, what is the maximum amplitude of the current in the purely capacitive branch?

7.13 Find the steady-state expression for v_o in the circuit shown if $i_g = 0.5 \cos 2000t$ A.

 ANSWER: $v_o = 30\sqrt{2} \cos(2000t + 45°)$ V.

 (c) At what finite frequency does the impedance of the interconnection become purely resistive?

 (d) What is the impedance at the frequency found in (c)?

 ANSWER: (a) $9 - j12$ Ω; (b) $21 + j3$ Ω; (c) 4 krad/s; (d) 15 Ω.

 ANSWER: 7.07 A.

 ANSWER: (a) 200 $\angle 36.87°$ mS; (b) 160 mS; (c) 120 mS; (d) 10 A.

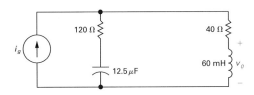

Source Transformations and Thévenin-Norton Equivalent Circuits

The source transformations introduced in Section 2.3 and the Thévenin-Norton equivalent circuits discussed in Section 3.9 are analytical techniques that also can be applied to frequency-domain circuits. We prove the validity of these techniques by following the same process used in Sections 3.9 and 3.10, except that we substitute impedance (Z) for resistance (R). Figure 7.20 shows a source-transformation equivalent circuit with the nomenclature of the frequency domain.

Figure 7.20 A source transformation in the frequency domain.

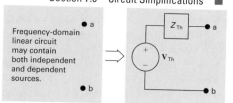

Figure 7.21 illustrates the frequency-domain version of a Thévenin equivalent circuit. Figure 7.22 shows the frequency-domain equivalent of a Norton equivalent circuit. The techniques for finding the Thévenin equivalent voltage and impedance are identical to those used for resistive circuits, except that the frequency-domain equivalent circuit involves the manipulation of complex quantities. The same holds for finding the Norton equivalent current and impedance.

Figure 7.21 The frequency-domain version of a Thévenin equivalent circuit.

Example 7.8 demonstrates the application of the source-transformation equivalent circuit to frequency-domain analysis. Example 7.9 illustrates the details of finding a Thévenin equivalent circuit in the frequency domain.

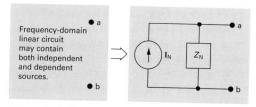

Figure 7.22 The frequency-domain version of a Norton equivalent circuit.

EXAMPLE 7.8

Use the concept of source transformation to find the phasor voltage $\mathbf{V}_0$ in the circuit shown in Fig. 7.23.

SOLUTION

We can replace the series combination of the voltage source $(40 \angle 0°)$ and the impedance of $1 + j3\ \Omega$ with the parallel combination of a current source and the $1 + j3\ \Omega$ impedance. The source current is

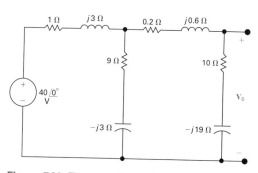

Figure 7.23 The circuit for Example 7.8.

$$\mathbf{I} = \frac{40}{1 + j3} = \frac{40}{10}(1 - j3) = 4 - j12 \text{ A}.$$

Thus we can modify the circuit shown in Fig. 7.23 to the one shown in Fig. 7.24. Note that the polarity reference of the 40 V source determines the reference direction for $\mathbf{I}$.

Next, we combine the two parallel branches into a single impedance,

$$Z = \frac{(1 + j3)(9 - j3)}{10} = 1.8 + j2.4\ \Omega,$$

which is in parallel with the current source of $4 - j12$ A. Another source transformation converts this parallel combination to a series combination consisting of a voltage source in series with the impedance of $1.8 + j2.4\ \Omega$. The voltage of the voltage source is

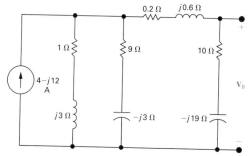

Figure 7.24 The first step in reducing the circuit shown in Fig. 7.23.

$$\mathbf{V} = (4 - j12)(1.8 + j2.4) = 36 - j12 \text{ V}.$$

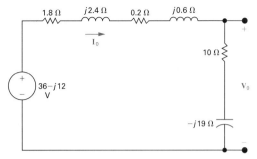

Figure 7.25 The second step in reducing the circuit shown in Fig. 7.23.

Using this source transformation, we redraw the circuit as Fig. 7.25. Note the polarity of the voltage source. We added the current $\mathbf{I}_0$ to the circuit to expedite the solution for $\mathbf{V}_0$.

Also note that we have reduced the circuit to a simple series circuit. We calculate the current $\mathbf{I}_0$ by dividing the voltage of the source by the total series impedance:

$$\mathbf{I}_0 = \frac{36 - j12}{12 - j16} = \frac{12(3 - j1)}{4(3 - j4)}$$

$$= \frac{39 + j27}{25} = 1.56 + j1.08 \text{ A}.$$

We now obtain the value of $\mathbf{V}_0$ by multiplying $\mathbf{I}_0$ by the impedance $10 - j19$:

$$\mathbf{V}_0 = (1.56 + j1.08)(10 - j19) = 36.12 - j18.84 \text{ V}$$

$$= 40.74 \ \underline{/-27.55°} \text{ V}.$$

EXAMPLE 7.9

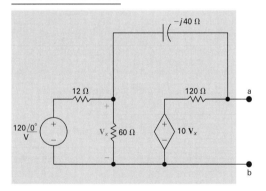

Figure 7.26 The circuit for Example 7.9.

Find the Thévenin equivalent circuit with respect to terminals a,b for the circuit shown in Fig. 7.26.

SOLUTION

We first determine the Thévenin equivalent voltage. This voltage is the open-circuit voltage appearing at terminals a,b. We choose the reference for the Thévenin voltage as positive at terminal a. We can make two source transformations relative to the 120 V, 12 Ω, and 60 Ω circuit elements to simplify this portion of the circuit. At the same time, these transformations must preserve the identity of the controlling voltage $\mathbf{V}_x$ because of the dependent voltage source.

We determine the two source transformations by first replacing the series combination of the 120 V source and 12 Ω resistor with a 10 A current source in parallel with 12 Ω. Next, we replace the parallel combination of the 12 and 60 Ω resistors with a single 10 Ω resistor. Finally, we replace the 10 A source in parallel with 10 Ω with a 100 V source in series with 10 Ω. Figure 7.27 shows the resulting circuit.

We added the current **I** to Fig. 7.27 to aid further discussion. Note that once we know the current **I**, we can compute the Thévenin voltage. We find **I** by summing the voltages around the closed path in the circuit shown in Fig. 7.27. Hence

$$100 = 10\mathbf{I} - j40\mathbf{I} + 120\mathbf{I} + 10\mathbf{V}_x = (130 - j40)\mathbf{I} + 10\mathbf{V}_x.$$

We relate the controlling voltage $\mathbf{V}_x$ to the current **I** by noting from Fig. 7.27 that

$$\mathbf{V}_x = 100 - 10\mathbf{I}.$$

Then,

$$\mathbf{I} = \frac{-900}{30 - j40} = 18 \ \angle{-126.87°} \ \text{A}.$$

We now calculate $\mathbf{V}_x$:

$$\mathbf{V}_x = 100 - 180 \ \angle{-126.87°} = 208 + j144 \ \text{V}.$$

Finally, we note from Fig. 7.27 that

$$\mathbf{V}_{\text{Th}} = 10\mathbf{V}_x + 120\mathbf{I}$$

$$= 2080 + j1440 + 120(18) \ \angle{-126.87°}$$

$$= 784 - j288 = 835.22 \ \angle{-20.17°} \ \text{V}.$$

To obtain the Thévenin impedance, we may use any of the techniques previously used to find the Thévenin resistance. We illustrate the test-source method in this example. Recall that in using this method, we deactivate all independent sources from the circuit and then apply either a test voltage source or a test current source to the terminals of interest. The ratio of the voltage to the current at the source is the Thévenin impedance. Figure 7.28 shows the result of applying this technique to the circuit shown in Fig. 7.26. Note that we chose a test voltage source $\mathbf{V}_T$. Also note that we deactivated the independent voltage source with an appropriate short-circuit and preserved the identity of $\mathbf{V}_x$. The branch currents $\mathbf{I}_a$ and $\mathbf{I}_b$ have been added to the circuit to simplify the calculation of $\mathbf{I}_T$. By straightforward applications of Kirchhoff's circuit laws, you should be able to verify the following

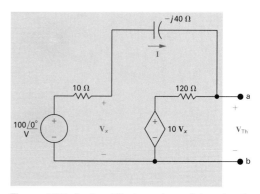

Figure 7.27 A simplified version of the circuit shown in Fig. 7.26.

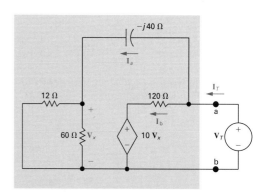

Figure 7.28 A circuit for calculating the Thévenin equivalent impedance.

relationships:

$$\mathbf{I}_a = \frac{\mathbf{V}_T}{10 - j40}, \quad \mathbf{V}_x = 10\mathbf{I}_a,$$

$$\mathbf{I}_b = \frac{\mathbf{V}_T - 10\mathbf{V}_x}{120}$$

$$= \frac{-\mathbf{V}_T(9 + j4)}{120(1 - j4)},$$

$$\mathbf{I}_T = \mathbf{I}_a + \mathbf{I}_b$$

$$= \frac{\mathbf{V}_T}{10 - j40}\left(1 - \frac{9 + j4}{12}\right)$$

$$= \frac{\mathbf{V}_T(3 - j4)}{12(10 - j40)},$$

$$Z_{\text{Th}} = \frac{\mathbf{V}_T}{\mathbf{I}_T} = 91.2 - j38.4 \ \Omega.$$

Figure 7.29 depicts the Thévenin equivalent circuit.

Figure 7.29 The Thévenin equivalent for the circuit shown in Fig. 7.26.

DRILL EXERCISES

7.14 Find the Thévenin equivalent with respect to terminals a,b in the circuit shown.

ANSWER: $\mathbf{V}_{\text{Th}} = \mathbf{V}_{ab} = 10 \ \underline{/45°}$ V; $Z_{\text{Th}} = 5 - j5 \ \Omega$.

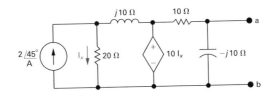

7.15 Find the Norton equivalent with respect to terminals a,b in the circuit shown.

ANSWER: $\mathbf{I}_N = 10 \ \underline{/-45°}$ A; $Z_{Th} = 1.6 + j3.2 \ \Omega$.

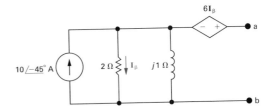

7.7 THE NODE-VOLTAGE METHOD

In Sections 3.2–3.4, we introduced the basic concepts of the node-voltage method of circuit analysis. The same concepts apply when we use the node-voltage method to analyze frequency-domain circuits. Example 7.10 illustrates the solution of such a circuit by the node-voltage technique. Drill Exercise 7.16 and many of the Chapter Problems give you an opportunity to use the node-voltage method to solve for steady-state sinusoidal responses.

EXAMPLE 7.10

Use the node-voltage method to find the branch currents $\mathbf{I}_a$, $\mathbf{I}_b$, and $\mathbf{I}_c$ in the circuit shown in Fig. 7.30.

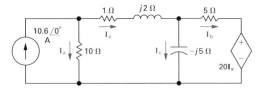

Figure 7.30 The circuit for Example 7.10.

SOLUTION

We can describe the circuit in terms of two node voltages because it contains three essential nodes. Four branches terminate at the essential node that stretches across the bottom of Fig. 7.30, so we use it as the reference node. The remaining two essential nodes are labeled 1 and 2, and the appropriate node voltages are designated $\mathbf{V}_1$ and $\mathbf{V}_2$. Figure 7.31 reflects the choice of reference node and the terminal labels.

Summing the currents away from node 1 yields

$$-10.6 + \frac{\mathbf{V}_1}{10} + \frac{\mathbf{V}_1 - \mathbf{V}_2}{1 + j2} = 0.$$

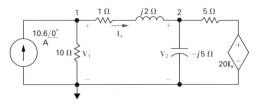

Figure 7.31 The circuit shown in Fig. 7.30, with the node voltages defined.

Multiplying by $1 + j2$ and collecting the coefficients of $\mathbf{V}_1$ and $\mathbf{V}_2$ generates the expression

$$\mathbf{V}_1(1.1 + j0.2) - \mathbf{V}_2 = 10.6 + j21.2.$$

Summing the currents away from node 2 gives

$$\frac{\mathbf{V}_2 - \mathbf{V}_1}{1 + j2} + \frac{\mathbf{V}_2}{-j5} + \frac{\mathbf{V}_2 - 20\mathbf{I}_x}{5} = 0.$$

The controlling current $\mathbf{I}_x$ is

$$\mathbf{I}_x = \frac{\mathbf{V}_1 - \mathbf{V}_2}{1 + j2}.$$

Substituting this expression for $\mathbf{I}_x$ into the node 2 equation, multiplying by $1 + j2$, and collecting coefficients of $\mathbf{V}_1$ and $\mathbf{V}_2$ produces the equation

$$-5\mathbf{V}_1 + (4.8 + j0.6)\mathbf{V}_2 = 0.$$

The solutions for $\mathbf{V}_1$ and $\mathbf{V}_2$ are

$$\mathbf{V}_1 = 68.40 - j16.80 \text{ V},$$

$$\mathbf{V}_2 = 68 - j26 \text{ V}.$$

Hence the branch currents are

$$\mathbf{I}_a = \frac{\mathbf{V}_1}{10} = 6.84 - j1.68 \text{ A},$$

$$\mathbf{I}_x = \frac{\mathbf{V}_1 - \mathbf{V}_2}{1 + j2} = 3.76 + j1.68 \text{ A},$$

$$\mathbf{I}_b = \frac{\mathbf{V}_2 - 20\mathbf{I}_x}{5} = -1.44 - j11.92 \text{ A},$$

$$\mathbf{I}_c = \frac{\mathbf{V}_2}{-j5} = 5.2 + j13.6 \text{ A}.$$

To check our work, we note that

$$\mathbf{I}_a + \mathbf{I}_x = 6.84 - j1.68 + 3.76 + j1.68$$

$$= 10.6 \text{ A},$$

$$\mathbf{I}_x = \mathbf{I}_b + \mathbf{I}_c = -1.44 - j11.92 + 5.2 + j13.6$$

$$= 3.76 + j1.68 \text{ A}.$$

DRILL EXERCISE

7.16 Use the node-voltage method to find the steady-state expression for $v(t)$ in the circuit shown. The sinusoidal sources are $i_s = 10\cos\omega t$ A and $v_s = 100\sin\omega t$ V, where $\omega = 50$ krad/s.

ANSWER: $v(t) = 31.62\cos(50{,}000t - 71.57°)$ V.

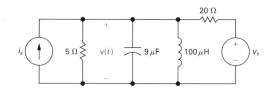

7.8 THE MESH-CURRENT METHOD

We can also use the mesh-current method to analyze frequency-domain circuits.   The procedures used in frequency-domain applications are the same as those used in analyzing resistive circuits. In Sections 3.5–3.7, we introduced the basic techniques of the mesh-current method; we demonstrate the extension of this method to frequency-domain circuits in Example 7.11.

EXAMPLE 7.11

Use the mesh-current method to find the voltages V_1, V_2, and V_3 in the circuit shown in Fig. 7.32.

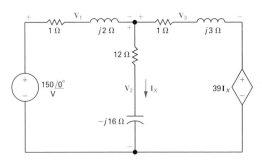

Figure 7.32 The circuit for Example 7.11.

SOLUTION

The circuit has two meshes and a dependent voltage source, so we must write two mesh-current equations and a constraint equation. The reference direction for the mesh currents I_1 and I_2 is clockwise, as shown in Fig. 7.33. Once we know I_1 and I_2, we can easily find the unknown voltages. Summing the voltages around mesh 1 gives

$$150 = (1 + j2)I_1 + (12 - j16)(I_1 - I_2),$$

or

$$150 = (13 - j14)I_1 - (12 - j16)I_2.$$

Summing the voltages around mesh 2 generates the equation

$$0 = (12 - j16)(I_2 - I_1) + (1 + j3)I_2 + 39I_x.$$

Figure 7.33 reveals that the controlling current I_x is the difference between I_1 and I_2; that is, the constraint is

$$I_x = I_1 - I_2.$$

Substituting this constraint into the mesh 2 equation and simplifying the resulting expression gives

$$0 = (27 + j16)I_1 - (26 + j13)I_2.$$

Solving for I_1 and I_2 yields

$$I_1 = -26 - j52 \text{ A},$$

$$I_2 = -24 - j58 \text{ A},$$

$$I_x = -2 + j6 \text{ A}.$$

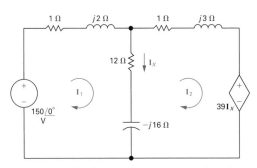

Figure 7.33 Mesh currents used to solve the circuit shown in Fig. 7.32.

The three voltages are

$$\mathbf{V}_1 = (1 + j2)\mathbf{I}_1 = 78 - j104 \text{ V},$$

$$\mathbf{V}_2 = (12 - j16)\mathbf{I}_x = 72 + j104 \text{ V},$$

$$\mathbf{V}_3 = (1 + j3)\mathbf{I}_2 = 150 - j130 \text{ V},$$

$$39\mathbf{I}_x = -78 + j234 \text{ V}.$$

We check these calculations by summing the voltages around closed paths:

$$-150 + \mathbf{V}_1 + \mathbf{V}_2 = -150 + 78 - j104 + 72 + j104 = 0,$$

$$-\mathbf{V}_2 + \mathbf{V}_3 + 39\mathbf{I}_x = -72 - j104 + 150 - j130 - 78 + j234 = 0,$$

$$-150 + \mathbf{V}_1 + \mathbf{V}_3 + 39\mathbf{I}_x = -150 + 78 - j104 + 150 - j130 - 78 + j234 = 0.$$

DRILL EXERCISE

7.17 Use the mesh-current method to find the phasor current **I** in the circuit shown.

ANSWER: $\mathbf{I} = 29 + j2 = 29.07 \ \angle 3.95° \text{ A}.$

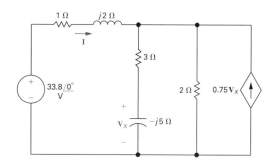

7.9 INSTANTANEOUS, AVERAGE, AND REACTIVE POWER

Instantaneous Power

We begin our investigation of sinusoidal power calculations with the familiar circuit in Fig. 7.34. Here, v and i are steady-state sinusoidal signals. Using the passive sign convention, the power at any instant of time is

$$p = vi. \tag{7.51}$$

Figure 7.34 The black box representation of a circuit used for calculating power.

This is **instantaneous power**. Remember that if the reference direction of the current is in the direction of the voltage rise, Eq. (7.51) must be written with a minus sign. Instantaneous power is measured in watts when the voltage is in volts and the current is in amperes. First, we write expressions for v and i:

$$v = V_m \cos(\omega t + \theta_v), \tag{7.52}$$

$$i = I_m \cos(\omega t + \theta_i), \tag{7.53}$$

where θ_v is the voltage phase angle, and θ_i is the current phase angle.

We are operating in the sinusoidal steady state, so we may choose any convenient reference for zero time. Engineers designing systems that transfer large blocks of power have found it convenient to use a zero time corresponding to the instant the current is passing through a positive maximum. This reference system requires a shift of both the voltage and current by θ_i. Thus Eqs. (7.52) and (7.53) become

$$v = V_m \cos(\omega t + \theta_v - \theta_i), \tag{7.54}$$

$$i = I_m \cos \omega t. \tag{7.55}$$

When we substitute Eqs. (7.54) and (7.55) into Eq. (7.51), the expression for the instantaneous power becomes

$$p = V_m I_m \cos(\omega t + \theta_v - \theta_i) \cos \omega t. \tag{7.56}$$

We could use Eq. (7.56) directly to find the average power; however, by simply applying a couple of trigonometric identities, we can put Eq. (7.56) into a much more informative form.

We begin with the trigonometric identity[2]

$$\cos \alpha \cos \beta = \frac{1}{2} \cos(\alpha - \beta) + \frac{1}{2} \cos(\alpha + \beta)$$

to expand Eq. (7.56); letting $\alpha = \omega t + \theta_v - \theta_i$ and $\beta = \omega t$ gives

$$p = \frac{V_m I_m}{2} \cos(\theta_v - \theta_i) + \frac{V_m I_m}{2} \cos(2\omega t + \theta_v - \theta_i). \tag{7.57}$$

Now use the trigonometric identity

$$\cos(\alpha + \beta) = \cos \alpha \cos \beta - \sin \alpha \sin \beta$$

to expand the second term on the right-hand side of Eq. (7.57), which gives

$$p = \frac{V_m I_m}{2} \cos(\theta_v - \theta_i) + \frac{V_m I_m}{2} \cos(\theta_v - \theta_i) \cos 2\omega t$$

$$- \frac{V_m I_m}{2} \sin(\theta_v - \theta_i) \sin 2\omega t. \tag{7.58}$$

---

[2] See entry 8 in Appendix C.

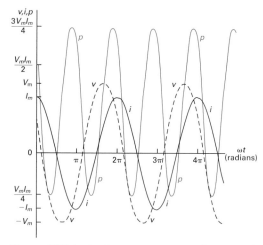

Figure 7.35 Instantaneous power, voltage, and current versus ωt for steady-state sinusoidal operation.

Figure 7.35 depicts a representative relationship among v, i, and p, based on the assumptions $\theta_v = 60°$ and $\theta_i = 0°$. You can see that the frequency of the instantaneous power is twice the frequency of the voltage or current. This observation also follows directly from the second two terms on the right-hand side of Eq. (7.58). Therefore, the instantaneous power goes through two complete cycles for every cycle of either the voltage or the current. Also note that the instantaneous power may be negative for a portion of each cycle, even if the network between the terminals is passive. In a completely passive network, negative power implies that energy stored in the inductors or capacitors is now being extracted. The fact that the instantaneous power varies with time in the sinusoidal steady-state operation of a circuit explains why some motor-driven appliances (such as refrigerators) experience vibration and require resilient motor mountings to prevent excessive vibration.

We are now ready to use Eq. (7.58) to find the average power at the terminals of the circuit represented by Fig. 7.34 and, at the same time, introduce the concept of reactive power.

Average and Reactive Power

We begin by noting that Eq. (7.58) has three terms, which we can rewrite as follows:

$$p = P + P \cos 2\omega t - Q \sin 2\omega t, \qquad (7.59)$$

where

$$P = \frac{V_m I_m}{2} \cos(\theta_v - \theta_i), \qquad (7.60)$$

$$Q = \frac{V_m I_m}{2} \sin(\theta_v - \theta_i). \qquad (7.61)$$

P is called the **average power**, and Q is called the **reactive power**. Average power is sometimes called **real power**, because it describes the power in a circuit that is transformed from electric to nonelectric energy. Although the two terms are interchangeable, we primarily use the term *average power* in this text.

It is easy to see why P is called the average power. The average power associated with sinusoidal signals is the average of the instantaneous power over one period, or, in equation form,

$$P = \frac{1}{T} \int_{t_0}^{t_0+T} p \, dt, \qquad (7.62)$$

where T is the period of the sinusoidal function. The limits on Eq. (7.62) imply that we can initiate the integration process at any convenient time t_0 but we must terminate the integration exactly one period later. (We could integrate over nT periods, where n is an integer, provided we multiplied the integral by $1/nT$.)

We could find the average power by substituting Eq. (7.59) directly into Eq. (7.62) and then performing the integration. But note that the average value of p is given by the first term on the right-hand side of Eq. (7.59), because the integral of both $\cos 2\omega t$ and $\sin 2\omega t$ over one period is zero. Thus the average power is given in Eq. (7.60).

We can develop a better understanding of all the terms in Eq. (7.59) and the relationships among them by examining the power in circuits that are purely resistive, purely inductive, or purely capacitive.

Power for Purely Resistive Circuits

If the circuit between the terminals is purely resistive, the voltage and current are in phase, which means that $\theta_v = \theta_i$. Equation (7.59) then reduces to

$$p = P + P \cos 2\omega t. \qquad (7.63)$$

The instantaneous power expressed in Eq. (7.63) is referred to as the **instantaneous real power**. Figure 7.36 shows a graph of Eq. (7.63) for a representative purely resistive circuit, assuming $\omega = 377$ rad/s. By definition, the average power, P, is the average of p over one period. Thus it is easy to see just by looking at the graph that $P = 1$ for this circuit. Note from Eq. (7.63) that the instantaneous real power can never be negative, which is also shown in Fig. 7.36. In other words, power cannot be extracted from a purely resistive network. Rather, all the electric energy is dissipated in the form of thermal energy.

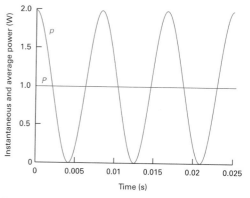

Figure 7.36 Instantaneous real power and average power for a purely resistive circuit.

Power for Purely Inductive Circuits

If the circuit between the terminals is purely inductive, the voltage and current are out of phase by precisely 90°. In particular, the current lags the voltage by 90° (that is, $\theta_i = \theta_v - 90°$); therefore $\theta_v - \theta_i = +90°$. The expression for the instantaneous power then reduces to

$$p = -Q \sin 2\omega t. \qquad (7.64)$$

In a purely inductive circuit, the average power is zero. Therefore no transformation of energy from electric to nonelectric form

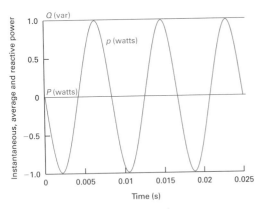

Figure 7.37 Instantaneous real power, average power, and reactive power for a purely inductive circuit.

takes place. The instantaneous power at the terminals in a purely inductive circuit is continually exchanged between the circuit and the source driving the circuit, at a frequency of 2ω. In other words, when p is positive, energy is being stored in the magnetic fields associated with the inductive elements, and when p is negative, energy is being extracted from the magnetic fields.

A measure of the power associated with purely inductive circuits is the reactive power Q. The name *reactive power* comes from the characterization of an inductor as a reactive element; its impedance is purely reactive. Note that average power P and reactive power Q carry the same dimension. To distinguish between average and reactive power, we use the units *watt* (W) for average power and **var** (*volt-amp reactive*, or VAR) for reactive power. Figure 7.37 plots the instantaneous power for a representative purely inductive circuit, assuming $\omega = 377$ rad/s.

Power for Purely Capacitive Circuits

If the circuit between the terminals is purely capacitive, the voltage and current are precisely 90° out of phase. In this case, the current leads the voltage by 90° (that is, $\theta_i = \theta_v + 90°$); thus, $\theta_v - \theta_i = -90°$. The expression for the instantaneous power then becomes

$$p = Q \sin 2\omega t. \tag{7.65}$$

Again, the average power is zero, so there is no transformation of energy from electric to nonelectric form. In a purely capacitive circuit, the power is continually exchanged between the source driving the circuit and the electric field associated with the capacitive elements. Figure 7.38 plots the instantaneous power for a representative purely capacitive circuit, assuming $\omega = 377$ rad/s.

Note that the decision to use the current as the reference leads to Q being positive for inductors (that is, $\theta_v - \theta_i = 90°$) and negative for capacitors (that is, $\theta_v - \theta_i = -90°$). Power engineers recognize this difference in the algebraic sign of Q by saying that inductors demand (or absorb) magnetizing vars, and capacitors furnish (or deliver) magnetizing vars. We say more about this convention later.

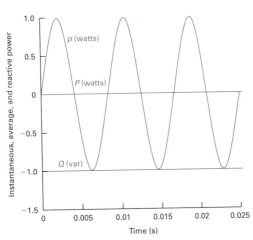

Figure 7.38 Instantaneous real power and average power for a purely capacitive circuit.

The Power Factor

The angle $\theta_v - \theta_i$ plays a role in the computation of both average and reactive power and is referred to as the **power factor angle**. The cosine of this angle is called the **power factor**, abbreviated pf, and the sine of this angle is called the **reactive factor**, abbreviated rf. Thus

$$pf = \cos(\theta_v - \theta_i), \tag{7.66}$$

$$rf = \sin(\theta_v - \theta_i). \tag{7.67}$$

Knowing the value of the power factor does not tell you the value of the power factor angle, because $\cos(\theta_v - \theta_i) = \cos(\theta_i - \theta_v)$. To completely describe this angle, we use the descriptive phrases **lagging power factor** and **leading power factor**. Lagging power factor implies that current lags voltage—hence an inductive load. Leading power factor implies that current leads voltage—hence a capacitive load. Both the power factor and the reactive factor are convenient quantities to use in describing electrical loads.

Example 7.12 illustrates the interpretation of P and Q on the basis of a numerical calculation.

EXAMPLE 7.12

(a) Calculate the average power and the reactive power at the terminals of the network shown in Fig. 7.39 if

$$v = 100 \cos(\omega t + 15°) \text{ V},$$

$$i = 4 \sin(\omega t - 15°) \text{ A}.$$

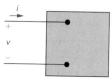

Figure 7.39 A pair of terminals used for calculating power.

(b) State whether the network inside the box is absorbing or delivering average power.

(c) State whether the network inside the box is absorbing or supplying magnetizing vars.

SOLUTION

(a) Because i is expressed in terms of the sine function, the first step in the calculation for P and Q is to rewrite i as a cosine function:

$$i = 4 \cos(\omega t - 105°) \text{ A}.$$

We now calculate P and Q directly from Eqs. (7.60) and (7.61). Thus

$$P = \frac{1}{2}(100)(4) \cos[15 - (-105)] = -100 \text{ W},$$

$$Q = \frac{1}{2}100(4) \sin[15 - (-105)] = 173.21 \text{ VAR}.$$

(b) Note from Fig. 7.39 the use of the passive sign convention. Because of this, the negative value of −100 W means that the network inside the box is delivering average power to the terminals.

(c) The passive sign convention means that, because Q is positive, the network inside the box is absorbing magnetizing vars at its terminals.

DRILL EXERCISES

7.18 For each of the following sets of voltage and current, calculate the real and reactive power in the line between networks A and B in the circuit shown. In each case, state whether the power flow is from A to B or vice versa. Also state whether magnetizing vars are being transferred from A to B or vice versa.

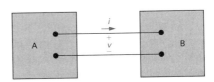

(a) $v = 100 \cos(\omega t - 45°)$ V; $i = 20 \cos(\omega t + 15°)$ A.

(b) $v = 100 \cos(\omega t - 45°)$ V; $i = 20 \cos(\omega t + 165°)$ A.

(c) $v = 100 \cos(\omega t - 45°)$ V; $i = 20 \cos(\omega t - 105°)$ A.

(d) $v = 100 \cos \omega t$ V; $i = 20 \cos(\omega t + 120°)$ A.

ANSWER: (a) $P = 500$ W (A to B), $Q = -866.03$ VAR (B to A); (b) $P = -866.03$ W (B to A), $Q = 500$ VAR (A to B); (c) $P = 500$ W (A to B), $Q = 866.03$ VAR (A to B); (d) $P = -500$ W (B to A), $Q = -866.03$ VAR (B to A).

7.19 Compute the power factor and the reactive factor for the network inside the box in Fig. 7.39, whose voltage and current are described in Example 7.12. *Hint:* Use $-i$ to calculate the power and reactive factors.

ANSWER: pf = 0.5 leading; rf = −0.866.

7.10 THE RMS VALUE AND POWER CALCULATIONS

In introducing the rms value of a sinusoidal voltage (or current) in Section 7.1, we mentioned that it would play an important role in power calculations. We can now discuss this role.

Assume that a sinusoidal voltage is applied to the terminals of a resistor, as shown in Fig. 7.40, and that we want to determine the average power delivered to the resistor. From Eq. (7.62),

$$P = \frac{1}{T} \int_{t_0}^{t_0+T} \frac{V_m^2 \cos^2(\omega t + \theta_v)}{R} \, dt$$

$$= \frac{1}{R} \left[\frac{1}{T} \int_{t_0}^{t_0+T} V_m^2 \cos^2(\omega t + \theta_v) \, dt \right]. \qquad (7.68)$$

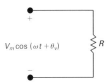

Figure 7.40 A sinusoidal voltage applied to the terminals of a resistor.

Comparing Eq. (7.68) with Eq. (7.5) reveals that the average power delivered to R is simply the rms value of the voltage squared divided by R, or

$$P = \frac{V_{\text{rms}}^2}{R}. \qquad (7.69)$$

If the resistor is carrying a sinusoidal current, say, $I_m \cos(\omega t + \theta_i)$, the average power delivered to the resistor is

$$P = I_{\text{rms}}^2 R. \qquad (7.70)$$

The rms value is also referred to as the **effective value** of the sinusoidal voltage (or current). The rms value has an interesting property: Given an equivalent resistive load, R, and an equivalent time period, T, the rms value of a sinusoidal source delivers the same energy to R as does a dc source of the same value. For example, a dc source of 100 V delivers the same energy in T seconds that a sinusoidal source of 100 V$_{\text{rms}}$ delivers, assuming equivalent load resistances (see Problem 7.62). Figure 7.41 demonstrates this equivalence. Energywise, the effect of the two sources is identical. This has led to the term *effective value* being used interchangeably with *rms value*.

The average power given by Eq. (7.60) and the reactive power given by Eq. (7.61) can be written in terms of effective values:

$$P = \frac{V_m I_m}{2} \cos(\theta_v - \theta_i)$$

$$= \frac{V_m}{\sqrt{2}} \frac{I_m}{\sqrt{2}} \cos(\theta_v - \theta_i)$$

$$= V_{\text{eff}} I_{\text{eff}} \cos(\theta_v - \theta_i); \qquad (7.71)$$

and, by similar manipulation,

$$Q = V_{\text{eff}} I_{\text{eff}} \sin(\theta_v - \theta_i). \qquad (7.72)$$

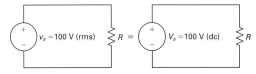

Figure 7.41 The effective value of v_s (100 V rms) delivers the same power to R as the dc voltage V_s (100 V dc).

The effective value of the sinusoidal signal in power calculations is so widely used that voltage and current ratings of circuits and equipment involved in power utilization are given in terms of rms values. For example, the voltage rating of residential electric wiring is often 240 V/120 V service. These voltage levels are the rms values of the sinusoidal voltages supplied by the utility company, which provides power at two voltage levels to accommodate low-voltage appliances (such as televisions) and higher-voltage appliances (such as electric ranges). Appliances such as electric lamps, irons, and toasters all carry rms ratings on their nameplates. For example, a 120 V, 100 W lamp has a resistance of $120^2/100$, or 144 Ω, and draws an rms current of 120/144, or 0.833 A. The peak value of the lamp current is $0.833\sqrt{2}$, or 1.18 A.

The phasor transform of a sinusoidal function may also be expressed in terms of the rms value. The magnitude of the rms phasor is equal to the rms value of the sinusoidal function. If a phasor is based on the rms value, we indicate this by either an explicit statement, a parenthetical "rms" adjacent to the phasor quantity, or the subscript "eff," as in Eq. (7.71).

In Example 7.13, we illustrate the use of rms values for calculating power.

EXAMPLE 7.13

(a) A sinusoidal voltage having a maximum amplitude of 625 V is applied to the terminals of a 50 Ω resistor. Find the average power delivered to the resistor.

(b) Repeat (a) by first finding the current in the resistor.

SOLUTION

(a) The rms value of the sinusoidal voltage is $625/\sqrt{2}$, or approximately 441.94 V. From Eq. (7.69), the average power delivered to the 50 Ω resistor is

$$P = \frac{(441.94)^2}{50} = 3906.25 \text{ W.}$$

(b) The maximum amplitude of the current in the resistor is 625/50, or 12.5 A. The rms value of the current is $12.5/\sqrt{2}$, or approximately 8.84 A. Hence the average power delivered to the resistor is

$$P = (8.84)^2 50 = 3906.25 \text{ W.}$$

DRILL EXERCISE

7.20 The periodic triangular current in Example 7.4, repeated here, has a peak value of 180 mA. Find the average power that this current delivers to a 5 kΩ resistor.

ANSWER: 54 W.

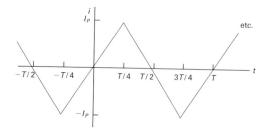

7.11 COMPLEX POWER AND POWER CALCULATIONS

Complex Power

Before proceeding to the various methods of calculating real and reactive power in circuits operating in the sinusoidal steady state, we need to introduce and define complex power. **Complex power** is the complex sum of real power and reactive power, or

$$S = P + jQ. \qquad (7.73)$$

As you will see, we can compute the complex power directly from the voltage and current phasors for a circuit. Equation (7.73) can then be used to compute the average power and the reactive power, because $P = \Re\{S\}$ and $Q = \Im\{S\}$.

Dimensionally, complex power is the same as average or reactive power. However, to distinguish complex power from either average or reactive power, we use the units **volt-amps** (VA). Thus we use volt-amps for complex power, watts for average power, and vars for reactive power, as summarized in Table 7.3.

TABLE 7.3 Three Power Quantities and Their Units

QUANTITY	UNITS
Complex power	volt-amps
Average power	watts
Reactive power	vars

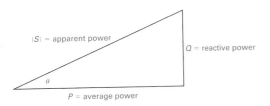

Figure 7.42 A power triangle.

Another advantage of using complex power is the geometric interpretation it provides. When working with Eq. (7.73), think of P, Q, and $|S|$ as the sides of a right triangle, as shown in Fig. 7.42. It is easy to show that the angle θ in the power triangle is the power factor angle $\theta_v - \theta_i$. For the right triangle shown in Fig. 7.42,

$$\tan \theta = \frac{Q}{P}. \tag{7.74}$$

But from the definitions of P and Q (Eqs. [7.60] and [7.61], respectively),

$$\frac{Q}{P} = \frac{(V_m I_m / 2) \sin(\theta_v - \theta_i)}{(V_m I_m / 2) \cos(\theta_v - \theta_i)}$$

$$= \tan(\theta_v - \theta_i). \tag{7.75}$$

Therefore, $\theta = \theta_v - \theta_i$. The geometric relations for a right triangle mean also that the four power triangle dimensions (the three sides and the power factor angle) can be determined if any two of the four are known.

The magnitude of complex power is referred to as **apparent power**. Specifically,

$$|S| = \sqrt{P^2 + Q^2}. \tag{7.76}$$

Apparent power, like complex power, is measured in volt-amps. The apparent power, or volt-amp, requirement of a device designed to convert electric energy to a nonelectric form is more important than the average power requirement. Although the average power represents the useful output of the energy-converting device, the apparent power represents the volt-amp capacity required to supply the average power. As you can see from the power triangle in Fig. 7.42, unless the power factor angle is $0°$ (that is, the device is purely resistive, pf = 1, and $Q = 0$), the volt-amp capacity required by the device is larger than the average power used by the device. As we will see in Example 7.16, it makes sense to operate devices at a power factor close to 1.

Many useful appliances (such as refrigerators, fans, air conditioners, fluorescent lighting fixtures, and washing machines) and most industrial loads operate at a lagging power factor. The power factor of these loads sometimes is corrected either by adding a capacitor to the device itself or by connecting capacitors across the line feeding the load; the latter method is often used for large industrial loads. Many of the Chapter Problems give you a chance to make some calculations that correct a lagging power factor load and improve the operation of a circuit.

Example 7.14 uses a power triangle to calculate several quantities associated with an electrical load.

EXAMPLE 7.14

An electrical load operates at 240 V rms. The load absorbs an average power of 8 kW at a lagging power factor of 0.8.

(a) Calculate the complex power of the load.

(b) Calculate the impedance of the load.

SOLUTION

(a) The power factor is described as lagging, so we know that the load is inductive and that the algebraic sign of the reactive power is positive. From the power triangle shown in Fig. 7.43,

$$P = |S| \cos \theta,$$

$$Q = |S| \sin \theta.$$

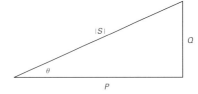

Figure 7.43 A power triangle.

Now, because $\cos \theta = 0.8$, $\sin \theta = 0.6$. Therefore

$$|S| = \frac{P}{\cos \theta} = \frac{8 \text{ kW}}{0.8} = 10 \text{ kVA},$$

$$Q = 10 \sin \theta = 6 \text{ kVAR},$$

and

$$S = 8 + j6 \text{ kVA}.$$

(b) From the computation of the complex power of the load, we see that $P = 8$ kW. Using Eq. (7.71),

$$P = V_{\text{eff}} I_{\text{eff}} \cos(\theta_v - \theta_i)$$

$$= (240) I_{\text{eff}} (0.8)$$

$$= 8000 \text{ W}.$$

Solving for I_{eff},

$$I_{\text{eff}} = 41.67 \text{ A}.$$

We already know the angle of the load impedance, because it is the power factor angle:

$$\theta = \cos^{-1}(0.8) = 36.87°.$$

We also know that θ is positive because the power factor is lagging, indicating an inductive load. We compute the magnitude of the load impedance from its definition as the ratio of the magnitude of the voltage to the magnitude of the current:

$$|Z| = \frac{|V_{\text{eff}}|}{|I_{\text{eff}}|} = \frac{240}{41.67} = 5.76.$$

Hence,

$$Z = 5.76 \;\angle{36.87°}\; \Omega = 4.608 + j3.456 \; \Omega.$$

Power Calculations

We are now ready to develop additional equations that can be used to calculate real, reactive, and complex power. We begin by combining Eqs. (7.60), (7.61), and (7.73) to get

$$S = \frac{V_m I_m}{2} \cos(\theta_v - \theta_i) + j \frac{V_m I_m}{2} \sin(\theta_v - \theta_i)$$

$$= \frac{V_m I_m}{2} [\cos(\theta_v - \theta_i) + j \sin(\theta_v - \theta_i)]$$

$$= \frac{V_m I_m}{2} e^{j(\theta_v - \theta_i)} = \frac{1}{2} V_m I_m \;\angle{(\theta_v - \theta_i)}. \tag{7.77}$$

If we use the effective values of the sinusoidal voltage and current, Eq. (7.77) becomes

$$S = V_{\text{eff}} I_{\text{eff}} \;\angle{(\theta_v - \theta_i)}. \tag{7.78}$$

Equations (7.77) and (7.78) are important relationships in power calculations because they show that if the phasor current and voltage are known at a pair of terminals, the complex power associated with that pair of terminals is either one-half the product of the voltage and the conjugate of the current, or the product of the rms phasor voltage and the conjugate of the rms phasor current. We can show this for the rms phasor voltage and current in Fig. 7.44 as follows:

$$S = V_{\text{eff}} I_{\text{eff}} \;\angle{(\theta_v - \theta_i)}$$

$$= V_{\text{eff}} I_{\text{eff}} e^{j(\theta_v - \theta_i)}$$

$$= V_{\text{eff}} e^{j\theta_v} I_{\text{eff}} e^{-j\theta_i}$$

$$= \mathbf{V}_{\text{eff}} \mathbf{I}_{\text{eff}}^{*}. \tag{7.79}$$

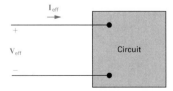

Figure 7.44 The phasor voltage and current associated with a pair of terminals.

Note that $\mathbf{I}_{\text{eff}}^* = I_{\text{eff}}e^{-j\theta_i}$ follows from Euler's identity and the trigonometric identities $\cos(-\theta) = \cos(\theta)$ and $\sin(-\theta) = -\sin(\theta)$:

$$I_{\text{eff}}e^{-j\theta_i} = I_{\text{eff}}\cos(-\theta_i) + jI_{\text{eff}}\sin(-\theta_i)$$

$$= I_{\text{eff}}\cos(\theta_i) - jI_{\text{eff}}\sin(\theta_i)$$

$$= \mathbf{I}_{\text{eff}}^*.$$

The same derivation technique could be applied to Eq. (7.77) to yield

$$S = \frac{1}{2}\mathbf{VI}^*. \tag{7.80}$$

Both Eqs. (7.79) and (7.80) are based on the passive sign convention. If the current reference is in the direction of the voltage rise across the terminals, we insert a minus sign on the right-hand side of each equation.

To illustrate the use of Eq. (7.80) in a power calculation, let's use the same circuit that we used in Example 7.12. Expressed in terms of the phasor representation of the terminal voltage and current,

$$\mathbf{V} = 100 \ \underline{/15°} \ \text{V},$$

$$\mathbf{I} = 4 \ \underline{/-105°} \ \text{A}.$$

Therefore

$$S = \frac{1}{2}(100 \ \underline{/15°})(4 \ \underline{/+105°}) = 200 \ \underline{/120°}$$

$$= -100 + j173.21 \ \text{VA}.$$

Once we calculate the complex power, we can read off both the real and reactive powers, because $S = P + jQ$. Thus

$$P = -100 \ \text{W},$$

$$Q = 173.21 \ \text{VAR}.$$

The interpretations of the algebraic signs on P and Q are identical to those given in the solution of Example 7.12.

Alternate Forms for Complex Power

Equations (7.79) and (7.80) have several useful variations. Here, we use the rms value form of the equations, because rms values are the most common type of representation for voltages and currents in power computations.

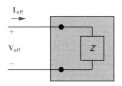

Figure 7.45 The general circuit of Fig. 7.44 replaced with an equivalent impedance.

The first variation of Eq. (7.79) is to replace the voltage with the product of the current times the impedance. In other words, we can always represent the circuit inside the box of Fig. 7.44 by an equivalent impedance, as shown in Fig. 7.45. Then,

$$\mathbf{V}_{\text{eff}} = Z\mathbf{I}_{\text{eff}}. \tag{7.81}$$

Substituting Eq. (7.81) into Eq. (7.79) yields

$$
\begin{aligned}
S &= Z\mathbf{I}_{\text{eff}}\mathbf{I}_{\text{eff}}^* \\
&= |\mathbf{I}_{\text{eff}}|^2 Z \\
&= |\mathbf{I}_{\text{eff}}|^2 (R + jX) \\
&= |\mathbf{I}_{\text{eff}}|^2 R + j|\mathbf{I}_{\text{eff}}|^2 X = P + jQ,
\end{aligned} \tag{7.82}
$$

from which

$$P = |\mathbf{I}_{\text{eff}}|^2 R = \frac{1}{2} I_m^2 R, \tag{7.83}$$

$$Q = |\mathbf{I}_{\text{eff}}|^2 X = \frac{1}{2} I_m^2 X. \tag{7.84}$$

In Eq. (7.84), X is the reactance of either the equivalent inductance or equivalent capacitance of the circuit. Recall from our earlier discussion of reactance that it is positive for inductive circuits and negative for capacitive circuits.

A second useful variation of Eq. (7.79) comes from replacing the current with the voltage divided by the impedance:

$$S = \mathbf{V}_{\text{eff}} \left(\frac{\mathbf{V}_{\text{eff}}}{Z} \right)^* = \frac{|\mathbf{V}_{\text{eff}}|^2}{Z^*} = P + jQ. \tag{7.85}$$

Note that if Z is a pure resistive element,

$$P = \frac{|\mathbf{V}_{\text{eff}}|^2}{R}, \tag{7.86}$$

and if Z is a pure reactive element,

$$Q = \frac{|\mathbf{V}_{\text{eff}}|^2}{X}. \tag{7.87}$$

In Eq. (7.87), X is positive for an inductor and negative for a capacitor.

The following examples demonstrate various power calculations in circuits operating in the sinusoidal steady state.

EXAMPLE 7.15

In the circuit shown in Fig. 7.46, a load having an impedance of $39 + j26\ \Omega$ is fed from a voltage source through a line having an impedance of $1 + j4\ \Omega$. The effective, or rms, value of the source voltage is 250 V.

(a) Calculate the load current $\mathbf{I}_L$ and voltage $\mathbf{V}_L$.

(b) Calculate the average and reactive power delivered to the load.

(c) Calculate the average and reactive power delivered to the line.

(d) Calculate the average and reactive power supplied by the source.

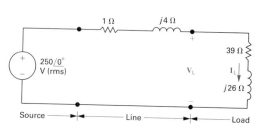

Figure 7.46 The circuit for Example 7.15.

SOLUTION

(a) The line and load impedances are in series across the voltage source, so the load current equals the voltage divided by the total impedance, or

$$\mathbf{I}_L = \frac{250\ \angle 0^\circ}{40 + j30} = 4 - j3 = 5\ \angle{-36.87^\circ}\ \text{A (rms)}.$$

Because the voltage is given in terms of its rms value, the current also is rms. The load voltage is the product of the load current and load impedance:

$$\mathbf{V}_L = (39 + j26)\mathbf{I}_L = 234 - j13$$

$$= 234.36\ \angle{-3.18^\circ}\ \text{V (rms)}.$$

(b) The average and reactive power delivered to the load can be computed using Eq. (7.79). Therefore

$$S = \mathbf{V}_L\mathbf{I}_L^* = (234 - j13)(4 + j3)$$

$$= 975 + j650\ \text{VA}.$$

Thus the load is absorbing an average power of 975 W and a reactive power of 650 VAR.

(c) The average and reactive power delivered to the line are most easily calculated from Eqs. (7.83) and (7.84) because the line current is known. Thus

$$P = (5)^2(1) = 25\ \text{W},$$

$$Q = (5)^2(4) = 100\ \text{VAR}.$$

Note that the reactive power associated with the line is positive because the line reactance is inductive.

(d) One way to calculate the average and reactive power delivered by the source is to add the complex power delivered to the line to that delivered to the load, or

$$S = 25 + j100 + 975 + j650$$

$$= 1000 + j750 \text{ VA}.$$

The complex power at the source can also be calculated from Eq. (7.79):

$$S_s = -250\mathbf{I}_L^*.$$

The minus sign is inserted in Eq. (7.79) whenever the current reference is in the direction of a voltage rise. Thus

$$S_s = -250(4 + j3) = -(1000 + j750) \text{ VA}.$$

The minus sign implies that both average power and magnetizing reactive power are being delivered by the source. Note that this result agrees with the previous calculation of S, as it must, because the source must furnish all the average and reactive power absorbed by the line and load.

EXAMPLE 7.16

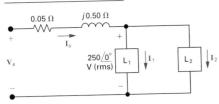

Figure 7.47 The circuit for Example 7.16.

The two loads in the circuit shown in Fig. 7.47 can be described as follows: Load 1 absorbs an average power of 8 kW at a leading power factor of 0.8. Load 2 absorbs 20 kVA at a lagging power factor of 0.6.

(a) Determine the power factor of the two loads in parallel.

(b) Determine the apparent power required to supply the loads, the magnitude of the current, $\mathbf{I}_s$, and the average power loss in the transmission line.

(c) Given that the frequency of the source is 60 Hz, compute the value of the capacitor that would correct the power factor to 1 if placed in parallel with the two loads. Recompute the values in (b) for the load with the corrected power factor.

SOLUTION

(a) All voltage and current phasors in this problem are assumed to represent effective values. Note from the circuit diagram in Fig. 7.47 that $\mathbf{I}_s = \mathbf{I}_1 + \mathbf{I}_2$. The total complex power absorbed by the two loads is

$$S = (250)\mathbf{I}_s^*$$
$$= (250)(\mathbf{I}_1 + \mathbf{I}_2)^*$$
$$= (250)\mathbf{I}_1^* + (250)\mathbf{I}_2^*$$
$$= S_1 + S_2.$$

We can sum the complex powers geometrically, using the power triangles for each load, as shown in Fig. 7.48. By hypothesis,

$$S_1 = 8000 - j\frac{8000(.6)}{(.8)}$$
$$= 8000 - j6000 \text{ VA},$$
$$S_2 = 20,000(.6) + j20,000(.8)$$
$$= 12,000 + j16,000 \text{ VA}.$$

It follows that

$$S = 20,000 + j10,000 \text{ VA},$$

and
$$\mathbf{I}_s^* = \frac{20,000 + j10,000}{250} = 80 + j40 \text{ A}.$$

Therefore

$$\mathbf{I}_s = 80 - j40 = 89.44 \angle{-26.57°} \text{ A}.$$

Thus the power factor of the combined load is

$$\text{pf} = \cos(0 + 26.57°)$$
$$= .8944 \text{ lagging}.$$

The power factor of the two loads in parallel is lagging because the net reactive power is positive.

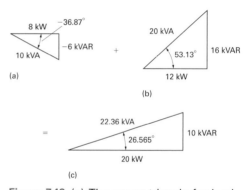

Figure 7.48 (a) The power triangle for load 1. (b) The power triangle for load 2. (c) The sum of the power triangles.

(b) The apparent power which must be supplied to these loads is

$$|S| = |20 + j10|$$
$$= 22.36 \text{ kVA}.$$

The magnitude of the current that supplies this apparent power is

$$|\mathbf{I}_s| = |80 - j40|$$
$$= 89.44 \text{ A}.$$

The average power lost in the line results from the current flowing through the line resistance:

$$P_{\text{line}} = |\mathbf{I}_s|^2 R = (89.44)^2 (0.05) = 400 \text{ W}.$$

Note that the power supplied totals $20{,}000 + 400 = 20{,}400$ W, even though the loads require a total of only 20,000 W.

(c) As we can see from the power triangle in Fig. 7.48(c), we can correct the power factor to 1 if we place a capacitor in parallel with the existing loads such that the capacitor supplies 10 kVAR of magnetizing reactive power. The value of the capacitor is calculated as follows. First, find the capacitive reactance from Eq. (7.87):

$$X = \frac{|V_{\text{eff}}|^2}{Q}$$
$$= \frac{(250)^2}{-10{,}000}$$
$$= -6.25 \ \Omega.$$

Recall that the reactive impedance of a capacitor is $-1/\omega C$, and $\omega = 2\pi(60) = 376.99$ rad/s, if the source frequency is 60 Hz. Thus,

$$C = \frac{-1}{\omega X} = \frac{-1}{(376.99)(-6.25)} = 424.4 \ \mu\text{F}.$$

The addition of the capacitor as the third load is represented in geometric form as the sum of the two power triangles shown in Fig. 7.49. When the power factor is 1, the apparent power and the average power are the same, as seen from the power triangle in Fig. 7.49(c). Therefore, the apparent power once the power factor has been corrected is

$$|S| = P = 20 \text{ kVA}.$$

The magnitude of the current that supplies this apparent power is

$$|\mathbf{I}_s| = \frac{20{,}000}{250}$$

$$= 80 \text{ A}.$$

The average power lost in the line is thus reduced to

$$P_{\text{line}} = |\mathbf{I}_s|^2 R = (80)^2(0.05) = 320 \text{ W}.$$

Now, the power supplied totals $20{,}000 + 320 = 20{,}320$ W. Note that the addition of the capacitor has reduced the line loss from 400 W to 320 W.

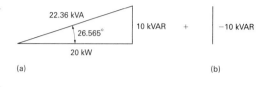

(a)   (b)

(c)

Figure 7.49 (a) The sum of the power triangles for loads 1 and 2. (b) The power triangle for a 424.4 μF capacitor at 60 Hz. (c) The sum of the power triangles in (a) and (b).

EXAMPLE 7.17

(a) Calculate the total average and reactive power delivered to each impedance in the circuit shown in Fig. 7.50.

(b) Calculate the average and reactive powers associated with each source in the circuit.

(c) Verify that the average power delivered equals the average power absorbed, and that the magnetizing reactive power delivered equals the magnetizing reactive power absorbed.

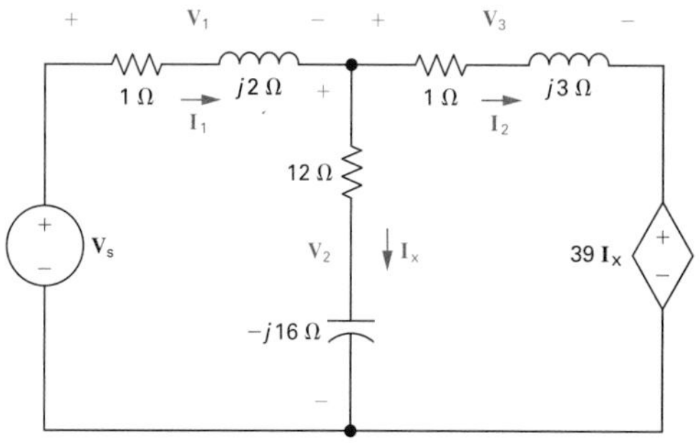

SOLUTION

(a) The complex power delivered to the $(1 + j2)$ Ω impedance is

$$S_1 = \frac{1}{2}\mathbf{V}_1\mathbf{I}_1^* = P_1 + jQ_1$$

$$= \frac{1}{2}(78 - j104)(-26 + j52)$$

$$= \frac{1}{2}(3380 + j6760)$$

$$= 1690 + j3380 \text{ VA}.$$

$\mathbf{V}_s = 150\underline{/0°} \text{ V}$

$\mathbf{V}_1 = (78 - j104) \text{ V}$ $\mathbf{I}_1 = (-26 - j52) \text{ A}$

$\mathbf{V}_2 = (72 + j104) \text{ V}$ $\mathbf{I}_x = (-2 + j6) \text{ A}$

$\mathbf{V}_3 = (150 - j130) \text{ V}$ $\mathbf{I}_2 = (-24 - j58) \text{ A}$

Figure 7.50 The circuit, with solution, for Example 7.17.

Thus this impedance is absorbing an average power of 1690 W and a reactive power of 3380 VAR. The complex power delivered to the $(12 - j16)$ Ω impedance is

$$S_2 = \frac{1}{2}\mathbf{V}_2\mathbf{I}_x^* = P_2 + jQ_2$$

$$= \frac{1}{2}(72 + j104)(-2 - j6)$$

$$= 240 - j320 \text{ VA.}$$

Therefore the impedance in the vertical branch is absorbing 240 W and delivering 320 VAR. The complex power delivered to the $(1 + j3)$ Ω impedance is

$$S_3 = \frac{1}{2}\mathbf{V}_3\mathbf{I}_2^* = P_3 + jQ_3$$

$$= \frac{1}{2}(150 - j130)(-24 + j58)$$

$$= 1970 + j5910 \text{ VA.}$$

This impedance is absorbing 1970 W and 5910 VAR.

(b) The complex power associated with the independent voltage source is

$$S_s = -\frac{1}{2}\mathbf{V}_s\mathbf{I}_1^* = P_s + jQ_s$$

$$= -\frac{1}{2}(150)(-26 + j52)$$

$$= 1950 - j3900 \text{ VA.}$$

Note that the independent voltage source is absorbing an average power of 1950 W and delivering 3900 VAR. The complex power associated with the current-controlled voltage source is

$$S_x = \frac{1}{2}(39\mathbf{I}_x)(\mathbf{I}_2^*) = P_x + jQ_x$$

$$= \frac{1}{2}(-78 + j234)(-24 + j58)$$

$$= -5850 - j5070 \text{ VA.}$$

Both average power and magnetizing reactive power are being delivered by the dependent source.

(c) The total power absorbed by the passive impedances and the
     independent voltage source is

$$P_{\text{absorbed}} = P_1 + P_2 + P_3 + P_s = 5850 \text{ W}.$$

The dependent voltage source is the only circuit element delivering average power. Thus

$$P_{\text{delivered}} = 5850 \text{ W}.$$

Magnetizing reactive power is being absorbed by the two
horizontal branches. Thus

$$Q_{\text{absorbed}} = Q_1 + Q_3 = 9290 \text{ VAR}.$$

Magnetizing reactive power is being delivered by the independent voltage source, the capacitor in the vertical impedance
branch, and the dependent voltage source. Therefore

$$Q_{\text{delivered}} = 9290 \text{ VAR}.$$

DRILL EXERCISES

7.21 The load impedance in the circuit shown is shunted
by a capacitor having a capacitive reactance of
$-52 \ \Omega$. Calculate:

(a) the rms phasors $\mathbf{V}_L$ and $\mathbf{I}_L$

(b) the average power and magnetizing reactive
power absorbed by the $(39 + j26) \ \Omega$ load
impedance

(c) the average power and magnetizing reactive power absorbed by the $(1 + j4) \ \Omega$ line
impedance

(d) the average power and magnetizing reactive
power delivered by the source

(e) the magnetizing reactive power delivered by
the shunting capacitor

ANSWER: (a) 252.20 $\angle{-4.54°}$ V (rms),
5.38 $\angle{-38.23°}$ A (rms); (b) 1129.09 W,
752.73 VAR; (c) 23.52 W, 94.09 VAR;
(d) 1152.62 W, −376.36 VAR; (e) 1223.18 VAR.

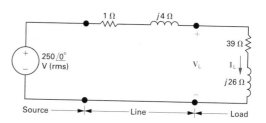

7.22 The rms voltage at the terminals of a load is 250 V. The load is absorbing an average power of 40 kW and delivering a magnetizing reactive power of 30 kVAR. Derive two equivalent impedance models of the load.

7.23 Find the phasor voltage $\mathbf{V}_s$ (rms) in the circuit shown if loads L_1 and L_2 are absorbing 15 kVA at 0.6 pf lagging and 6 kVA at 0.8 pf leading, respectively. Express $\mathbf{V}_s$ in polar form.

ANSWER: 251.64 $\angle 15.91°$ V(rms).

ANSWER: 1 Ω in series with −0.75 Ω of capacitive reactance; 1.5625 Ω in parallel with −2.083 Ω of capacitive reactance.

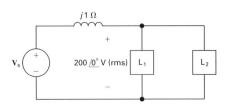

Practical Perspective

Household Distribution Circuit

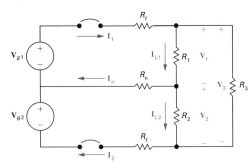

Figure 7.51 Three-wire household distribution circuit.

Let us return to the household distribution circuit introduced at the beginning of the chapter. We will modify the circuit slightly by adding resistance to each conductor to simulate more accurately the residential wiring conductors. The modified circuit is shown in Fig. 7.51. In this circuit the resistor R_3 is used to model a 240 V appliance (such as an electric range) and the resistor R_1 and R_2 are used to model 120 V appliances (such as a lamp, toaster, and iron). The resistor R_n is used to model the neutral conductor and the resistor R_ℓ is used to model what electricians refer to as the hot conductors of the circuit.

In the example that follows, we will show why the neutral conductor, i.e., the R_n conductor in Fig. 7.51, never contains an overcurrent protective device such as a circuit breaker or fuse.

EXAMPLE 7.18

(a) Show that if $\mathbf{V}_{g1} = \mathbf{V}_{g2}$ and $R_1 = R_2$ that $\mathbf{I}_n = 0$ and $\mathbf{V}_1 = \mathbf{V}_2$.

(b) Calculate $\mathbf{I}_n$, $\mathbf{V}_1$ and $\mathbf{V}_2$ if $R_\ell = 0.04$ Ω; $R_n = 0.08$ Ω; $R_1 = 60$ Ω; $R_2 = 300$ Ω; $R_3 = 20$ Ω; and $\mathbf{V}_{g1} = \mathbf{V}_{g2} = 120 \angle 0°$ V(rms).

(c) Repeat (b) if the neutral conductor is open.

(d) On the basis of the calculations in (b) and (c) explain why the neutral conductor is never fused in such a manner that it can open while the hot conductors are energized.

SOLUTION

(a) Let $\mathbf{I}_a$ be the clockwise mesh current in the $\mathbf{V}_{g1}-R_\ell-R_1-R_n$ mesh. Let $\mathbf{I}_b$ be the clockwise mesh current in the $\mathbf{V}_{g2}-R_n-R_2-R_\ell$ mesh. Let $\mathbf{I}_c$ be the clockwise mesh current in the $R_1-R_3-R_2$ mesh. Then the three mesh-current equations are

$$\mathbf{V}_{g1} = (R_1 + R_\ell + R_n)\mathbf{I}_a - R_n\mathbf{I}_b - R_1\mathbf{I}_c,$$

$$\mathbf{V}_{g2} = -R_n\mathbf{I}_a + (R_2 + R_\ell + R_n)\mathbf{I}_b - R_2\mathbf{I}_c,$$

and

$$0 = -R_1\mathbf{I}_a - R_2\mathbf{I}_b + (R_1 + R_2 + R_3)\mathbf{I}_c.$$

For convenience let

$$R_a = R_1 + R_n + R_\ell$$

$$R_b = R_2 + R_n + R_\ell$$

and

$$R_c = R_1 + R_2 + R_3.$$

With these substitutions the three simultaneous equations become

$$\mathbf{V}_{g1} = R_a\mathbf{I}_a - R_n\mathbf{I}_b - R_1\mathbf{I}_c,$$

$$\mathbf{V}_{g2} = -R_n\mathbf{I}_a + R_b\mathbf{I}_b - R_2\mathbf{I}_c,$$

$$0 = -R_1\mathbf{I}_a - R_2\mathbf{I}_b + R_c\mathbf{I}_c.$$

Using Cramer's method, the solutions for $\mathbf{I}_a$, $\mathbf{I}_b$, and $\mathbf{I}_c$ are

$$\mathbf{I}_a = \frac{\mathbf{V}_{g1}(R_b R_c - R_2^2) + \mathbf{V}_{g2}(R_n R_c + R_1 R_2)}{\Delta},$$

$$\mathbf{I}_b = \frac{\mathbf{V}_{g1}(R_n R_c + R_1 R_2) + \mathbf{V}_{g2}(R_a R_c - R_1^2)}{\Delta},$$

$$\mathbf{I}_c = \frac{\mathbf{V}_{g1}(R_2 R_n + R_1 R_b) + \mathbf{V}_{g2}(R_a R_2 + R_n R_1)}{\Delta},$$

where Δ is the characteristic determinate, i.e.,

$$\Delta = \begin{vmatrix} R_a & -R_n & -R_1 \\ -R_n & R_b & -R_2 \\ -R_1 & -R_2 & R_c \end{vmatrix}.$$

The current in the neutral conductor $\mathbf{I}_n$ is equal to $\mathbf{I}_a - \mathbf{I}_b$. If $\mathbf{V}_{g1} = \mathbf{V}_{g2}$, the expression for $\mathbf{I}_n$ reduces to

$$\mathbf{I}_n = \frac{\mathbf{V}_{g1}}{\Delta}\left[R_c(R_b - R_a) + (R_1 + R_2)(R_1 - R_2)\right].$$

Now if $R_1 = R_2$, then $R_a = R_b$, and since $\Delta \neq 0$, $\mathbf{I}_n = 0$. Therefore, if a three-wire distribution circuit is balanced (i.e., $R_1 = R_2$ and $\mathbf{V}_{g1} = \mathbf{V}_{g2}$), the current in the neutral conductor is zero.

We also note that when the three-wire distribution circuit is balanced, the load voltages $\mathbf{V}_1$ and $\mathbf{V}_2$ are equal. This can be verified by noting $\mathbf{I}_{L1} = (\mathbf{I}_a - \mathbf{I}_c)$; $\mathbf{I}_{L2} = (\mathbf{I}_b - \mathbf{I}_c)$; and $\mathbf{I}_{L1} = \mathbf{I}_{L2}$ when $R_1 = R_2$ and $\mathbf{V}_{g1} = \mathbf{V}_{g2}$.

(b) Using the equations developed in part (a), we have

$$\mathbf{I}_n = 1.59 \; \angle 0° \; \text{A(rms)},$$

$$\mathbf{V}_1 = 119.32 \; \angle 0° \; \text{V(rms)},$$

and

$$\mathbf{V}_2 = 119.63 \; \angle 0° \; \text{V(rms)}.$$

(c) With the neutral conductor open, the solution reduces to solving two simultaneous mesh-current equations. The results are

$$\mathbf{I}_n = 0 \quad \text{(by hypothesis)},$$

$$\mathbf{V}_1 = 39.83 \; \angle 0° \; \text{V(rms)},$$

and

$$\mathbf{V}_2 = 199.16 \; \angle 0° \; \text{V(rms)}.$$

(d) Note that when the neutral conductor is in service, $\mathbf{V}_1$ and $\mathbf{V}_2$ are very nearly equal to the rated voltage of the distribution circuit (120 V in this example) in spite of the fact that R_2 is five times larger than R_1. If the neutral conductor is taken out of service, the voltages on each side become very unbalanced. With $\mathbf{V}_2$ rising to 199.16 V, appliances rated at 120 V would become damaged. Therefore overcurrent protection is never installed in the neutral conductor of the three-wire distribution circuit.

SUMMARY

- The general equation for a **sinusoidal source** is

$$v = V_m \cos(\omega t + \phi) \text{ (voltage source),}$$

or

$$i = I_m \cos(\omega t + \phi) \text{ (current source),}$$

where V_m (or I_m) is the maximum amplitude, ω is the frequency, and ϕ is the phase angle.

- The frequency, ω, of a sinusoidal response is the same as the frequency of the sinusoidal source driving the circuit. The amplitude and phase angle of the response are usually different from those of the source.

- The best way to find the steady-state voltages and currents in a circuit driven by sinusoidal sources is to perform the analysis in the frequency domain. The following mathematical transforms allow us to move between the time and frequency domains.

  - The phasor transform (from the time domain to the frequency domain):

$$\mathbf{V} = V_m e^{j\phi} = \mathcal{P}\{V_m \cos(\omega t + \phi)\}.$$

  - The inverse phasor transform (from the frequency domain to the time domain):

$$\mathcal{P}^{-1}\{V_m e^{j\phi}\} = \Re\{V_m e^{j\phi} e^{j\omega t}\}.$$

- When working with sinusoidally varying signals, remember that voltage leads current by 90° at the terminals of an inductor, and current leads voltage by 90° at the terminals of a capacitor.

TABLE 7.4   **Impedance and Related Values**

ELEMENT	IMPEDANCE (Z)	REACTANCE	ADMITTANCE (Y)	SUSCEPTANCE
Resistor	R (resistance)	—	G (conductance)	—
Capacitor	$j(-1/\omega C)$	$-1/\omega C$	$j\omega C$	ωC
Inductor	$j\omega L$	ωL	$j(-1/\omega L)$	$-1/\omega L$

- **Impedance** (Z) plays the same role in the frequency domain as resistance, inductance, and capacitance play in the time domain. Specifically, the relationship between phasor current and phasor voltage for resistors, inductors, and capacitors is

$$\mathbf{V} = Z\mathbf{I},$$

where the reference direction for $\mathbf{I}$ obeys the passive sign convention. The reciprocal of impedance is **admittance** (Y), so another way to express the current-voltage relationship for resistors, inductors, and capacitors in the frequency domain is

$$\mathbf{I} = Y\mathbf{V}.$$

- **Instantaneous power** is the product of the instantaneous terminal voltage and current, or $p = \pm vi$. The positive sign is used when the reference direction for the current is from the positive to the negative reference polarity of the voltage. The frequency of the instantaneous power is twice the frequency of the voltage (or current).

- **Average power** is the average value of the instantaneous power over one period. It is the power converted from electric to nonelectric form and vice versa. This conversion is the reason that average power is also referred to as real power. Average power, with the passive sign convention, is expressed as

$$P = \frac{1}{2} V_m I_m \cos(\theta_v - \theta_i)$$

$$= V_{\text{eff}} I_{\text{eff}} \cos(\theta_v - \theta_i).$$

- **Reactive power** is the electric power exchanged between the magnetic field of an inductor and the source that drives it or between the electric field of a capacitor and the source that drives it. Reactive power is never converted to nonelectric power. Reactive power, with the passive sign convention, is expressed as

$$Q = \frac{1}{2} V_m I_m \sin(\theta_v - \theta_i)$$

$$= V_{\text{eff}} I_{\text{eff}} \sin(\theta_v - \theta_i).$$

Both average power and reactive power can be expressed in terms of either peak (V_m, I_m) or effective (V_{eff}, I_{eff}) current and voltage. Effective values are widely used in both household and industrial applications. *Effective value* and *rms value* are interchangeable terms for the same value.

- The **power factor** is the cosine of the phase angle between the voltage and the current:

$$\text{pf} = \cos(\theta_v - \theta_i).$$

The terms *lagging* and *leading* added to the description of the power factor indicate whether the current is lagging or leading the voltage and thus whether the load is inductive or capacitive.

- The **reactive factor** is the sine of the phase angle between the voltage and the current:

$$\text{rf} = \sin(\theta_v - \theta_i).$$

- **Complex power** is the complex sum of the real and reactive powers, or

$$S = P + jQ$$

$$= \frac{1}{2}\mathbf{VI}^* = \mathbf{V}_{eff}\mathbf{I}_{eff}^*$$

$$= |\mathbf{I}_{eff}|^2 Z = \frac{|\mathbf{V}_{eff}|^2}{Z^*}.$$

- **Apparent power** is the magnitude of the complex power:

$$|S| = \sqrt{P^2 + Q^2}.$$

- The **watt** is used as the unit for both instantaneous and real power.

- The **var** (volt amp reactive, or VAR) is used as the unit for reactive power.

- The **volt-amp** (VA) is used as the unit for complex and apparent power.

PROBLEMS

7.1. A sinusoidal current is zero at $t = 150$ μs and increasing at a rate of $2 \times 10^4 \pi$ A/s. The maximum amplitude of the current is 10 A.

    (a) What is the frequency of i in radians per second?

    (b) What is the expression for i?

7.2. In a single graph, sketch $v = 60\cos(\omega t + \phi)$ versus ωt for $\phi = -60°$, $-30°$, $0°$, $+30°$, and $60°$.

    (a) State whether the voltage function is shifting to the right or left as ϕ becomes more positive.

    (b) What is the direction of shift if ϕ changes from 0 to $-30°$?

7.3. Consider the sinusoidal voltage

$$v = 170\cos(120\pi t - 60°) \text{ V}.$$

    (a) What is the maximum amplitude of the voltage?

    (b) What is the frequency in hertz?

    (c) What is the frequency in radians per second?

    (d) What is the phase angle in radians?

    (e) What is the phase angle in degrees?

    (f) What is the period in milliseconds?

    (g) What is the first time after $t = 0$ that $v = 170$ V?

    (h) The sinusoidal function is shifted 125/18 ms to the left along the time axis. What is the expression for $v(t)$?

    (i) What is the minimum number of milliseconds that the function must be shifted to the right if the expression for $v(t)$ is 170 sin $120\pi t$ V?

    (j) What is the minimum number of milliseconds that the function must be shifted to the left if the expression for $v(t)$ is 170 cos $120\pi t$ V?

7.4. At $t = -250/6$ μs, a sinusoidal voltage is known to be zero and going positive. The voltage is next zero at $t = 1250/6$ μs. It is also known that the voltage is 75 V at $t = 0$.

    (a) What is the frequency of v in hertz?

    (b) What is the expression for v?

7.5. **(a)** Verify that Eq. (7.9) is the solution of Eq. (7.8). This can be done by substituting Eq. (7.9) into the left-hand side of Eq. (7.8) and then noting that it equals the right-hand side for all values of $t > 0$. At $t = 0$, Eq. (7.9) should reduce to the initial value of the current.

    (b) Because the transient component vanishes as time elapses and because our solution must satisfy the differential equation for all values of t, the steady-state component, by itself, must also satisfy the differential equation. Verify this observation by showing that the steady-state component of Eq. (7.9) satisfies Eq. (7.8).

7.6. Use the concept of the phasor to combine the following sinusoidal functions into a single trigonometric expression:

  (a) $y = 100\cos(300t + 45°) + 500\cos(300t - 60°)$,

  (b) $y = 250\cos(377t + 30°) - 150\sin(377t + 140°)$,

  (c) $y = 60\cos(100t + 60°) - 120\sin(100t - 125°) + 100\cos(100t + 90°)$, and

  (d) $y = 100\cos(\omega t + 40°) + 100\cos(\omega t + 160°) + 100\cos(\omega t - 80°)$

7.7. Show that

$$\int_{t_o}^{t_o+T} V_m^2 \cos^2(\omega t + \phi)\, dt = \frac{V_m^2 T}{2}.$$

7.8. The rms value of the sinusoidal voltage supplied to the convenience outlet of a U.S. home is 120 V. What is the maximum value of the voltage at the outlet?

 7.9. A 20 Ω resistor and a 1 μF capacitor are connected in parallel. This parallel combination is also in parallel with the series combination of a 1 Ω resistor and an 40 μH inductor. These three parallel branches are driven by a sinusoidal current source whose current is $20\cos(50,000t - 20°)$ A.

  (a) Draw the frequency-domain equivalent circuit.

  (b) Reference the voltage across the current source as a rise in the direction of the source current, and find the phasor voltage.

  (c) Find the steady-state expression for $v(t)$.

 7.10. A 400 Ω resistor, an 87.5 mH inductor, and a 312.5 nF capacitor are connected in series. The series-connected elements are energized by a sinusoidal voltage source whose voltage is $500\cos(8000t + 60°)$ V.

  (a) Draw the frequency-domain equivalent circuit.

  (b) Reference the current in the direction of the voltage rise across the source, and find the phasor current.

  (c) Find the steady-state expression for $i(t)$.

7.11. A 50 Hz sinusoidal voltage with a maximum amplitude of 340 V at $t = 0$ is applied across the terminals of an inductor. The maximum amplitude of the steady-state current in the inductor is 8.5 A.

(a) What is the frequency of the inductor current?

(b) What is the phase angle of the voltage?

(c) What is the phase angle of the current?

(d) What is the inductive reactance of the inductor?

(e) What is the inductance of the inductor in millihenrys?

(f) What is the impedance of the inductor?

7.12. A 40 kHz sinusoidal voltage has zero phase angle and a maximum amplitude of 2.5 mV. When this voltage is applied across the terminals of a capacitor, the resulting steady-state current has a maximum amplitude of 125.67 μA.

(a) What is the frequency of the current in radians per second?

(b) What is the phase angle of the current?

(c) What is the capacitive reactance of the capacitor?

(d) What is the capacitance of the capacitor in microfarads?

(e) What is the impedance of the capacitor?

 7.13. The circuit in Fig. P7.13 is operating in the sinusoidal steady state. Find the steady-state expression for $v_o(t)$ if $v_g = 64\cos 8000t$ V.

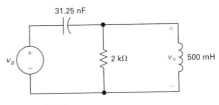

Figure P7.13

 7.14. Find the steady-state expression for $i_o(t)$ in the circuit in Fig. P7.14 if $v_s = 750\cos 5000t$ mV.

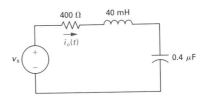

Figure P7.14

7.15. The circuit shown in Fig. P7.15 is operating in the sinusoidal steady state. The capacitor is adjusted until the current i_g is in phase with the sinusoidal voltage v_g.

    (a) Specify the values of capacitance in microfarads if $v_g = 250 \cos 1000t$ V.

    (b) Give the steady-state expressions for i_g when C has the values found in (a).

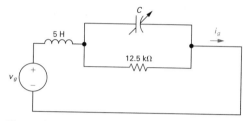

Figure P7.15

7.16. The frequency of the sinusoidal voltage source in the circuit in Fig. P7.16 is adjusted until the current i_o is in phase with v_g.

    (a) Find the frequency in hertz.

    (b) Find the steady-state expression for i_o (at the frequency found in [a]) if $v_g = 10 \cos \omega t$ V.

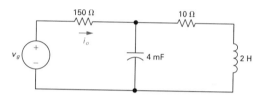

Figure P7.16

7.17. The circuit shown in Fig. P7.17 is operating in the sinusoidal steady state. Find the value of ω if

$$i_o = 100 \sin(\omega t + 81.87°) \text{ mA},$$

$$v_g = 50 \cos(\omega t - 45°) \text{ V}.$$

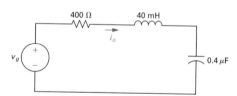

Figure P7.17

7.18. Find Z_{ab} in the circuit shown in Fig. P7.18 when the circuit is operating at a frequency of 1.6 Mrad/s.

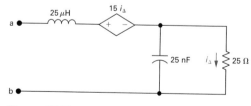

Figure P7.18

7.19. Find the admittance Y_{ab} in the circuit seen in Fig. P7.19. Express Y_{ab} in both polar and rectangular form. Give the value of Y_{ab} in millisiemens.

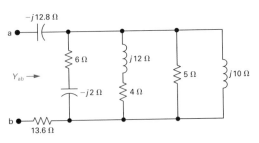

Figure P7.19

7.20. (a) The frequency of the source voltage in the circuit in Fig. P7.20 is adjusted until i_g is in phase with v_g. What is the value of ω in radians per second?

(b) If $v_g = 45 \cos \omega t$ V (where ω is the frequency found in [a]), what is the steady-state expression for v_o?

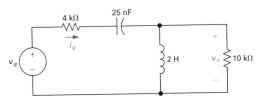

Figure P7.20

7.21. (a) The source voltage in the circuit in Fig. P7.21 is $v_g = 96 \cos 10{,}000t$ V. Find the values of L such that i_g is in phase with v_g when the circuit is operating in the steady state.

(b) For the values of L found in (a), find the steady-state expressions for i_g.

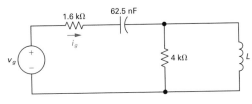

Figure P7.21

7.22. (a) For the circuit shown in Fig. P7.22, find the frequency (in radians per second) at which the impedance Z_{ab} is purely resistive.

(b) Find the value of Z_{ab} at the frequency of (a).

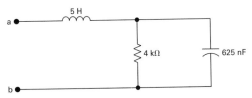

Figure P7.22

7.23. Find the impedance Z_{ab} in the circuit seen in Fig. P7.23. Express Z_{ab} in both polar and rectangular form.

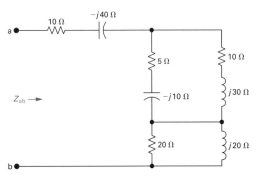

Figure P7.23

7.24. The frequency of the sinusoidal current source in the circuit in Fig. P7.24 is adjusted until v_o is in phase with i_g.

(a) What is the value of ω in radians per second?

(b) If $i_g = 2.5 \cos \omega t$ mA (where ω is the frequency found in [a]), what is the steady-state expression for v_o?

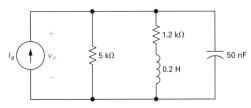

Figure P7.24

7.25. (a) Show that, at a given frequency ω, the circuits in Fig. P7.25(a) and (b) will have the same impedance between the terminals a,b if

$$R_1 = \frac{\omega^2 L_2^2 R_2}{R_2^2 + \omega^2 L_2^2},$$

$$L_1 = \frac{R_2^2 L_2}{R_2^2 + \omega^2 L_2^2}.$$

(b) Find the values of resistance and inductance that when connected in series will have the same impedance at 20 krad/s as that of a 50 kΩ resistor connected in parallel with a 2.5 H inductor.

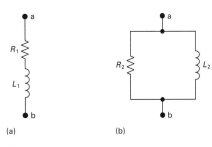

Figure P7.25

7.26. **(a)** Show that at a given frequency ω, the circuits in Fig. P7.25(a) and (b) will have the same impedance between the terminals a,b if

$$R_2 = \frac{R_1^2 + \omega^2 L_1^2}{R_1},$$

$$L_2 = \frac{R_1^2 + \omega^2 L_1^2}{\omega^2 L_1}.$$

(*Hint:* The two circuits will have the same impedance if they have the same admittance.)

(b) Find the values of resistance and inductance that when connected in parallel will have the same impedance at 10 krad/s as a 5 kΩ resistor connected in series with a 0.5 H inductor.

7.27. **(a)** Show that at a given frequency ω, the circuits in Fig. P7.27(a) and (b) will have the same impedance between the terminals a,b if

$$R_1 = \frac{R_2}{1 + \omega^2 R_2^2 C_2^2},$$

$$C_1 = \frac{1 + \omega^2 R_2^2 C_2^2}{\omega^2 R_2^2 C_2}.$$

(b) Find the values of resistance and capacitance that when connected in series will have the same impedance at 80 krad/s as that of a 500 Ω resistor connected in parallel with a 25 nF capacitor.

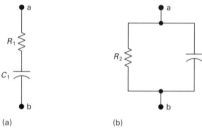

(a) (b)

Figure P7.27

7.28. **(a)** Show that at a given frequency ω, the circuits in Fig. P7.27(a) and (b) will have the same impedance between the terminals a,b if

$$R_2 = \frac{1 + \omega^2 R_1^2 C_1^2}{\omega^2 R_1 C_1^2},$$

$$C_2 = \frac{C_1}{1 + \omega^2 R_1^2 C_1^2}.$$

(*Hint:* The two circuits will have the same impedance if they have the same admittance.)

(b) Find the values of resistance and capacitance that when connected in parallel will give the same impedance at 20 krad/s as that of a 2 kΩ resistor connected in series with a capacitance of 50 nF.

7.29. The expressions for the steady-state voltage and current at the terminals of the circuit seen in Fig. P7.29 are

$$v_g = 150\cos(8000\pi t + 20°) \text{ V},$$

$$i_g = 30\sin(8000\pi t + 38°) \text{ A}.$$

(a) What is the impedance seen by the source?

(b) By how many microseconds is the current out of phase with the voltage?

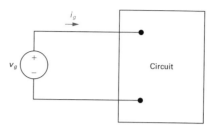

Figure P7.29

7.30. Use the concept of current division to find the steady-state expression for i_o in the circuit in Fig. P7.30 if $i_g = 125\cos 500t$ mA.

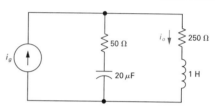

Figure P7.30

7.31. Use the concept of voltage division to find the steady-state expression for $v_o(t)$ in the circuit in Fig. P7.31 if $v_g = 75\cos 5000t$ V.

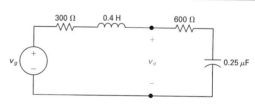

Figure P7.31

7.32. Find the steady-state expression for $v_o(t)$ in the circuit seen in Fig. P7.32 by using the technique of source transformations. The sinusoidal voltage sources are

$$v_1 = 240\cos(4000t + 53.13°) \text{ V},$$

$$v_2 = 96\sin 4000t \text{ V}.$$

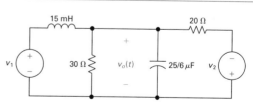

Figure P7.32

7.33. The phasor current $\mathbf{I}_b$ in the circuit shown in Fig. P7.33 is $5\ \angle 45°$ A.

(a) Find $\mathbf{I}_a$, $\mathbf{I}_c$, and $\mathbf{V}_g$.

(b) If $\omega = 800$ rad/s, write the expressions for $i_a(t)$, $i_c(t)$, and $v_g(t)$.

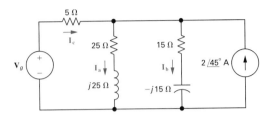

Figure P7.33

7.34. The circuit in Fig. P7.34 is operating in the sinusoidal steady state. Find $v_o(t)$ if $i_s(t) = 15\cos 8000t$ mA.

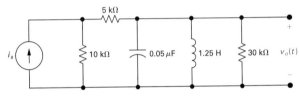

Figure P7.34

7.35. Find $\mathbf{I}_b$ and Z in the circuit shown in Fig. P7.35 if $\mathbf{V}_g = 60\ \angle 0°$ V and $\mathbf{I}_a = 5\ \angle -90°$ A.

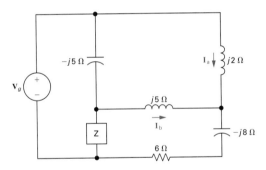

Figure P7.35

7.36. Find the Thévenin impedance seen looking into the terminals a,b of the circuit in Fig. P7.36 if the frequency of operation is 25 krad/s.

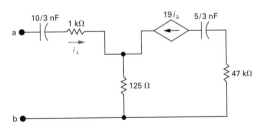

Figure P7.36

7.37. The sinusoidal voltage source in the circuit in Fig. P7.37 is developing a voltage equal to $247.49\cos(1000t + 45°)$ V.

    (a) Find the Thévenin voltage with respect to the terminals a,b.

    (b) Find the Thévenin impedance with respect to the terminals a,b.

    (c) Draw the Thévenin equivalent.

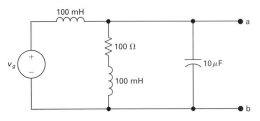

Figure P7.37

7.38. Find the Norton equivalent circuit with respect to the terminals a,b for the circuit shown in Fig. P7.38 when $\mathbf{V}_s = 25 \angle 0°$ V.

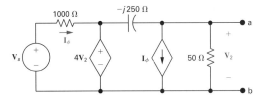

Figure P7.38

7.39. Find the Norton equivalent circuit with respect to the terminals a,b for the circuit shown in Fig. P7.39.

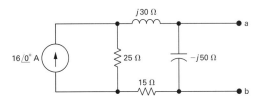

Figure P7.39

7.40. The device in Fig. P7.40 is represented in the frequency domain by a Norton equivalent. When an inductor having an impedance of $j100$ Ω is connected across the device, the value of $\mathbf{V}_0$ is $100 \angle 120°$ mV. When a capacitor having an impedance of $-j100$ Ω is connected across the device, the value of $\mathbf{I}_0$ is $-3 \angle 210°$ mA. Find the Norton current $\mathbf{I}_n$ and the Norton impedance Z_n.

Figure P7.40

7.41. Find the Thévenin equivalent circuit with respect to the terminals a,b for the circuit shown in Fig. P7.41.

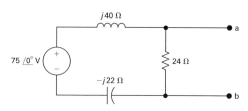

Figure P7.41

7.42. Find the Thévenin equivalent circuit with respect to the terminals a,b for the circuit shown in Fig. P7.42.

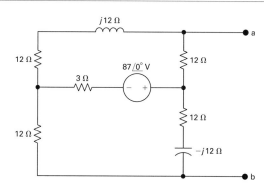

Figure P7.42

7.43. Find the Thévenin equivalent circuit with respect to the terminals a,b of the circuit shown in Fig. P7.43.

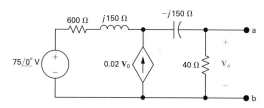

Figure P7.43

7.44. The circuit shown in Fig. P7.44 is operating at a frequency of 10 krad/s. Assume β is real and lies between -50 and $+50$, that is, $-50 \le \beta \le 50$.

   (a) Find the value of β so that the Thévenin impedance looking into the terminals a,b is purely resistive.

   (b) What is the value of the Thévenin impedance for the β found in (a)?

   (c) Can β be adjusted so that the Thévenin impedance equals $5 + j5\ \Omega$? If so, what is the value of β?

   (d) For what values of β will the Thévenin impedance be capacitive?

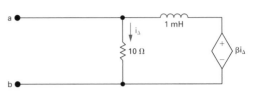

Figure P7.44

7.45. Use the mesh-current method to find the phasor current $\mathbf{I}_g$ in the circuit shown in Fig. P7.45.

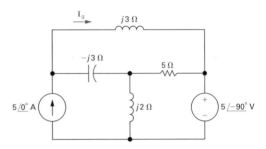

Figure P7.45

7.46. Use the node-voltage method to find the phasor voltage across the capacitor in the circuit in Fig. P7.45. Assume the voltage is positive at the left-hand terminal of the capacitor.

7.47. Use the node-voltage method to find $\mathbf{V}_o$ in the circuit in Fig. P7.47.

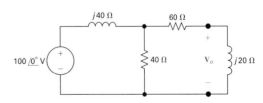

Figure P7.47

7.48. Use the mesh-current method to find the steady-state expression for v_o in the circuit seen in Fig. P7.48 if v_g equals $72 \cos 5000t$ V.

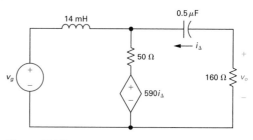

Figure P7.48

7.49. Use the node-voltage method to find the steady-state expression for $v_o(t)$ in the circuit in Fig. P7.49 if

$$v_{g1} = 10\cos(5000t + 53.13°) \text{ V},$$

$$v_{g2} = 8\sin 5000t \text{ V}.$$

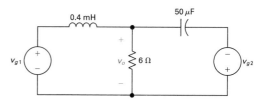

Figure P7.49

7.50. Use source transformations to find the steady-state expression for $v_o(t)$ in the circuit in Fig. P7.49.

7.51. Use the mesh-current method to find the steady-state expression for $v_o(t)$ in the circuit in Fig. P7.49.

7.52. Use the mesh-current method to find the branch currents $\mathbf{I}_a$, $\mathbf{I}_b$, $\mathbf{I}_c$, and $\mathbf{I}_d$ in the circuit shown in Fig. P7.52.

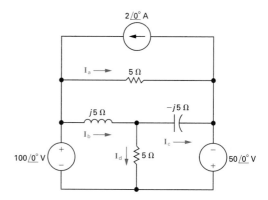

Figure P7.52

7.53. Find the value of Z in the circuit seen in Fig. P7.53 if $\mathbf{V}_g = 100 - j50$ V, $\mathbf{I}_g = 20 + j30$ A, and $\mathbf{V}_1 = 40 + j30$ V.

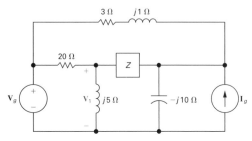

Figure P7.53

7.54. Use the mesh-current method to find the steady-state expression for $i_o(t)$ in the circuit in Fig. P7.54 if

$$v_a = 60 \cos 40{,}000t \text{ V},$$

$$v_b = 90 \sin(40{,}000t + 180°) \text{ V}.$$

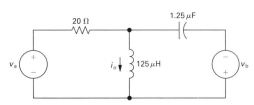

Figure P7.54

7.55. **(a)** For the circuit shown in Fig. P7.55, find the steady-state expression for v_o if $i_g = 5 \cos(8 \times 10^5 t)$ A.

(b) By how many nanoseconds does v_o lag i_g?

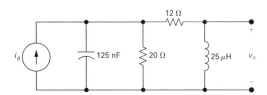

Figure P7.55

7.56. Use the node-voltage method to find $\mathbf{V}_o$ and $\mathbf{I}_o$ in the circuit seen in Fig. P7.56.

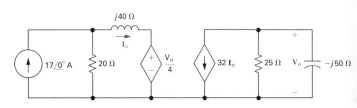

Figure P7.56

7.57. Use the node-voltage method to find the phasor voltage $\mathbf{V}_o$ in the circuit shown in Fig. P7.57. Express the voltage in both polar and rectangular form.

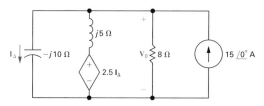

Figure P7.57

 7.58. Find the steady-state expressions for the branch currents i_a and i_b in the circuit seen in Fig. P7.58 if $v_a = 100 \sin 10{,}000t$ V and $v_b = 500 \cos 10{,}000t$ V.

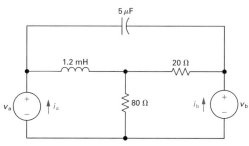

Figure P7.58

7.59. Use the principle of superposition to find the steady-state expression for the voltage $v_o(t)$ in the circuit in Fig. P7.49.

7.60. The following sets of values for v and i pertain to the circuit seen in Fig. 7.34. For each set of values, calculate P and Q and state whether the circuit inside the box is absorbing or delivering (1) average power and (2) magnetizing vars.

(a) $v = 340 \cos(\omega t + 60°)$ V, $i = 20 \cos(\omega t + 15°)$ A.

(b) $v = 75 \cos(\omega t - 15°)$ V, $i = 16 \cos(\omega t + 60°)$ A.

(c) $v = 625 \cos(\omega t + 40°)$ V, $i = 4 \sin(\omega t + 240°)$ A.

(d) $v = 180 \sin(\omega t + 220°)$ V, $i = 10 \cos(\omega t + 20°)$ A.

7.61. Show that the maximum value of the instantaneous power given by Eq. (7.59) is $P + \sqrt{P^2 + Q^2}$ and that the minimum value is $P - \sqrt{P^2 + Q^2}$.

7.62. A dc voltage equal to V_{dc} V is applied to a resistor of R Ω. A sinusoidal voltage equal to v_s V is also applied to a resistor of R Ω. Show that the dc voltage will deliver the same amount of energy in T seconds (where T is the period of the sinusoidal voltage) as the sinusoidal voltage provided V_{dc} equals the rms value of v_s. (*Hint:* Equate the two expressions for the energy delivered to the resistor.)

7.63. **(a)** Find the rms value of the periodic voltage shown in Fig. P7.63.

(b) If this voltage is applied to the terminals of a 12 Ω resistor, what is the average power dissipated in the resistor?

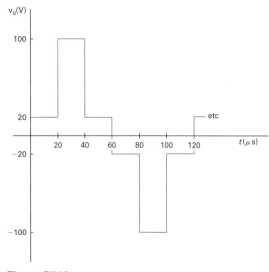

Figure P7.63

7.64. Find the rms value of the periodic current shown in Fig. P7.64.

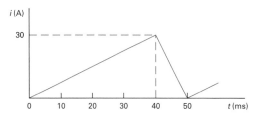

Figure P7.64

7.65. The periodic current shown in Fig. P7.64 dissipates an average power of 24 kW in a resistor. What is the value of the resistor?

7.66. The op amp in the circuit shown in Fig. P7.66 is ideal. Calculate the average power delivered to the 1000 Ω resistor when $v_g = 4 \cos 5000t$ V.

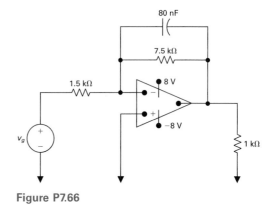

Figure P7.66

7.67. Find the average power, the reactive power, and the apparent power absorbed by the load in the circuit in Fig. P7.67 if i_g equals $30 \cos 100t$ mA.

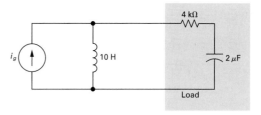

Figure P7.67

7.68. Find the average power dissipated in the 20 Ω resistor in the circuit seen in Fig. P7.68 if $i_g = 15 \cos 10,000t$ A.

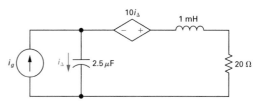

Figure P7.68

7.69. The three loads in the circuit in Fig. P7.69 can be described as follows: Load 1 is a 240 Ω resistor in series with an inductive reactance of 70 Ω; load 2 is a capacitive reactance of 120 Ω in series with a 160 Ω resistor; and load 3 is a 30 Ω resistor in series with a capacitive reactance of 40 Ω. The frequency of the voltage source is 60 Hz.

  (a) Give the power factor and reactive factor of each load.

  (b) Give the power factor and reactive factor of the composite load seen by the voltage source.

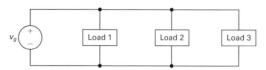

Figure P7.69

7.70. The voltage $\mathbf{V}_g$ in the frequency-domain circuit shown in Fig. P7.70 is 340 $\angle 0°$ V (rms).

  (a) Find the average and reactive power delivered by the voltage source.

  (b) Is the voltage source absorbing or delivering average power?

  (c) Is the voltage source absorbing or delivering magnetizing vars?

  (d) Find the average and reactive powers associated with each impedance branch in the circuit.

  (e) Check the balance between delivered and absorbed average power.

  (f) Check the balance between delivered and absorbed magnetizing vars.

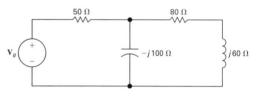

Figure P7.70

7.71. Find the average power delivered by the ideal current source in the circuit in Fig. P7.71 if $i_g = 30\cos 25{,}000t$ mA.

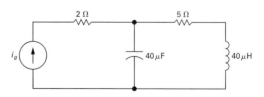

Figure P7.71

7.72. A load consisting of a 480 Ω resistor in parallel with a 5/9 μF capacitor is connected across the terminals of a sinusoidal voltage source v_g, where $v_g = 240 \cos 5000t$ V.

(a) What is the peak value of the instantaneous power delivered by the source?

(b) What is the peak value of the instantaneous power absorbed by the source?

(c) What is the average power delivered to the load?

(d) What is the reactive power delivered to the load?

(e) Does the load absorb or generate magnetizing vars?

(f) What is the power factor of the load?

(g) What is the reactive factor of the load?

7.73. **(a)** Find the average power, the reactive power, and the apparent power supplied by the voltage source in the circuit in Fig. P7.73 if $v_g = 50 \cos 10^5 t$ V.

(b) Check your answer in (a) by showing $P_{\text{dev}} = \sum P_{\text{abs}}$.

(c) Check your answer in (a) by showing $Q_{\text{dev}} = \sum Q_{\text{abs}}$.

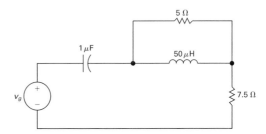

Figure P7.73

7.74. The two loads shown in Fig. P7.74 can be described as follows: Load 1 absorbs an average power of 24.96 kW and 47.04 kVAR magnetizing reactive power; load 2 has an impedance of $5 - j5$ Ω. The voltage at the terminals of the loads is $480\sqrt{2} \cos 120\pi t$ V.

(a) Find the rms value of the source voltage.

(b) By how many microseconds is the load voltage out of phase with the source voltage?

(c) Does the load voltage lead or lag the source voltage?

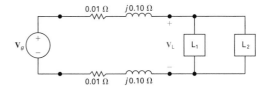

Figure P7.74

7.75. Three loads are connected in parallel across a 2400 V (rms) line, as shown in Fig. P7.75. Load 1 absorbs 18 kW and 24 kVAR. Load 2 absorbs 60 kVA at 0.6 pf lead. Load 3 absorbs 18 kW at unity power factor.

(a) Find the impedance that is equivalent to the three parallel loads.

(b) Find the power factor of the equivalent load as seen from the line's input terminals.

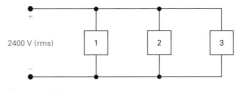

Figure P7.75

7.76. The three loads in Problem 7.75 are fed from a line having a series impedance $0.2 + j1.6 \, \Omega$, as shown in Fig. P7.76.

(a) Calculate the rms value of the voltage ($\mathbf{V}_s$) at the sending end of the line.

(b) Calculate the average and reactive powers associated with the line impedance.

(c) Calculate the average and reactive powers at the sending end of the line.

(d) Calculate the efficiency (η) of the line if the efficiency is defined as

$$\eta = (P_{\text{load}}/P_{\text{sending end}}) \times 100.$$

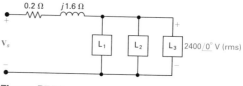

Figure P7.76

7.77. The three parallel loads in the circuit shown in Fig. P7.77 can be described as follows: Load 1 is absorbing an average power of 24 kW and 18 kVAR of magnetizing vars; load 2 is absorbing an average power of 48 kW and generating 30 kVAR of magnetizing reactive power; load 3 consists of a 60 Ω resistor in parallel with an inductive reactance of 480 Ω. Find the rms magnitude and the phase angle of $\mathbf{V}_g$ if $\mathbf{V}_o = 2400 \; \underline{/0°}$ V (rms).

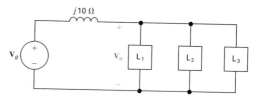

Figure P7.77

7.78. Two 660 V (rms) loads are connected in parallel. The two loads draw a total average power of 52,800 W at a power factor of 0.80 leading. One of the loads draws 40 kVA at a power factor of 0.96 lagging. What is the power factor of the other load?

7.79. **(a)** Find the average power dissipated in the line in Fig. P7.79.

(b) Find the capacitive reactance that when connected in parallel with the load will make the load look purely resistive.

(c) What is the equivalent impedance of the load in (b)?

(d) Find the average power dissipated in the line when the capacitive reactance is connected across the load.

(e) Express the power loss in (d) as a percentage of the power loss found in (a).

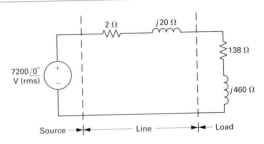

Figure P7.79

7.80. The three loads in the circuit shown in Fig. P7.80 are $S_1 = 5 + j2$ kVA, $S_2 = 3.75 + j1.5$ kVA, and $S_3 = 8 + j0$ kVA.

(a) Calculate the complex power delivered by each voltage source, $\mathbf{V}_{g1}$ and $\mathbf{V}_{g2}$.

(b) Verify that the total real and reactive power delivered by the sources equals the total real and reactive power absorbed by the network.

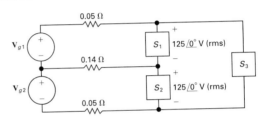

Figure P7.80

7.81. The three loads in the circuit seen in Fig. P7.81 are described as follows: Load 1 is absorbing 1.8 kW and 600 VAR; load 2 is 1.5 kVA at a 0.8 pf lead; load 3 is a 12 Ω resistor in parallel with an inductor that has a reactance of 48 Ω.

(a) Calculate the average power and the magnetizing reactive power delivered by each source if $\mathbf{V}_{g1} = \mathbf{V}_{g2} = 120 \, \underline{/0°}$ V (rms).

(b) Check your calculations by showing your results are consistent with the requirements

$$\sum P_{\text{dev}} = \sum P_{\text{abs}}$$

$$\sum Q_{\text{dev}} = \sum Q_{\text{abs}}.$$

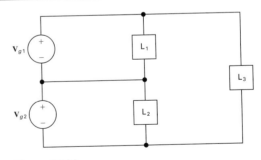

Figure P7.81

7.82. A group of small appliances on a 60 Hz system requires 25 kVA at 0.96 pf lagging when operated at 125 V (rms). The impedance of the feeder supplying the appliances is $0.006 + j0.048$ Ω. The voltage at the load end of the feeder is 125 V.

 (a) What is the rms magnitude of the voltage at the source end of the feeder?

 (b) What is the average power loss in the feeder?

 (c) What size capacitor (in microfarads) at the load end of the feeder is needed to improve the load power factor to unity?

 (d) After the capacitor is installed, what is the rms magnitude of the voltage at the source end of the feeder if the load voltage is maintained at 125 V?

 (e) What is the average power loss in the feeder for (d)?

◆ **7.83.** In the circuit in Fig. 7.51 $\mathbf{V}_{g1} = \mathbf{V}_{g2} = 120 \angle 0°$ V(rms). $R_\ell = 0.5$ Ω; $R_n = 1$ Ω; $R_1 = 30$ Ω; $R_2 = 300$ Ω; and $R_3 = 15$ Ω. Calculate: (a) $\mathbf{I}_n$; (b) $\mathbf{V}_1$; (c) $\mathbf{V}_2$; (d) $\mathbf{V}_3$; (e) the average power delivered to R_1, R_2, and R_3; (f) the average power delivered by the sinusoidal sources; and (g) the total average power dissipated in the circuit.

◆ **7.84.** Repeat Problem 7.83 if the neutral conductor is open.

Two-Stage RC Ladder

The two-stage RC-ladder network shown in the accompanying figure is used to model the interconnection between driving and load gates on an integrated circuit. After we have introduced the Laplace transform approach to circuit analysis, we will show how it can be used to find the solution for v_1 and v_2 (as functions of time) after the switch has been closed.

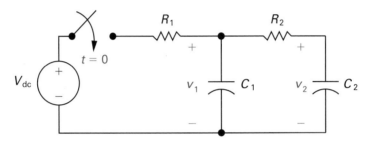

Two-stage RC-ladder network.

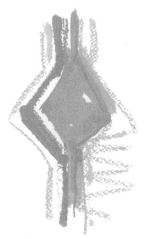

8 Introduction to the Laplace Transform

Chapter Contents

We now introduce a powerful analytical technique that is widely used to study the behavior of linear, lumped-parameter circuits. The method is based on the Laplace transform, which we define mathematically in Section 8.1. Before doing so, we need to explain why another analytical technique is needed. First, we wish to consider the transient behavior of circuits whose describing equations consist of more than a single node-voltage or mesh-current differential equation. In other words, we want to consider multiple-node and multiple-mesh circuits that are described by sets of linear differential equations.

Second, we wish to determine the transient response of circuits whose signal sources vary in ways more complicated than the simple dc level jumps considered in Chapters 5 and 6. Third, we can use the Laplace transform to introduce the concept of the transfer function as a tool for analyzing the steady-state sinusoidal response of a circuit when the frequency of the sinusoidal source is varied. We discuss the transfer function in Chapter 9. Finally, we wish to relate, in a systematic fashion, the time-domain behavior of a circuit to its frequency-domain behavior. Using the Laplace transform will provide a broader understanding of circuit functions.

In this chapter, we introduce the Laplace transform, discuss its pertinent characteristics, and present a systematic method for transforming from the frequency domain to the time domain.

8.1 DEFINITION OF THE LAPLACE TRANSFORM

The **Laplace transform** of a function is given by the expression

$$\mathcal{L}\{f(t)\} = \int_0^\infty f(t)e^{-st}\, dt, \qquad (8.1)$$

where the symbol $\mathcal{L}\{f(t)\}$ is read "the Laplace transform of $f(t)$."

The Laplace transform of $f(t)$ is also denoted $F(s)$; that is,

$$F(s) = \mathcal{L}\{f(t)\}. \qquad (8.2)$$

This notation emphasizes that when the integral in Eq. (8.1) has been evaluated, the resulting expression is a function of s. In our applications, t represents the time domain, and, because the exponent of e in the integral of Eq. (8.1) must be dimensionless, s must have the dimension of reciprocal time, or frequency. The Laplace transform transforms the problem from the time domain to the frequency domain. After obtaining the frequency-domain expression for the unknown, we inverse-transform it back to the time domain.

If the idea behind the Laplace transform seems foreign, consider another familiar mathematical transform. Logarithms are used to change a multiplication or division problem, such as A = BC, into a simpler addition or subtraction problem: log A = log BC = log B + log C. Antilogs are used to carry out the inverse process. The phasor is another transform; as we know from Chapter 7, it converts a sinusoidal signal into a complex number for easier, algebraic computation of circuit values. After determining the phasor value of a signal, we transform it back to its time-domain expression. Both of these examples point out the essential feature of mathematical transforms: They are designed to create a new domain to make the mathematical manipulations easier. After finding the unknown in the new domain, we inverse-transform it back to the original domain. In circuit analysis, we use the Laplace transform to transform a set of integrodifferential equations from the time domain to a set of algebraic equations in the frequency domain. We therefore simplify the solution for an unknown quantity to the manipulation of a set of algebraic equations.

Before we illustrate some of the important properties of the Laplace transform, some general comments are in order. First, note that the integral in Eq. (8.1) is improper because the upper limit is infinite. Thus we are confronted immediately with the question of whether the integral converges. In other words, does a given $f(t)$ have a Laplace transform? Obviously, the functions of primary interest in engineering analysis have Laplace transforms; otherwise we would not be interested in the transform. In linear circuit analysis, we excite circuits with sources that have Laplace transforms. Excitation functions such as t^t or e^{t^2}, which do not have Laplace transforms, are of no interest here.

Second, because the lower limit on the integral is zero, the Laplace transform ignores $f(t)$ for negative values of t. Put another way, $F(s)$ is determined by the behavior of $f(t)$ only for positive values of t. To emphasize that the lower limit is zero, Eq. (8.1) is frequently referred to as the **one-sided**, or **unilateral**, **Laplace transform**. In the two-sided, or bilateral, Laplace transform, the lower limit is $-\infty$. We do not use the bilateral form here; hence $F(s)$ is understood to be the one-sided transform.

Another point regarding the lower limit concerns the situation when $f(t)$ has a discontinuity at the origin. If $f(t)$ is continuous at the origin—as, for example, in Fig. 8.1(a)—$f(0)$ is not ambiguous. However, if $f(t)$ has a finite discontinuity at the origin—as, for example, in Fig. 8.1(b)—the question arises as to whether the Laplace transform integral should include or exclude the discontinuity. In other words, should we make the lower limit 0^- and include the discontinuity, or should we exclude the discontinuity by making the lower limit 0^+? (We use the notation 0^- and 0^+ to denote values of t just to the left and right of the origin, respectively.) Actually we may choose either, as long as we are consistent. For reasons to be explained later, we choose 0^- as the lower limit.

Because we are using 0^- as the lower limit, we note immediately that the integration from 0^- to 0^+ is zero. The only exception is when the discontinuity at the origin is an impulse function, a situation we discuss in Section 8.3. The important point now is that the two functions shown in Fig. 8.1 have the same unilateral Laplace transform because there is no impulse function at the origin.

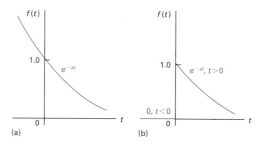

Figure 8.1 A continuous and discontinuous function at the origin. (a) $f(t)$ is continuous at the origin. (b) $f(t)$ is discontinuous at the origin.

The one-sided Laplace transform ignores $f(t)$ for $t < 0^-$. What happens prior to 0^- is accounted for by the initial conditions. Thus we use the Laplace transform to predict the response to a disturbance that occurs after initial conditions have been established.

In the discussion that follows, we divide the Laplace transforms into two types: functional and operational. A **functional transform** is the Laplace transform of a specific function, such as $\sin \omega t$, t, e^{-at}, and so on. An **operational transform** defines a general mathematical property of the Laplace transform, such as finding the transform of the derivative of $f(t)$. Before considering functional and operational transforms, however, we need to introduce the step and impulse functions.

8.2 THE STEP FUNCTION

In our introduction to the step response of RL and RC circuits (Sec. 5.5) we pointed out that a digital logic circuit switches between high and low dc voltage levels as it processes information. We can incorporate these abrupt voltage changes into the Laplace transform approach to circuit analysis by introducing the step function. The step function in turn leads to the concept of the impulse function and, as we shall see, we will find the impulse function very useful in the work that follows.

Figure 8.2 illustrates the step function. It is zero for $t < 0$. The symbol for the step function is $Ku(t)$. Thus, the mathematical definition of the **step function** is

Figure 8.2 The step function.

$$Ku(t) = 0, \quad t < 0,$$

$$Ku(t) = K, \quad t > 0. \tag{8.3}$$

If K is 1, the function defined by Eq. (8.3) is the **unit step**.

The step function is not defined at $t = 0$. In situations where we need to define the transition between 0^- and 0^+, we assume that it is linear and that

$$Ku(0) = 0.5K. \tag{8.4}$$

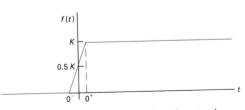

Figure 8.3 The linear approximation to the step function.

As before, 0^- and 0^+ represent symmetric points arbitrarily close to the left and right of the origin. Figure 8.3 illustrates the linear transition from 0^- to 0^+.

A discontinuity may occur at some time other than $t = 0$; for example, in sequential switching. A step that occurs at $t = a$ is expressed as $Ku(t - a)$. Thus

$$Ku(t - a) = 0, \quad t < a,$$

$$Ku(t - a) = K, \quad t > a. \tag{8.5}$$

If $a > 0$, the step occurs to the right of the origin, and if $a < 0$, the step occurs to the left of the origin. Figure 8.4 illustrates Eq. (8.5). Note that the step function is 0 when the argument $t - a$ is negative, and it is K when the argument is positive.

A step function equal to K for $t < a$ is written as $Ku(a - t)$. Thus

$$Ku(a - t) = K, \quad t < a,$$

$$Ku(a - t) = 0, \quad t > a. \tag{8.6}$$

The discontinuity is to the left of the origin when $a < 0$. Equation (8.6) is shown in Fig. 8.5.

One application of the step function is to use it to write the mathematical expression for a function that is nonzero for a finite duration but is defined for all positive time. One example useful in circuit analysis is a finite-width pulse, which we can create by adding two step functions. The function $K[u(t - 1) - u(t - 3)]$ has the value K for $1 < t < 3$ and the value 0 everywhere else, so it is a finite-width pulse of height K initiated at $t = 1$ and terminated at $t = 3$. In defining this pulse using step functions, it is helpful to think of the step function $u(t - 1)$ as "turning on" the constant value K at $t = 1$, and the step function $-u(t - 3)$ as "turning off" the constant value K at $t = 3$. We use step functions to turn on and turn off linear functions at desired times in Example 8.1.

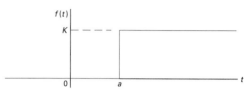

Figure 8.4 A step function occurring at $t = a$ when $a > 0$.

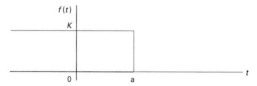

Figure 8.5 A step function $Ku(a-t)$ for $a > 0$.

EXAMPLE 8.1

Use step functions to write an expression for the function illustrated in Fig. 8.6.

SOLUTION

The function shown in Fig. 8.6 is made up of linear segments with break points at 0, 1, 3, and 4 s. To construct this function, we must add and subtract linear functions of the proper slope. We use the step function to initiate and terminate these linear

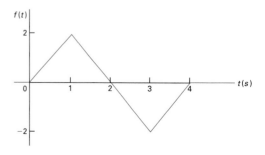

Figure 8.6 The function for Example 8.1.

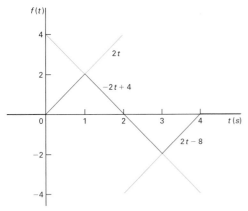

segments at the proper times. In other words, we use the step function to turn on and turn off a straight line with the following equations: $+2t$, on at $t = 0$, off at $t = 1$; $-2t + 4$, on at $t = 1$, off at $t = 3$; and $+2t - 8$, on at $t = 3$, off at $t = 4$. These straight line segments and their equations are shown in Fig. 8.7. The expression for $f(t)$ is

$$f(t) = 2t[u(t) - u(t-1)] + (-2t + 4)[u(t-1) - u(t-3)]$$
$$+ (2t - 8)[u(t-3) - u(t-4)].$$

Figure 8.7 Definition of the three line segments turned on and off with step functions to form the function shown in Fig. 8.6.

DRILL EXERCISE

8.1 Use step functions to write the expression for each function shown.

ANSWER: (a) $f(t) = 5t[u(t) - u(t-2)] + 10[u(t-2) - u(t-6)] + (-5t + 40)[u(t-6) - u(t-8)]$;
(b) $f(t) = 10\sin(\pi t)[u(t) - u(t-2)]$;
(c) $f(t) = 4t[u(t) - u(t-5)]$.

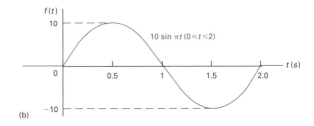

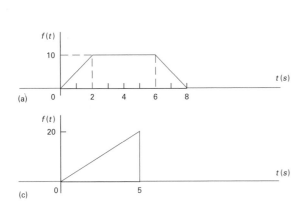

8.3 THE IMPULSE FUNCTION

When we have a finite discontinuity in a function, such as that illustrated in Fig. 8.1(b), the derivative of the function is not defined at the point of the discontinuity. The concept of an impulse function[1] enables us to define the derivative at a discontinuity, and thus to define the Laplace transform of that derivative. An **impulse** is a signal of infinite amplitude and zero duration. Such signals don't exist in nature, but some circuit signals come very close to approximating this definition, so we find a mathematical model of an impulse useful. Impulsive voltages and currents occur in circuit analysis either because of a switching operation or because the circuit is excited by an impulsive source. We will analyze these situations in Chapter 9, but here we focus on defining the impulse function generally.

To define the derivative of a function at a discontinuity, we first assume that the function varies linearly across the discontinuity, as shown in Fig. 8.8, where we observe that as $\epsilon \to 0$, an abrupt discontinuity occurs at the origin. When we differentiate the function, the derivative between $-\epsilon$ and $+\epsilon$ is constant at a value of $1/2\epsilon$. For $t > \epsilon$, the derivative is $-ae^{-a(t-\epsilon)}$. Figure 8.9 shows these observations graphically. As ϵ approaches zero, the value of $f'(t)$ between $\pm\epsilon$ approaches infinity. At the same time, the duration of this large value is approaching zero. Furthermore, the area under $f'(t)$ between $\pm\epsilon$ remains constant as $\epsilon \to 0$. In this example, the area is unity. As ϵ approaches zero, we say that the function between $\pm\epsilon$ approaches a **unit impulse function**, denoted $\delta(t)$. Thus the derivative of $f(t)$ at the origin approaches a unit impulse function as ϵ approaches zero, or

$$f'(0) \to \delta(t) \quad \text{as } \epsilon \to 0.$$

If the area under the impulse function curve is other than unity, the impulse function is denoted $K\delta(t)$, where K is the area. K is often referred to as the **strength** of the impulse function.

To summarize, an impulse function is created from a variable-parameter function whose parameter approaches zero. The variable-parameter function must exhibit the following three characteristics as the parameter approaches zero:

1. The amplitude approaches infinity.
2. The duration of the function approaches zero.
3. The area under the variable-parameter function is constant as the parameter changes.

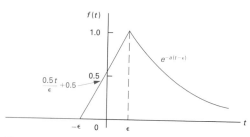

Figure 8.8 A magnified view of the discontinuity in Fig. 8.1(b), assuming a linear transition between $-\epsilon$ and $+\epsilon$.

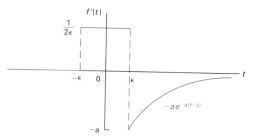

Figure 8.9 The derivative of the function shown in Fig. 8.8.

---

[1] The impulse function is also known as the Dirac delta function.

Many different variable-parameter functions have the afore-mentioned characteristics. In Fig. 8.8, we used a linear function $f(t) = 0.5t/\epsilon + 0.5$. Another example of a variable-parameter function is the exponential function:

$$f(t) = \frac{K}{2\epsilon} e^{-|t|/\epsilon}. \tag{8.7}$$

As ϵ approaches zero, the function becomes infinite at the origin and at the same time decays to zero in an infinitesimal length of time. Figure 8.10 illustrates the character of $f(t)$ as $\epsilon \to 0$. To show that an impulse function is created as $\epsilon \to 0$, we must also show that the area under the function is independent of ϵ. Thus,

$$\text{Area} = \int_{-\infty}^{0} \frac{K}{2\epsilon} e^{t/\epsilon} \, dt + \int_{0}^{\infty} \frac{K}{2\epsilon} e^{-t/\epsilon} \, dt$$

$$= \frac{K}{2\epsilon} \cdot \frac{e^{t/\epsilon}}{1/\epsilon} \Big|_{-\infty}^{0} + \frac{K}{2\epsilon} \cdot \frac{e^{-t/\epsilon}}{-1/\epsilon} \Big|_{0}^{\infty}$$

$$= \frac{K}{2} + \frac{K}{2} = K, \tag{8.8}$$

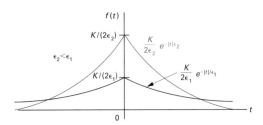

Figure 8.10 A variable-parameter function used to generate an impulse function.

which tells us that the area under the curve is constant and equal to K units. Therefore, as $\epsilon \to 0$, $f(t) \to K\delta(t)$.

Mathematically, the **impulse function** is defined

$$\int_{-\infty}^{\infty} K\delta(t) \, dt = K; \tag{8.9}$$

$$\delta(t) = 0, \quad t \neq 0. \tag{8.10}$$

Equation (8.9) states that the area under the impulse function is constant. This area represents the strength of the impulse. Equation (8.10) states that the impulse is zero everywhere except at $t = 0$. An impulse that occurs at $t = a$ is denoted $K\delta(t - a)$.

The graphic symbol for the impulse function is an arrow. The strength of the impulse is given parenthetically next to the head of the arrow. Figure 8.11 shows the impulses $K\delta(t)$ and $K\delta(t-a)$.

An important property of the impulse function is the **sifting property**, which is expressed as

$$\int_{-\infty}^{\infty} f(t)\delta(t - a) \, dt = f(a), \tag{8.11}$$

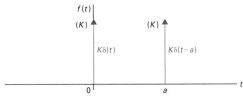

Figure 8.11 A graphic representation of the impulse $K\delta(t)$ and $K\delta(t - a)$.

where the function $f(t)$ is assumed to be continuous at $t = a$; that is, at the location of the impulse. Equation (8.11) shows that the impulse function sifts out everything except the value of $f(t)$ at $t = a$. The validity of Eq. (8.11) follows from noting that $\delta(t - a)$ is zero everywhere except at $t = a$, and hence the integral can be written

$$I = \int_{-\infty}^{\infty} f(t)\delta(t - a)\, dt = \int_{a-\epsilon}^{a+\epsilon} f(t)\delta(t - a)\, dt. \qquad (8.12)$$

But because $f(t)$ is continuous at a, it takes on the value $f(a)$ as $t \to a$, so

$$I = \int_{a-\epsilon}^{a+\epsilon} f(a)\delta(t-a)\, dt = f(a) \int_{a-\epsilon}^{a+\epsilon} \delta(t-a)\, dt = f(a). \quad (8.13)$$

We use the sifting property of the impulse function to find its Laplace transform:

$$\mathcal{L}\{\delta(t)\} = \int_{0^-}^{\infty} \delta(t)e^{-st}\, dt = \int_{0^-}^{\infty} \delta(t)\, dt = 1, \qquad (8.14)$$

which is an important Laplace transform pair that we make good use of in circuit analysis.

We can also define the derivatives of the impulse function and the Laplace transform of these derivatives. We discuss the first derivative, along with its transform and then state the result for the higher-order derivatives.

The function illustrated in Fig. 8.12(a) generates an impulse function as $\epsilon \to 0$. Figure 8.12(b) shows the derivative of this impulse-generating function, which is defined as the derivative of the impulse $[\delta'(t)]$ as $\epsilon \to 0$. The derivative of the impulse function sometimes is referred to as a moment function, or unit doublet.

To find the Laplace transform of $\delta'(t)$, we simply apply the defining integral to the function shown in Fig. 8.12(b) and, after integrating, let $\epsilon \to 0$. Then

$$\mathcal{L}\{\delta'(t)\} = \lim_{\epsilon \to 0} \left[\int_{-\epsilon}^{0^-} \frac{1}{\epsilon^2}e^{-st}\, dt + \int_{0^+}^{\epsilon} \left(-\frac{1}{\epsilon^2}\right)e^{-st}\, dt \right]$$

$$= \lim_{\epsilon \to 0} \frac{e^{s\epsilon} + e^{-s\epsilon} - 2}{s\epsilon^2}$$

$$= \lim_{\epsilon \to 0} \frac{se^{s\epsilon} - se^{-s\epsilon}}{2\epsilon s}$$

$$= \lim_{\epsilon \to 0} \frac{s^2 e^{s\epsilon} + s^2 e^{-s\epsilon}}{2s}$$

$$= s. \qquad (8.15)$$

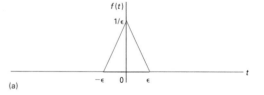

(a)

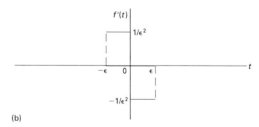

(b)

Figure 8.12 The first derivative of the impulse function. (a) The impulse-generating function used to define the first derivative of the impulse. (b) The first derivative of the impulse-generating function that approaches $\delta'(t)$ as $\epsilon \to 0$.

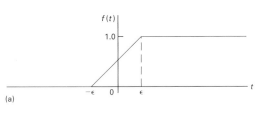

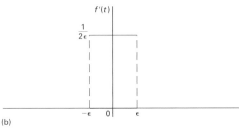

Figure 8.13 The impulse function as the derivative of the step function: (a) $f(t) \to u(t)$ as $\epsilon \to 0$; and (b) $f'(t) \to \delta(t)$ as $\epsilon \to 0$.

In deriving Eq. (8.15), we had to use l'Hôpital's rule twice to evaluate the indeterminate form 0/0.

Higher-order derivatives may be generated in a manner similar to that used to generate the first derivative (see Problem 8.6), and the defining integral may then be used to find its Laplace transform. For the nth derivative of the impulse function, we find that its Laplace transform simply is s^n; that is,

$$\mathcal{L}\{\delta^{(n)}(t)\} = s^n. \tag{8.16}$$

Finally, an impulse function can be thought of as a derivative of a step function; that is,

$$\delta(t) = \frac{du(t)}{dt}. \tag{8.17}$$

Figure 8.13 presents the graphic interpretation of Eq. (8.17). The function shown in Fig. 8.13(a) approaches a unit step function as $\epsilon \to 0$. The function shown in Fig. 8.13(b)—the derivative of the function in Fig. 8.13(a)—approaches a unit impulse as $\epsilon \to 0$.

The impulse function is an extremely useful concept in circuit analysis, and we say more about it in the following chapters. We introduced the concept here so that we can include discontinuities at the origin in our definition of the Laplace transform.

DRILL EXERCISES

8.2 **(a)** Find the area under the function shown in Fig. 8.12(a).

(b) What is the duration of the function when $\epsilon = 0$?

(c) What is the magnitude of $f(0)$ when $\epsilon = 0$?

ANSWER: (a) 1; (b) 0; (c) ∞.

8.4 Find $f(t)$ if

$$f(t) = \frac{1}{2\pi} \int_{-\infty}^{\infty} F(\omega)e^{jt\omega}\,d\omega,$$

$$F(\omega) = \frac{4 + j\omega}{9 + j\omega}\pi\delta(\omega).$$

ANSWER: 2/9.

8.3 Evaluate the following integrals:

(a) $I = \int_{-1}^{3}(t^3 + 2)[\delta(t) + 8\delta(t - 1)]\,dt$.

(b) $I = \int_{-2}^{2} t^2[\delta(t) + \delta(t + 1.5) + \delta(t - 3)]\,dt$.

ANSWER: (a) 26; (b) 2.25.

8.4 FUNCTIONAL TRANSFORMS

A functional transform is simply the Laplace transform of a specified function of t. Because we are limiting our introduction to the unilateral, or one-sided, Laplace transform, we define all functions to be zero for $t < 0^-$.

We derived one functional transform pair in Section 8.3, where we showed that the Laplace transform of the unit impulse function equals 1; see Eq. (8.14). A second illustration is the unit step function of Fig. 8.13(a), where

$$\mathcal{L}\{u(t)\} = \int_{0^-}^{\infty} f(t)e^{-st}\, dt = \int_{0^+}^{\infty} 1e^{-st}\, dt$$

$$= \frac{e^{-st}}{-s}\bigg|_{0^+}^{\infty} = \frac{1}{s}. \tag{8.18}$$

Equation (8.18) shows that the Laplace transform of the unit step function is $1/s$.

The Laplace transform of the decaying exponential function shown in Fig. 8.14 is

$$\mathcal{L}\{e^{-at}\} = \int_{0^+}^{\infty} e^{-at} e^{-st}\, dt = \int_{0^+}^{\infty} e^{-(a+s)t}\, dt = \frac{1}{s+a}. \tag{8.19}$$

In deriving Eqs. (8.18) and (8.19), we used the fact that integration across the discontinuity at the origin is zero.

A third illustration of finding a functional transform is the sinusoidal function shown in Fig. 8.15. The expression for $f(t)$ for $t > 0^-$ is $\sin \omega t$; hence the Laplace transform is

$$\mathcal{L}\{\sin \omega t\} = \int_{0^-}^{\infty} (\sin \omega t)e^{-st}\, dt$$

$$= \int_{0^-}^{\infty} \left(\frac{e^{j\omega t} - e^{-j\omega t}}{2j}\right) e^{-st}\, dt$$

$$= \int_{0^-}^{\infty} \frac{e^{-(s-j\omega)t} - e^{-(s+j\omega)t}}{2j}\, dt$$

$$= \frac{1}{2j}\left(\frac{1}{s - j\omega} - \frac{1}{s + j\omega}\right)$$

$$= \frac{\omega}{s^2 + \omega^2}. \tag{8.20}$$

Table 8.1 gives an abbreviated list of Laplace transform pairs. It includes the functions of most interest in an introductory course on circuit applications.

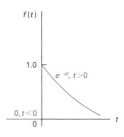

Figure 8.14 A decaying exponential function.

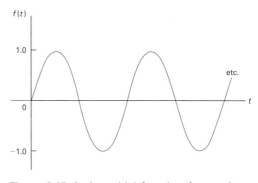

Figure 8.15 A sinusoidal function for $t > 0$.

TABLE 8.1   An Abbreviated List of Laplace Transform Pairs

TYPE	$f(t)(t > 0^-)$	$F(s)$
(impulse)	$\delta(t)$	1
(step)	$u(t)$	$\dfrac{1}{s}$
(ramp)	t	$\dfrac{1}{s^2}$
(exponential)	e^{-at}	$\dfrac{1}{s+a}$
(sine)	$\sin \omega t$	$\dfrac{\omega}{s^2+\omega^2}$
(cosine)	$\cos \omega t$	$\dfrac{s}{s^2+\omega^2}$
(damped ramp)	te^{-at}	$\dfrac{1}{(s+a)^2}$
(damped sine)	$e^{-at}\sin \omega t$	$\dfrac{\omega}{(s+a)^2+\omega^2}$
(damped cosine)	$e^{-at}\cos \omega t$	$\dfrac{s+a}{(s+a)^2+\omega^2}$

DRILL EXERCISE

8.5 Use the defining integral to

   (a) prove that the Laplace transform of t is $1/s^2$

   (b) find the Laplace transform of $\sinh \beta t$

ANSWER: (a) derivation; (b) $\beta/(s^2 - \beta^2)$.

8.5 OPERATIONAL TRANSFORMS

Operational transforms indicate how mathematical operations performed on either $f(t)$ or $F(s)$ are converted into the opposite domain. The operations of primary interest are (1) multiplication by a constant; (2) addition (subtraction); (3) differentiation; (4) integration; (5) translation in the time domain; (6) translation in the frequency domain; and (7) scale changing.

Multiplication by a Constant

From the defining integral, if

$$\mathcal{L}\{f(t)\} = F(s),$$

then

$$\mathcal{L}\{Kf(t)\} = KF(s). \qquad (8.21)$$

Thus, multiplication of $f(t)$ by a constant corresponds to multiplying $F(s)$ by the same constant.

Addition (Subtraction)

Addition (subtraction) in the time domain translates into addition (subtraction) in the frequency domain. Thus if

$$\mathcal{L}\{f_1(t)\} = F_1(s),$$

$$\mathcal{L}\{f_2(t)\} = F_2(s),$$

$$\mathcal{L}\{f_3(t)\} = F_3(s),$$

then

$$\mathcal{L}\{f_1(t) + f_2(t) - f_3(t)\} = F_1(s) + F_2(s) - F_3(s), \qquad (8.22)$$

which is derived by simply substituting the algebraic sum of time-domain functions into the defining integral.

Differentiation

Differentiation in the time domain corresponds to multiplying $F(s)$ by s and then subtracting the initial value of $f(t)$—that is, $f(0^-)$—from this product:

$$\mathcal{L}\left\{\frac{df(t)}{dt}\right\} = sF(s) - f(0^-), \qquad (8.23)$$

which is obtained directly from the definition of the Laplace transform, or

$$\mathcal{L}\left\{\frac{df(t)}{dt}\right\} = \int_{0^-}^{\infty} \left[\frac{df(t)}{dt}\right] e^{-st}\, dt. \qquad (8.24)$$

We evaluate the integral in Eq. (8.24) by integrating by parts. Letting $u = e^{-st}$ and $dv = [df(t)/dt]\, dt$ yields

$$\mathcal{L}\left\{\frac{df(t)}{dt}\right\} = e^{-st} f(t)\Big|_{0^-}^{\infty} - \int_{0^-}^{\infty} f(t)(-se^{-st}\, dt). \qquad (8.25)$$

Because we are assuming that $f(t)$ is Laplace transformable, the evaluation of $e^{-st} f(t)$ at $t = \infty$ is zero. Therefore the right-hand side of Eq. (8.25) reduces to

$$-f(0^-) + s \int_{0^-}^{\infty} f(t)e^{-st} \, dt = sF(s) - f(0^-).$$

This observation completes the derivation of Eq. (8.23). It is an important result because it states that differentiation in the time domain reduces to an algebraic operation in the s domain.

We determine the Laplace transform of higher-order derivatives by using Eq. (8.23) as the starting point. For example, to find the Laplace transform of the second derivative of $f(t)$, we first let

$$g(t) = \frac{df(t)}{dt}. \tag{8.26}$$

Now we use Eq. (8.23) to write

$$G(s) = sF(s) - f(0^-). \tag{8.27}$$

But because

$$\frac{dg(t)}{dt} = \frac{d^2 f(t)}{dt^2},$$

we write

$$\mathcal{L}\left\{\frac{dg(t)}{dt}\right\} = \mathcal{L}\left\{\frac{d^2 f(t)}{dt^2}\right\} = sG(s) - g(0^-). \tag{8.28}$$

Combining Eqs. (8.26), (8.27), and (8.28) gives

$$\mathcal{L}\left\{\frac{d^2 f(t)}{dt^2}\right\} = s^2 F(s) - sf(0^-) - \frac{df(0^-)}{dt}. \tag{8.29}$$

We find the Laplace transform of the nth derivative by successively applying the preceding process, which leads to the general result

$$\mathcal{L}\left\{\frac{d^n f(t)}{dt^n}\right\} = s^n F(s) - s^{n-1} f(0^-) - s^{n-2} \frac{df(0^-)}{dt}$$
$$- s^{n-3} \frac{d^2 f(0^-)}{dt^2} - \cdots - \frac{d^{n-1} f(0^-)}{dt^{n-1}}. \tag{8.30}$$

Integration

Integration in the time domain corresponds to dividing by s in the s domain. As before, we establish the relationship by the defining integral:

$$\mathcal{L}\left\{\int_{0^-}^{t} f(x)\,dx\right\} = \int_{0^-}^{\infty}\left[\int_{0^-}^{t} f(x)\,dx\right]e^{-st}\,dt. \qquad (8.31)$$

We evaluate the integral on the right-hand side of Eq. (8.31) by integrating by parts, first letting

$$u = \int_{0^-}^{t} f(x)\,dx,$$

$$dv = e^{-st}\,dt.$$

Then

$$du = f(t)\,dt$$

$$v = -\frac{e^{-st}}{s}.$$

The integration-by-parts formula yields

$$\mathcal{L}\left\{\int_{0^-}^{t} f(x)\,dx\right\} = -\frac{e^{-st}}{s}\int_{0^-}^{t} f(x)\,dx\,\Bigg|_{0^-}^{\infty} + \int_{0^-}^{\infty}\frac{e^{-st}}{s}f(t)\,dt.$$
$$(8.32)$$

The first term on the right-hand side of Eq. (8.32) is zero at both the upper and lower limits. The evaluation at the lower limit obviously is zero, whereas the evaluation at the upper limit is zero because we are assuming that $f(t)$ has a Laplace transform. The second term on the right-hand side of Eq. (8.32) is $F(s)/s$; therefore

$$\mathcal{L}\left\{\int_{0^-}^{t} f(x)\,dx\right\} = \frac{F(s)}{s}, \qquad (8.33)$$

which reveals that the operation of integration in the time domain is transformed to the algebraic operation of multiplying by $1/s$ in the s domain. Equations (8.33) and (8.30) form the basis of the earlier statement that the Laplace transform translates a set of integrodifferential equations into a set of algebraic equations.

Translation in the Time Domain

If we start with any function $f(t)u(t)$, we can represent the same function, translated in time by the constant a, as $f(t-a)u(t-a)$.[2] Translation in the time domain corresponds to multiplication by an exponential in the frequency domain. Thus

$$\mathcal{L}\{f(t-a)u(t-a)\} = e^{-as}F(s), \quad a > 0. \tag{8.34}$$

For example, when we know that

$$\mathcal{L}\{tu(t)\} = \frac{1}{s^2},$$

Eq. (8.34) lets us write the Laplace transform of $(t-a)u(t-a)$ directly:

$$\mathcal{L}\{(t-a)u(t-a)\} = \frac{e^{-as}}{s^2}.$$

The proof of Eq. (8.34) follows from the defining integral:

$$\mathcal{L}\{f(t-a)u(t-a)\} = \int_{0^-}^{\infty} u(t-a)f(t-a)e^{-st}\,dt$$

$$= \int_{a}^{\infty} f(t-a)e^{-st}\,dt. \tag{8.35}$$

In writing Eq. (8.35), we took advantage of $u(t-a) = 1$ for $t > a$. Now we change the variable of integration. Specifically, we let $x = t - a$. Then $x = 0$ when $t = a$, $x = \infty$ when $t = \infty$, and $dx = dt$. Thus we write the integral in Eq. (8.35) as

$$\mathcal{L}\{f(t-a)u(t-a)\} = \int_{0}^{\infty} f(x)e^{-s(x+a)}\,dx$$

$$= e^{-sa}\int_{0}^{\infty} f(x)e^{-sx}\,dx$$

$$= e^{-as}F(s),$$

which is what we set out to prove.

---

[2] Note that throughout we multiply any arbitrary function $f(t)$ by the unit step function $u(t)$ to ensure that the resulting function is defined for all positive time.

Translation in the Frequency Domain

Translation in the frequency domain corresponds to multiplication by an exponential in the time domain:

$$\mathcal{L}\{e^{-at} f(t)\} = F(s + a), \qquad (8.36)$$

which follows from the defining integral. The derivation of Eq. (8.36) is left to Problem 8.15.

We may use the relationship in Eq. (8.36) to derive new transform pairs. Thus, knowing that

$$\mathcal{L}\{\cos \omega t\} = \frac{s}{s^2 + \omega^2},$$

we use Eq. (8.36) to deduce that

$$\mathcal{L}\{e^{-at} \cos \omega t\} = \frac{s + a}{(s + a)^2 + \omega^2}.$$

Scale Changing

The scale-change property gives the relationship between $f(t)$ and $F(s)$ when the time variable is multiplied by a positive constant:

$$\mathcal{L}\{f(at)\} = \frac{1}{a} F\left(\frac{s}{a}\right), \qquad a > 0, \qquad (8.37)$$

the derivation of which is left to Problem 8.19. The scale-change property is particularly useful in experimental work, especially where time-scale changes are made to facilitate building a model of a system.

We use Eq. (8.37) to formulate new transform pairs. Thus, knowing that

$$\mathcal{L}\{\cos t\} = \frac{s}{s^2 + 1},$$

we deduce from Eq. (8.37) that

$$\mathcal{L}\{\cos \omega t\} = \frac{1}{\omega} \frac{s/\omega}{(s/\omega)^2 + 1} = \frac{s}{s^2 + \omega^2}.$$

Table 8.2 gives an abbreviated list of operational transforms. Some entries were not discussed in this section, but you will become more familiar with them by working Problems 8.20 and 8.21.

TABLE 8.2 **An Abbreviated List of Operational Transforms.**

OPERATION	$f(t)$	$F(s)$
Multiplication by a constant	$Kf(t)$	$KF(s)$
Addition/subtraction	$f_1(t) + f_2(t) - f_3(t) + \cdots$	$F_1(s) + F_2(s) - F_3(s) + \cdots$
First derivative (time)	$\dfrac{df(t)}{dt}$	$sF(s) - f(0^-)$
Second derivative (time)	$\dfrac{d^2 f(t)}{dt^2}$	$s^2 F(s) - sf(0^-) - \dfrac{df(0^-)}{dt}$
nth derivative (time)	$\dfrac{d^n f(t)}{dt^n}$	$s^n F(s) - s^{n-1} f(0^-) - s^{n-2}\dfrac{df(0^-)}{dt}$ $- s^{n-3}\dfrac{df^2(0^-)}{dt^2} - \cdots - \dfrac{d^{n-1} f(0^-)}{dt^{n-1}}$
Time integral	$\displaystyle\int_0^t f(x)\, dx$	$\dfrac{F(s)}{s}$
Translation in time	$f(t - a)u(t - a), a > 0$	$e^{-as} F(s)$
Translation in frequency	$e^{-at} f(t)$	$F(s + a)$
Scale changing	$f(at), a > 0$	$\dfrac{1}{a} F\left(\dfrac{s}{a}\right)$
First derivative (s)	$t f(t)$	$-\dfrac{dF(s)}{ds}$
nth derivative (s)	$t^n f(t)$	$(-1)^n \dfrac{d^n F(s)}{ds^n}$
s integral	$\dfrac{f(t)}{t}$	$\displaystyle\int_s^\infty F(u)\, du$

DRILL EXERCISE

8.6 Use the appropriate operational transform from Table 8.2 to find the Laplace transform of each function: (a) $t^2 e^{-at}$; (b) $\frac{d}{dt}(e^{-at} \sinh \beta t u(t))$; (c) $t \cos \omega t$.

ANSWER: (a) $\dfrac{2}{(s+a)^3}$; (b) $\dfrac{\beta s}{(s+a)^2 - \beta^2}$;

(c) $\dfrac{s^2 - \omega^2}{(s^2 + \omega^2)^2}$.

8.6 APPLYING THE LAPLACE TRANSFORM

We now illustrate how to use the Laplace transform to solve the ordinary integrodifferential equations that describe the behavior of lumped-parameter circuits. Consider the circuit shown in

Fig. 8.16. We assume that no initial energy is stored in the circuit at the instant when the switch, which is shorting the dc current source, is opened. The problem is to find the time-domain expression for $v(t)$ when $t \geq 0$.

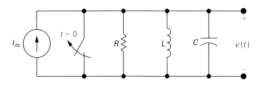

Figure 8.16 A parallel RLC circuit.

We begin by writing the integrodifferential equation that $v(t)$ must satisfy. We need only a single node-voltage equation to describe the circuit. Summing the currents away from the top node in the circuit generates the equation:

$$\frac{v(t)}{R} + \frac{1}{L} \int_0^t v(x)\,dx + C\frac{dv(t)}{dt} = I_{dc}u(t). \qquad (8.38)$$

Note that in writing Eq. (8.38), we indicated the opening of the switch in the step jump of the source current from zero to I_{dc}.

After deriving the integrodifferential equations (in this example, just one), we transform the equations to the s domain. We will not go through the steps of the transformation in detail, because in Chapter 9 we will discover how to bypass them and generate the s-domain equations directly. Briefly though, we use three operational transforms and one functional transform on Eq. (8.38) to obtain

$$\frac{V(s)}{R} + \frac{1}{L}\frac{V(s)}{s} + C[sV(s) - v(0^-)] = I_{dc}\left(\frac{1}{s}\right), \qquad (8.39)$$

an algebraic equation in which $V(s)$ is the unknown variable. We are assuming that the circuit parameters R, L, and C, as well as the source current I_{dc}, are known; the initial voltage on the capacitor $v(0^-)$ is zero because the initial energy stored in the circuit is zero. Thus we have reduced the problem to solving an algebraic equation.

Next we solve the algebraic equations (again, just one in this case) for the unknowns. Solving Eq. (8.39) for $V(s)$ gives

$$V(s)\left(\frac{1}{R} + \frac{1}{sL} + sC\right) = \frac{I_{dc}}{s},$$

$$V(s) = \frac{I_{dc}/C}{s^2 + (1/RC)s + (1/LC)}. \qquad (8.40)$$

To find $v(t)$ we must inverse-transform the expression for $V(s)$. We denote this inverse operation

$$v(t) = \mathcal{L}^{-1}\{V(s)\}. \qquad (8.41)$$

The next step in the analysis is to find the inverse transform of the s-domain expression; this is the subject of Section 8.7. In that section we also present a final, critical step: checking the validity of the resulting time-domain expression. The need for such checking is not unique to the Laplace transform; conscientious and prudent engineers always test any derived solution to be sure it makes sense in terms of known system behavior.

Simplifying the notation now is advantageous. We do so by dropping the parenthetical t in time-domain expressions and the parenthetical s in frequency-domain expressions. We use lowercase letters for all time-domain variables, and we represent the corresponding s-domain variables with uppercase letters. Thus

$$\mathcal{L}\{v\} = V \quad \text{or} \quad v = \mathcal{L}^{-1}\{V\},$$

$$\mathcal{L}\{i\} = I \quad \text{or} \quad i = \mathcal{L}^{-1}\{I\},$$

$$\mathcal{L}\{f\} = F \quad \text{or} \quad f = \mathcal{L}^{-1}\{F\},$$

and so on.

DRILL EXERCISE

8.7 In the circuit shown in Fig. 8.16, the dc current source is replaced with a sinusoidal source that delivers a current of $5 \cos 10t$ A. The circuit components are $R = 1 \ \Omega$, $C = 25$ mF, and $L = 625$ mH. Find the numerical expression for V.

ANSWER: $V = \dfrac{200s^2}{(s^2 + 40s + 64)(s^2 + 100)}.$

8.7 INVERSE TRANSFORMS

The expression for $V(s)$ in Eq. (8.40) is a **rational** function of s—that is, one that can be expressed in the form of a ratio of two polynomials in s such that no nonintegral powers of s appear in the polynomials. In fact, for linear, lumped-parameter circuits whose component values are constant, the s-domain expressions for the unknown voltages and currents are always rational functions of s. (You may verify this observation by working Problems 8.22, 8.24, 8.30, and 8.32.) If we can inverse-transform rational functions of s, we can solve for the time-domain expressions for the voltages and currents. The purpose of this section is to present a straightforward and systematic technique for finding the inverse transform of a rational function.

In general, we need to find the inverse transform of a function that has the form

$$F(s) = \frac{N(s)}{D(s)} = \frac{a_n s^n + a_{n-1} s^{n-1} + \cdots + a_1 s + a_0}{b_m s^m + b_{m-1} s^{m-1} + \cdots + b_1 s + b_0}. \quad (8.42)$$

The coefficients a and b are real constants, and the exponents m and n are positive integers. The ratio $N(s)/D(s)$ is called a **proper rational function** if $m > n$, and an **improper rational function** if $m \leq n$. Only a proper rational function can be expanded as a sum of partial fractions. This restriction poses no problem, as we show at the end of this section.

Partial Fraction Expansion: Proper Rational Functions

A proper rational function is expanded into a sum of partial fractions by writing a term or a series of terms for each root of $D(s)$. Thus $D(s)$ must be in factored form before we can make a partial fraction expansion. For each distinct root of $D(s)$, a single term appears in the sum of partial fractions. For each multiple root of $D(s)$ of multiplicity r, the expansion contains r terms. For example, in the rational function

$$\frac{s+6}{s(s+3)(s+1)^2},$$

the denominator has four roots. Two of these roots are distinct—namely, at $s = 0$ and $s = -3$. A multiple root of multiplicity 2 occurs at $s = -1$. Thus the partial fraction expansion of this function takes the form

$$\frac{s+6}{s(s+3)(s+1)^2} \equiv \frac{K_1}{s} + \frac{K_2}{s+3} + \frac{K_3}{(s+1)^2} + \frac{K_4}{s+1}. \quad (8.43)$$

The key to the partial fraction technique for finding inverse transforms lies in recognizing the $f(t)$ corresponding to each term in the sum of partial fractions. From Table 8.1 you should be able to verify that

$$\mathcal{L}^{-1} \left\{ \frac{s+6}{s(s+3)(s+1)^2} \right\}$$

$$= (K_1 + K_2 e^{-3t} + K_3 t e^{-t} + K_4 e^{-t}) u(t). \quad (8.44)$$

All that remains is to establish a technique for determining the coefficients ($K_1, K_2, K_3, \ldots$) generated by making a partial fraction expansion. There are four general forms this problem can take. Specifically, the roots of $D(s)$ are either (1) real and distinct; (2) complex and distinct; (3) real and repeated; or (4) complex and repeated. Before we consider each situation in turn, a few general comments are in order.

We used the identity sign $\equiv$ in Eq. (8.43) to emphasize that expanding a rational function into a sum of partial fractions establishes an identical equation. Thus both sides of the equation must be the same for all values of the variable s. Also, the identity relationship must hold when both sides are subjected to the same mathematical operation. These characteristics are pertinent to determining the coefficients, as we will see.

Be sure to verify that the rational function is proper. This check is important, because nothing in the procedure for finding the various K's will alert you to nonsense results if the rational function is improper. We present a procedure for checking the Ks, but you can avoid wasted effort by forming the habit of asking yourself, "Is $F(s)$ a proper rational function?"

Partial Fraction Expansion: Distinct Real Roots of D(s)

We first consider determining the coefficients in a partial fraction expansion when all the roots of $D(s)$ are real and distinct. To find a K associated with a term that arises because of a distinct root of $D(s)$, we multiply both sides of the identity by a factor equal to the denominator beneath the desired K. Then, when we evaluate both sides of the identity at the root corresponding to the multiplying factor, the right-hand side is always the desired K, and the left-hand side is always its numerical value. For example,

$$F(s) = \frac{96(s+5)(s+12)}{s(s+8)(s+6)} \equiv \frac{K_1}{s} + \frac{K_2}{s+8} + \frac{K_3}{s+6}. \quad (8.45)$$

To find the value of K_1, we multiply both sides by s and then evaluate both sides at $s = 0$:

$$\left.\frac{96(s+5)(s+12)}{(s+8)(s+6)}\right|_{s=0} \equiv K_1 + \left.\frac{K_2 s}{s+8}\right|_{s=0} + \left.\frac{K_3 s}{s+6}\right|_{s=0},$$

or

$$\frac{96(5)(12)}{8(6)} \equiv K_1 = 120. \quad (8.46)$$

To find the value of K_2, we multiply both sides by $s + 8$ and then evaluate both sides at $s = -8$:

$$\left.\frac{96(s + 5)(s + 12)}{s(s + 6)}\right|_{s=-8}$$

$$\equiv \left.\frac{K_1(s + 8)}{s}\right|_{s=-8} + K_2 + \left.\frac{K_3(s + 8)}{(s + 6)}\right|_{s=-8},$$

or

$$\frac{96(-3)(4)}{(-8)(-2)} = K_2 = -72. \qquad (8.47)$$

Then K_3 is

$$\left.\frac{96(s + 5)(s + 12)}{s(s + 8)}\right|_{s=-6} = K_3 = 48. \qquad (8.48)$$

From Eq. (8.45) and the K values obtained,

$$\frac{96(s + 5)(s + 12)}{s(s + 8)(s + 6)} \equiv \frac{120}{s} + \frac{48}{s + 6} - \frac{72}{s + 8}. \qquad (8.49)$$

At this point, testing the result to protect against computational errors is a good idea. As we already mentioned, a partial fraction expansion creates an identity; thus both sides of Eq. (8.49) must be the same for all s values. The choice of test values is completely open; hence we choose values that are easy to verify. For example, in Eq. (8.49), testing at either -5 or -12 is attractive because in both cases the left-hand side reduces to zero. Choosing -5 yields

$$\frac{120}{-5} + \frac{48}{1} - \frac{72}{3} = -24 + 48 - 24 = 0,$$

whereas testing -12 gives

$$\frac{120}{-12} + \frac{48}{-6} - \frac{72}{-4} = -10 - 8 + 18 = 0.$$

Confident now that the numerical values of the various Ks are correct, we proceed to find the inverse transform:

$$\mathcal{L}^{-1}\left\{\frac{96(s + 5)(s + 12)}{s(s + 8)(s + 6)}\right\} = (120 + 48e^{-6t} - 72e^{-8t})u(t). \quad (8.50)$$

DRILL EXERCISES

8.8 Find $f(t)$ if

$$F(s) = \frac{6s^2 + 26s + 26}{(s+1)(s+2)(s+3)}.$$

ANSWER: $f(t) = (3e^{-t} + 2e^{-2t} + e^{-3t})u(t)$.

8.9 Find $f(t)$ if

$$F(s) = \frac{7s^2 + 63s + 134}{(s+3)(s+4)(s+5)}.$$

ANSWER: $f(t) = (4e^{-3t} + 6e^{-4t} - 3e^{-5t})u(t)$.

Partial Fraction Expansion: Distinct Complex Roots of D(s)

The only difference between finding the coefficients associated with distinct complex roots and finding those associated with distinct real roots is that the algebra in the former involves complex numbers. We illustrate by expanding the rational function:

$$F(s) = \frac{100(s+3)}{(s+6)(s^2 + 6s + 25)}. \tag{8.51}$$

We begin by noting that $F(s)$ is a proper rational function. Next we must find the roots of the quadratic term $s^2 + 6s + 25$:

$$s^2 + 6s + 25 = (s + 3 - j4)(s + 3 + j4). \tag{8.52}$$

With the denominator in factored form, we proceed as before:

$$\frac{100(s+3)}{(s+6)(s^2 + 6s + 25)}$$

$$\equiv \frac{K_1}{s+6} + \frac{K_2}{s+3-j4} + \frac{K_3}{s+3+j4}. \tag{8.53}$$

To find K_1, K_2, and K_3, we use the same process as before:

$$K_1 = \frac{100(s+3)}{s^2 + 6s + 25}\bigg|_{s=-6} = \frac{100(-3)}{25} = -12, \tag{8.54}$$

$$K_2 = \frac{100(s+3)}{(s+6)(s+3+j4)}\bigg|_{s=-3+j4} = \frac{100(j4)}{(3+j4)(j8)}$$

$$= 6 - j8 = 10 \angle{-53.13°}, \tag{8.55}$$

$$K_3 = \frac{100(s+3)}{(s+6)(s+3-j4)}\bigg|_{s=-3-j4} = \frac{100(-j4)}{(3-j4)(-j8)}$$

$$= 6 + j8 = 10 \angle{53.13°}. \tag{8.56}$$

Then

$$\frac{100(s+3)}{(s+6)(s^2+6s+25)} = \frac{-12}{s+6} + \frac{10\ \angle-53.13°}{s+3-j4}$$

$$+ \frac{10\ \angle 53.13°}{s+3+j4}. \qquad (8.57)$$

Again, we need to make some observations. First, in physically realizable circuits, complex roots always appear in conjugate pairs. Second, the coefficients associated with these conjugate pairs are themselves conjugates. Note, for example, that K_3 (Eq. [8.56]) is the conjugate of K_2 (Eq. [8.55]). Thus for complex conjugate roots, you actually need to calculate only one of the two coefficients.

Before inverse-transforming Eq. (8.57), we check the partial fraction expansion numerically. Testing at -3 is attractive because the left-hand side reduces to zero at this value:

$$F(s) = \frac{-12}{3} + \frac{10\ \angle-53.13°}{-j4} + \frac{10\ \angle 53.13°}{j4}$$

$$= -4 + 2.5\ \angle 36.87° + 2.5\ \angle-36.87°$$

$$= -4 + 2.0 + j1.5 + 2.0 - j1.5 = 0.$$

We now proceed to inverse-transform Eq. (8.57):

$$\mathcal{L}^{-1}\left\{\frac{100(s+3)}{(s+6)(s^2+6s+25)}\right\} = (-12e^{-6t} + 10e^{-j53.13°}e^{-(3-j4)t}$$

$$+ 10e^{j53.13°}e^{-(3+j4)t})u(t). \,(8.58)$$

In general, having the function in the time domain contain imaginary components is undesirable. Fortunately, because the terms involving imaginary components always come in conjugate pairs, we can eliminate the imaginary components simply by adding the pairs:

$$10e^{-j53.13°}e^{-(3-j4)t} + 10e^{j53.13°}e^{-(3+j4)t}$$

$$= 10e^{-3t}\left(e^{j(4t-53.13°)} + e^{-j(4t-53.13°)}\right)$$

$$= 20e^{-3t}\cos(4t - 53.13°), \qquad (8.59)$$

which enables us to simplify Eq. (8.58):

$$\mathcal{L}^{-1}\left\{\frac{100(s+3)}{(s+6)(s^2+6s+25)}\right\}$$

$$= [-12e^{-6t} + 20e^{-3t}\cos(4t - 53.13°)]u(t). \qquad (8.60)$$

Because distinct complex roots appear frequently in lumped-parameter linear circuit analysis, we need to summarize these results with a new transform pair. Whenever $D(s)$ contains distinct complex roots—that is, factors of the form $(s + \alpha - j\beta)(s + \alpha + j\beta)$—a pair of terms of the form

$$\frac{K}{s + \alpha - j\beta} + \frac{K^*}{s + \alpha + j\beta} \tag{8.61}$$

appears in the partial fraction expansion, where the partial fraction coefficient is, in general, a complex number. In polar form,

$$K = |K|e^{j\theta} = |K| \; \underline{/\theta°}, \tag{8.62}$$

where $|K|$ denotes the magnitude of the complex coefficient. Then

$$K^* = |K|e^{-j\theta} = |K| \; \underline{/-\theta°}. \tag{8.63}$$

The complex conjugate pair in Eq. (8.61) always inverse-transforms as

$$\mathcal{L}^{-1}\left\{ \frac{K}{s + \alpha - j\beta} + \frac{K^*}{s + \alpha + j\beta} \right\}$$
$$= 2|K|e^{-\alpha t}\cos(\beta t + \theta). \tag{8.64}$$

In applying Eq. (8.64) it is important to note that K is defined as the coefficient associated with the denominator term $s + \alpha - j\beta$, and K^* is defined as the coefficient associated with the denominator $s + \alpha + j\beta$.

DRILL EXERCISES

8.10 Find $f(t)$ if

$$F(s) = 10(s^2 + 119)/[(s + 5)(s^2 + 10s + 169)].$$

ANSWER: $f(t) = (10e^{-5t} - 8.33e^{-5t}\sin 12t)u(t)$.

8.11 Find $v(t)$ in Drill Exercise 8.7.

ANSWER: $v(t) = [4.98\cos(10t - 5.14°) + 0.15e^{-1.67t} - 5.11e^{-38.33t}]u(t)$ V.

Partial Fraction Expansion: Repeated Real Roots of D(s)

To find the coefficients associated with the terms generated by a multiple root of multiplicity r, we multiply both sides of the identity by the multiple root raised to its rth power. We find the K appearing over the factor raised to the rth power by evaluating both sides of the identity at the multiple root. To find the remaining $(r-1)$ coefficients, we differentiate both sides of the identity $(r-1)$ times. At the end of each differentiation, we evaluate both sides of the identity at the multiple root. The right-hand side is always the desired K, and the left-hand side is always its numerical value. For example,

$$\frac{100(s+25)}{s(s+5)^3} = \frac{K_1}{s} + \frac{K_2}{(s+5)^3} + \frac{K_3}{(s+5)^2} + \frac{K_4}{s+5}. \quad (8.65)$$

We find K_1 as previously described; that is,

$$K_1 = \left. \frac{100(s+25)}{(s+5)^3} \right|_{s=0} = \frac{100(25)}{125} = 20. \quad (8.66)$$

To find K_2, we multiply both sides by $(s+5)^3$ and then evaluate both sides at -5:

$$\left. \frac{100(s+25)}{s} \right|_{s=-5} = \left. \frac{K_1(s+5)^3}{s} \right|_{s=-5} + K_2 + K_3(s+5)|_{s=-5}$$

$$+ \left. K_4(s+5)^2 \right|_{s=-5}, \quad (8.67)$$

$$\frac{100(20)}{(-5)} = K_1 \times 0 + K_2 + K_3 \times 0 + K_4 \times 0$$

$$= K_2 = -400. \quad (8.68)$$

To find K_3 we first must multiply both sides of Eq. (8.65) by $(s+5)^3$. Next we differentiate both sides once with respect to s and then evaluate at $s = -5$:

$$\left. \frac{d}{ds}\left[\frac{100(s+25)}{s} \right] \right|_{s=-5} = \left. \frac{d}{ds}\left[\frac{K_1(s+5)^3}{s} \right] \right|_{s=-5}$$

$$+ \frac{d}{ds}[K_2]_{s=-5}$$

$$+ \frac{d}{ds}[K_3(s+5)]_{s=-5}$$

$$+ \frac{d}{ds}[K_4(s+5)^2]_{s=-5}, \quad (8.69)$$

$$100\left[\frac{s-(s+25)}{s^2} \right]_{s=-5} = K_3 = -100. \quad (8.70)$$

To find K_4 we first multiply both sides of Eq. (8.65) by $(s + 5)^3$. Next we differentiate both sides twice with respect to s and then evaluate both sides at $s = -5$. After simplifying the first derivative, the second derivative becomes

$$100\frac{d}{ds}\left[-\frac{25}{s^2}\right]_{s=-5} = K_1\frac{d}{ds}\left[\frac{(s+5)^2(2s-5)}{s^2}\right]_{s=-5}$$

$$+ 0 + \frac{d}{ds}[K_3]_{s=-5} + \frac{d}{ds}[2K_4(s+5)]_{s=-5},$$

or

$$-40 = 2K_4. \tag{8.71}$$

Solving Eq. (8.71) for K_4 gives

$$K_4 = -20. \tag{8.72}$$

Then

$$\frac{100(s+25)}{s(s+5)^3} = \frac{20}{s} - \frac{400}{(s+5)^3} - \frac{100}{(s+5)^2} - \frac{20}{s+5}. \tag{8.73}$$

At this point we can check our expansion by testing both sides of Eq. (8.73) at $s = -25$. Noting that both sides of Eq. (8.73) equal zero when $s = -25$ gives us confidence in the correctness of the partial fraction expansion. The inverse transform of Eq. (8.73) yields

$$\mathcal{L}^{-1}\left\{\frac{100(s+25)}{s(s+5)^3}\right\}$$

$$= [20 - 200t^2e^{-5t} - 100te^{-5t} - 20e^{-5t}]u(t). \tag{8.74}$$

DRILL EXERCISE

8.12 Find $f(t)$ if $F(s) = (4s^2 + 7s + 1)/[s(s+1)^2]$.

ANSWER: $f(t) = (1 + 2te^{-t} + 3e^{-t})u(t)$.

Partial Fraction Expansion: Repeated Complex Roots of D(s)

We handle repeated complex roots in the same way that we did repeated real roots; the only difference is that the algebra involves complex numbers. Recall that complex roots always appear in conjugate pairs and that the coefficients associated with a conjugate pair are also conjugates, so that only half the Ks need to be evaluated. For example,

$$F(s) = \frac{768}{(s^2 + 6s + 25)^2}.\qquad(8.75)$$

After factoring the denominator polynomial, we write

$$F(s) = \frac{768}{(s + 3 - j4)^2(s + 3 + j4)^2}$$

$$= \frac{K_1}{(s + 3 - j4)^2} + \frac{K_2}{s + 3 - j4}$$

$$+ \frac{K_1^*}{(s + 3 + j4)^2} + \frac{K_2^*}{s + 3 + j4}.\qquad(8.76)$$

Now we need to evaluate only K_1 and K_2, because K_1^* and K_2^* are conjugate values. The value of K_1 is

$$K_1 = \frac{768}{(s + 3 + j4)^2}\bigg|_{s=-3+j4}$$

$$= \frac{768}{(j8)^2} = -12 = 12\ \underline{/180°}.\qquad(8.77)$$

The value of K_2 is

$$K_2 = \frac{d}{ds}\left[\frac{768}{(s + 3 + j4)^2}\right]_{s=-3+j4}$$

$$= -\frac{2(768)}{(s + 3 + j4)^3}\bigg|_{s=-3+j4}$$

$$= -\frac{2(768)}{(j8)^3}$$

$$= -j3 = 3\ \underline{/-90°}.\qquad(8.78)$$

From Eqs. (8.77) and (8.78),

$$K_1^* = -12 = 12\ \underline{/180°}\qquad(8.79)$$

$$K_2^* = j3 = 3\ \underline{/90°}.\qquad(8.80)$$

We now group the partial fraction expansion by conjugate terms to obtain

$$F(s) = \left[\frac{-12}{(s+3-j4)^2} + \frac{-12}{(s+3+j4)^2} \right]$$
$$+ \left(\frac{3\ \angle-90°}{s+3-j4} + \frac{3\ \angle 90°}{s+3+j4} \right). \qquad (8.81)$$

We now write the inverse transform of $F(s)$:

$$f(t) = [-24te^{-3t}\cos 4t + 6e^{-3t}\cos(4t - 90°)]u(t). \qquad (8.82)$$

Note that if $F(s)$ has a real root a of multiplicity r in its denominator, the term in a partial fraction expansion is of the form

$$\frac{K}{(s+a)^r}.$$

The inverse transform of this term is

$$\mathcal{L}^{-1}\left\{ \frac{K}{(s+a)^r} \right\} = \frac{Kt^{r-1}e^{-at}}{(r-1)!}u(t). \qquad (8.83)$$

If $F(s)$ has a complex root of $\alpha + j\beta$ of multiplicity r in its denominator, the term in partial fraction expansion is the conjugate pair

$$\frac{K}{(s+\alpha-j\beta)^r} + \frac{K^*}{(s+\alpha+j\beta)^r}.$$

The inverse transform of this pair is

$$\mathcal{L}^{-1}\left\{ \frac{K}{(s+\alpha-j\beta)^r} + \frac{K^*}{(s+\alpha+j\beta)^r} \right\}$$
$$= \left[\frac{2|K|t^{r-1}}{(r-1)!}e^{-\alpha t}\cos(\beta t + \theta) \right]u(t). \qquad (8.84)$$

Equations (8.83) and (8.84) are the key to being able to inverse-transform any partial fraction expansion by inspection. One further note regarding these two equations: In most circuit analysis problems, r is seldom greater than 2. Therefore, the inverse transform of a rational function can be handled with four transform pairs. Table 8.3 lists these pairs.

DRILL EXERCISE

8.13 Find $f(t)$ if $F(s) = 40/(s^2 + 4s + 5)^2$.

ANSWER: $f(t) = (-20te^{-2t}\cos t + 20e^{-2t}\sin t)u(t)$.

TABLE 8.3   **Four Useful Transform Pairs**

PAIR NUMBER	NATURE OF ROOTS	$F(S)$	$f(t)$
1	Distinct real	$\dfrac{K}{s+a}$	$Ke^{-at}u(t)$
2	Repeated real	$\dfrac{K}{(s+a)^2}$	$Kte^{-at}u(t)$
3	Distinct complex	$\dfrac{K}{s+\alpha-j\beta}+\dfrac{K^*}{s+\alpha+j\beta}$	$2|K|e^{-\alpha t}\cos(\beta t+\theta)u(t)$
4	Repeated complex	$\dfrac{K}{(s+\alpha-j\beta)^2}+\dfrac{K^*}{(s+\alpha+j\beta)^2}$	$2t|K|e^{-\alpha t}\cos(\beta t+\theta)u(t)$

Note: In pairs 1 and 2, K is a real quantity, whereas in pairs 3 and 4, K is the complex quantity $|K| \; \angle\theta$.

Partial Fraction Expansion: Improper Rational Functions

We conclude the discussion of partial fraction expansions by returning to an observation made at the beginning of this section, namely, that improper rational functions pose no serious problem in finding inverse transforms. An improper rational function can always be expanded into a polynomial plus a proper rational function. The polynomial is then inverse-transformed into impulse functions and derivatives of impulse functions. The proper rational function is inverse-transformed by the techniques outlined in this section. To illustrate the procedure, we use the function

$$F(s) = \frac{s^4 + 13s^3 + 66s^2 + 200s + 300}{s^2 + 9s + 20}. \qquad (8.85)$$

Dividing the denominator into the numerator until the remainder is a proper rational function gives

$$F(s) = s^2 + 4s + 10 + \frac{30s + 100}{s^2 + 9s + 20}, \qquad (8.86)$$

where the term $(30s + 100)/(s^2 + 9s + 20)$ is the remainder.

Next we expand the proper rational function into a sum of partial fractions:

$$\frac{30s + 100}{s^2 + 9s + 20} = \frac{30s + 100}{(s + 4)(s + 5)} = \frac{-20}{s + 4} + \frac{50}{s + 5}. \qquad (8.87)$$

Substituting Eq. (8.87) into Eq. (8.86) yields

$$F(s) = s^2 + 4s + 10 - \frac{20}{s + 4} + \frac{50}{s + 5}. \qquad (8.88)$$

Now we can inverse-transform Eq. (8.88) by inspection. Hence

$$f(t) = \frac{d^2\delta(t)}{dt^2} + 4\frac{d\delta(t)}{dt} + 10\delta(t) - (20e^{-4t} - 50e^{-5t})u(t). \quad (8.89)$$

DRILL EXERCISES

8.14 Find $f(t)$ if

$$F(s) = (5s^2 + 29s + 32)/[(s+2)(s+4)].$$

ANSWER: $f(t) = 5\delta(t) - (3e^{-2t} - 2e^{-4t})u(t).$

8.15 Find $f(t)$ if

$$F(s) = (2s^3 + 8s^2 + 2s - 4)/(s^2 + 5s + 4).$$

ANSWER: $f(t) = 2\frac{d\delta(t)}{dt} - 2\delta(t) + 4e^{-4t}u(t).$

8.8 POLES AND ZEROS OF F(s)

The rational function of Eq. (8.42) also may be expressed as the ratio of two factored polynomials. In other words, we may write $F(s)$ as

$$F(s) = \frac{K(s+z_1)(s+z_2)\cdots(s+z_n)}{(s+p_1)(s+p_2)\cdots(s+p_m)}, \quad (8.90)$$

where K is the constant a_n/b_m. For example, we may also write the function

$$F(s) = \frac{8s^2 + 120s + 400}{2s^4 + 20s^3 + 70s^2 + 100s + 48}$$

as

$$F(s) = \frac{8(s^2 + 15s + 50)}{2(s^4 + 10s^3 + 35s^2 + 50s + 24)}$$

$$= \frac{4(s+5)(s+10)}{(s+1)(s+2)(s+3)(s+4)}. \quad (8.91)$$

The roots of the denominator polynomial, that is, $-p_1, -p_2, -p_3, \ldots, -p_m$, are called the **poles of** $F(s)$; they are the values of s at which $F(s)$ becomes infinitely large. In the function described by Eq. (8.91), the poles of $F(s)$ are $-1, -2, -3,$ and -4.

The roots of the numerator polynomial, that is, $-z_1$, $-z_2$, $-z_3$, ..., $-z_n$, are called the **zeros of** $F(s)$; they are the values of s at which $F(s)$ becomes zero. In the function described by Eq. (8.91), the zeros of $F(s)$ are -5 and -10.

In what follows, you may find that being able to visualize the poles and zeros of $F(s)$ as points on a complex s plane is helpful. A complex plane is needed because the roots of the polynomials may be complex. In the complex s plane, we use the horizontal axis to plot the real values of s and the vertical axis to plot the imaginary values of s.

As an example of plotting the poles and zeros of $F(s)$, consider the function

$$F(s) = \frac{10(s+5)(s+3-j4)(s+3+j4)}{s(s+10)(s+6-j8)(s+6+j8)}. \qquad (8.92)$$

The poles of $F(s)$ are at 0, -10, $-6+j8$, and $-6-j8$. The zeros are at -5, $-3+j4$, and $-3-j4$. Figure 8.17 shows the poles and zeros plotted on the s plane, where X's represent poles and O's represent zeros.

Note that the poles and zeros for Eq. (8.90) are located in the finite s plane. $F(s)$ can also have either an rth-order pole or an rth-order zero at infinity. For example, the function described by Eq. (8.91) has a second-order zero at infinity, because for large values of s the function reduces to $4/s^2$, and $F(s) = 0$ when $s = \infty$. In this text, we are interested in the poles and zeros located in the finite s plane. Therefore, when we refer to the poles and zeros of a rational function of s, we are referring to the finite poles and zeros.

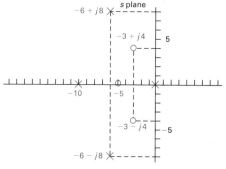

Figure 8.17 Plotting poles and zeros on the s plane.

8.9   INITIAL- AND FINAL-VALUE THEOREMS

The initial- and final-value theorems are useful because they enable us to determine from $F(s)$ the behavior of $f(t)$ at 0 and ∞. Hence we can check the initial and final values of $f(t)$ to see if they conform with known circuit behavior, before actually finding the inverse transform of $F(s)$.

The initial-value theorem states that

$$\lim_{t \to 0^+} f(t) = \lim_{s \to \infty} sF(s), \qquad (8.93)$$

and the final-value theorem states that

$$\lim_{t \to \infty} f(t) = \lim_{s \to 0} sF(s). \qquad (8.94)$$

The initial-value theorem is based on the assumption that $f(t)$ contains no impulse functions. In Eq. (8.94), we must add the restriction that the theorem is valid only if the poles of $F(s)$, except for a first-order pole at the origin, lie in the left half of the s plane.

To prove Eq. (8.93), we start with the operational transform of the first derivative:

$$\mathcal{L}\left\{\frac{df}{dt}\right\} = sF(s) - f(0^-) = \int_{0^-}^{\infty} \frac{df}{dt} e^{-st}\, dt. \qquad (8.95)$$

Now we take the limit as $s \to \infty$:

$$\lim_{s\to\infty}[sF(s) - f(0^-)] = \lim_{s\to\infty}\int_{0^-}^{\infty} \frac{df}{dt} e^{-st}\, dt. \qquad (8.96)$$

Observe that the right-hand side of Eq. (8.96) may be written as

$$\lim_{s\to\infty}\left(\int_{0^-}^{0^+} \frac{df}{dt} e^{0}\, dt + \int_{0^+}^{\infty} \frac{df}{dt} e^{-st}\, dt\right).$$

As $s \to \infty$, $(df/dt)e^{-st} \to 0$; hence the second integral vanishes in the limit. The first integral reduces to $f(0^+) - f(0^-)$, which is independent of s. Thus the right-hand side of Eq. (8.96) becomes

$$\lim_{s\to\infty}\int_{0^-}^{\infty} \frac{df}{dt} e^{-st}\, dt = f(0^+) - f(0^-). \qquad (8.97)$$

Because $f(0^-)$ is independent of s, the left-hand side of Eq. (8.96) may be written

$$\lim_{s\to\infty}[sF(s) - f(0^-)] = \lim_{s\to\infty}[sF(s)] - f(0^-). \qquad (8.98)$$

From Eqs. (8.97) and (8.98),

$$\lim_{s\to\infty} sF(s) = f(0^+) = \lim_{t\to 0^+} f(t),$$

which completes the proof of the initial-value theorem.

The proof of the final-value theorem also starts with Eq. (8.95). Here we take the limit as $s \to 0$:

$$\lim_{s\to 0}[sF(s) - f(0^-)] = \lim_{s\to 0}\left(\int_{0^-}^{\infty} \frac{df}{dt} e^{-st}\, dt\right). \qquad (8.99)$$

The integration is with respect to t and the limit operation is with respect to s, so the right-hand side of Eq. (8.99) reduces to

$$\lim_{s\to 0}\left(\int_{0^-}^{\infty} \frac{df}{dt} e^{-st}\, dt\right) = \int_{0^-}^{\infty} \frac{df}{dt}\, dt. \qquad (8.100)$$

Because the upper limit on the integral is infinite, this integral may also be written as a limit process:

$$\int_{0^-}^{\infty} \frac{df}{dt}\, dt = \lim_{t \to \infty} \int_{0^-}^{t} \frac{df}{dy}\, dy, \qquad (8.101)$$

where we use y as the symbol of integration to avoid confusion with the upper limit on the integral. Carrying out the integration process yields

$$\lim_{t \to \infty} [f(t) - f(0^-)] = \lim_{t \to \infty} [f(t)] - f(0^-). \qquad (8.102)$$

Substituting Eq. (8.102) into Eq. (8.99) gives

$$\lim_{s \to 0} [sF(s)] - f(0^-) = \lim_{t \to \infty} [f(t)] - f(0^-). \qquad (8.103)$$

Because $f(0^-)$ cancels, Eq. (8.103) reduces to the final-value theorem, namely,

$$\lim_{s \to 0} sF(s) = \lim_{t \to \infty} f(t).$$

The final-value theorem is useful only if $f(\infty)$ exists. This condition is true only if all the poles of $F(s)$, except for a simple pole at the origin, lie in the left half of the s plane.

The Application of Initial- and Final-Value Theorems

To illustrate the application of the initial- and final-value theorems, we apply them to a function we used to illustrate partial fraction expansions. Consider the transform pair given by Eq. (8.60). The initial-value theorem gives

$$\lim_{s \to \infty} sF(s) = \lim_{s \to \infty} \frac{100s^2[1 + (3/s)]}{s^3[1 + (6/s)][1 + (6/s) + (25/s^2)]} = 0,$$

$$\lim_{t \to 0^+} f(t) = [-12 + 20\cos(-53.13°)](1) = -12 + 12 = 0.$$

The final-value theorem gives

$$\lim_{s \to 0} sF(s) = \lim_{s \to 0} \frac{100s(s + 3)}{(s + 6)(s^2 + 6s + 25)} = 0,$$

$$\lim_{t \to \infty} f(t) = \lim_{t \to \infty} [-12e^{-6t} + 20e^{-3t}\cos(4t - 53.13°)]u(t) = 0.$$

In applying the theorems to Eq. (8.60), we already had the time-domain expression and were merely testing our understanding. But the real value of the initial- and final-value theorems lies in being able to test the s-domain expressions before working out the inverse transform. For example, consider the expression for $V(s)$ given by Eq. (8.40). Although we cannot calculate $v(t)$ until the circuit parameters are specified, we can check to see if $V(s)$ predicts the correct values of $v(0^+)$ and $v(\infty)$. We know from the statement of the problem that generated $V(s)$ that $v(0^+)$ is zero. We also know that $v(\infty)$ must be zero, because the ideal inductor is a perfect short circuit across the dc current source. Finally, we know that the poles of $V(s)$ must lie in the left half of the s plane because R, L, and C are positive constants. Hence the poles of $sV(s)$ also lie in the left half of the s plane.

Applying the initial-value theorem yields

$$\lim_{s\to\infty} sV(s) = \lim_{s\to\infty} \frac{s(I_{dc}/C)}{s^2[1 + 1/(RCs) + 1/(LCs^2)]} = 0.$$

Applying the final-value theorem gives

$$\lim_{s\to 0} sV(s) = \lim_{s\to 0} \frac{s(I_{dc}/C)}{s^2 + (s/RC) + (1/LC)} = 0.$$

The derived expression for $V(s)$ correctly predicts the initial and final values of $v(t)$.

DRILL EXERCISES

8.16 Use the initial- and final-value theorems to find the initial and final values of $f(t)$ in Drill Exercises 8.9, 8.12, and 8.13.

ANSWER: 7, 0; 4, 1; and 0, 0.

8.17 **(a)** Use the initial-value theorem to find the initial value of v in Drill Exercise 8.7.

(b) Can the final-value theorem be used to find the steady-state value of v? Why?

ANSWER: (a) 0; (b) no, because V has a pair of poles on the imaginary axis.

Practical Perspective

Two-Stage RC Ladder

Let us return to the two-stage RC-ladder network introduced at the beginning of this chapter. We have redrawn the circuit in Fig. 8.18, where we have identified v_1 and v_2 as unknown node voltages. We are interested in how fast the voltages v_1 and v_2 rise in response to the application of a dc voltage which represents a logical high. We assume the initial voltage on each capacitor is zero at the instant the logical high is applied.

We begin our analysis by writing the node-voltage differential equations that describe the circuit, namely

$$C_1 \frac{dv_1}{dt} + \frac{v_1 - V_{dc}}{R_1} + \frac{v_1 - v_2}{R_2} = 0,$$

and

$$C_2 \frac{dv_2}{dt} + \frac{v_2 - v_1}{R_2} = 0.$$

Before Laplace-transforming the equations, we rearrange them as follows

$$R_1 R_2 C_1 \frac{dv_1}{dt} + (R_1 + R_2)v_1 - R_1 v_2 = R_2 V_{dc}$$

and

$$R_2 C_2 \frac{dv_2}{dt} + v_2 - v_1 = 0.$$

Now we transform the equations to the s-domain using the fact that $v_1(0^-) = v_2(0^-) = 0$. Also note that the Laplace transform of the constant $V_{dc} R_2$ is $V_{dc} R_2/s$. Thus we can write

$$[R_1 R_2 C_1 s + (R_1 + R_2)] V_1 - R_1 V_2 = \frac{V_{dc} R_2}{s},$$

$$-V_1 + [R_2 C_2 s + 1]V_2 = 0.$$

These two s domain equations can be solved for V_1 and V_2, yielding

$$V_1 = \frac{V_{dc}}{R_1 C_1} \cdot \frac{[s + (1/R_2 C_2)]}{s \left\{ s^2 + \frac{[R_1 C_1 + (R_1 + R_2)C_2]s}{R_1 C_1 R_2 C_2} + \frac{1}{R_1 C_1 R_2 C_2} \right\}}$$

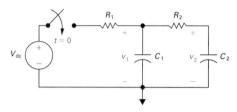

Figure 8.18 Two-stage RC-ladder network.

and

$$V_2 = \frac{V_{dc}}{R_1 C_1 R_2 C_2} \cdot \frac{1}{s \left\{ s^2 + \frac{[R_1 C_1 + (R_1 + R_2) C_2]s}{R_1 C_1 R_2 C_2} + \frac{1}{R_1 C_1 R_2 C_2} \right\}}.$$

The reader can become aquainted with the numerical solutions for v_1 and v_2 by working Problems 8.43 and 8.44.

SUMMARY

- The **Laplace transform** is a tool for converting time-domain equations into frequency-domain equations, according to the following general definition:

$$\mathcal{L}\{f(t)\} = \int_0^\infty f(t) e^{-st}\, dt = F(s),$$

where $f(t)$ is the time-domain expression, and $F(s)$ is the frequency-domain expression.

- A **functional transform** is the Laplace transform of a specific function. Important functional transform pairs are summarized in Tables 8.1 and 8.3.

- **Operational transforms** define the general mathematical properties of the Laplace transform. Important operational transform pairs are summarized in Table 8.2.

- The **step function** $Ku(t)$ describes a function that experiences a discontinuity from one constant level to another at some point in time. K is the magnitude of the jump; if $K = 1$, $Ku(t)$ is the **unit step function**.

- The **impulse function** $K\delta(t)$ is defined

$$\int_{-\infty}^{\infty} K\delta(t)\, dt = K,$$

$$\delta(t) = 0, \quad t \neq 0.$$

K is the strength of the impulse; if $K = 1$, $K\delta(t)$ is the **unit impulse function**.

- In linear lumped-parameter circuits, $F(s)$ is a rational function of s.

- If $F(s)$ is a proper rational function, the inverse transform is found by a partial fraction expansion.

- If $F(s)$ is an improper rational function, it can be inverse-transformed by first expanding it into a sum of a polynomial and a proper rational function.

- $F(s)$ can be expressed as the ratio of two factored polynomials. The roots of the denominator are called **poles** and are plotted as X's on the complex s plane. The roots of the numerator are called **zeros** and are plotted as O's on the complex s plane.

- The initial-value theorem states that

$$\lim_{t \to 0^+} f(t) = \lim_{s \to \infty} sF(s).$$

The theorem assumes that $f(t)$ contains no impulse functions.

- The final-value theorem states that

$$\lim_{t \to \infty} f(t) = \lim_{s \to 0^+} sF(s).$$

The theorem is valid only if the poles of $F(s)$, except for a first-order pole at the origin, lie in the left half of the s plane.

- The initial- and final-value theorems allow us to predict the initial and final values of $f(t)$ from an s-domain expression.

PROBLEMS

8.1. Use step functions to write the expression for each of the functions shown in Fig. P8.1.

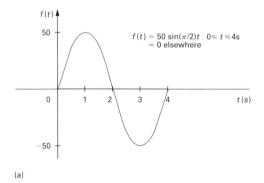

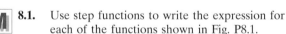

(a)

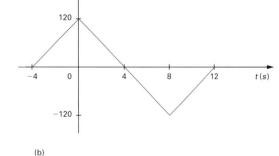

(b)

Figure P8.1

8.2. Step functions can be used to define a **window** function. Thus $u(t-1) - u(t-4)$ defines a window 1 unit high and 3 units wide located on the time axis between 1 and 4.

A function $f(t)$ is defined as follows:

$$f(t) = 0, \quad t \le 0$$
$$= 30t, \quad 0 \le t \le 2s$$
$$= 60, \quad 2s \le t \le 4s$$
$$= 60\cos\left(\frac{\pi}{4}t - \pi\right), \quad 4s \le t \le 8s;$$
$$= 30t - 300, \quad 8s \le t \le 10s$$
$$= 0, \quad 10s \le t \le \infty.$$

(a) Sketch $f(t)$ over the interval $-2s \le t \le 12s$.

(b) Use the concept of the window function to write an expression for $f(t)$.

8.3. Make a sketch of $f(t)$ for $-25s \le t \le 25s$ when $f(t)$ is given by the following expression: $f(t) = -(20t + 400)u(t + 20) + (40t + 400)u(t + 10) + (400 - 40t)u(t - 10) + (20t - 400)u(t - 20)$.

8.4. Explain why the following function generates an impulse function as $\epsilon \to 0$:

$$f(t) = \frac{\epsilon/\pi}{\epsilon^2 + t^2}, \quad -\infty \le t \le \infty.$$

8.5. In Section 8.3, we used the sifting property of the impulse function to show that $\mathcal{L}\{\delta(t)\} = 1$. Show that we can obtain the same result by finding the Laplace transform of the rectangular pulse that exists between $\pm\epsilon$ in Fig. 8.9 and then finding the limit of this transform as $\epsilon \to 0$.

8.6. The triangular pulses shown in Fig. P8.6 are equivalent to the rectangular pulses in Fig. 8.12(b), because they both enclose the same area $(1/\epsilon)$ and they both approach infinity proportional to $1/\epsilon^2$ as $\epsilon \to 0$. Use this triangular-pulse representation for $\delta'(t)$ to find the Laplace transform of $\delta''(t)$.

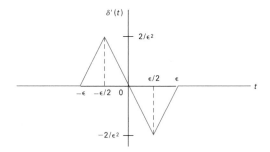

Figure P8.6

8.7. **(a)** Show that

$$\int_{-\infty}^{\infty} f(t)\delta'(t-a)\,dt = -f'(a).$$

(*Hint:* Integrate by parts.)

(b) Use the formula in (a) to show that

$$\mathcal{L}\{\delta'(t)\} = s.$$

8.8. Find the Laplace transform of each of the following functions:

(a) $f(t) = te^{-at}$;

(b) $f(t) = \sin \omega t$;

(c) $f(t) = \sin(\omega t + \theta)$;

(d) $f(t) = \cosh t$;

(e) $f(t) = \cosh(t + \theta)$.

8.9. Find the Laplace transform (when $\epsilon \to 0$) of the derivative of the exponential function illustrated in Fig. 8.8, using each of the following two methods:

(a) First differentiate the function and then find the transform of the resulting function.

(b) Use the operational transform given by Eq. (8.23).

8.10. Show that

$$\mathcal{L}\{\delta^{(n)}(t)\} = s^n.$$

8.11. **(a)** Find the Laplace transform of te^{-at}.

(b) Use the operational transform given by Eq. (8.23) to find the Laplace transform of
$$\frac{d}{dt}(te^{-at}u(t)).$$

(c) Check your result in part (b) by first differentiating and then transforming the resulting expression.

8.12. **(a)** Find $\mathcal{L}\left\{\dfrac{d}{dt}\sin \omega t\, u(t)\right\}$.

(b) Find $\mathcal{L}\left\{\dfrac{d}{dt}\cos \omega t\, u(t)\right\}$.

(c) Find $\mathcal{L}\left\{\dfrac{d^3}{dt^3}t^2u(t)\right\}$.

(d) Check the results of parts (a), (b), and (c) by first differentiating and then transforming.

8.13. **(a)** Find the Laplace transform of

$$\int_{0-}^{t} x\,dx$$

by first integrating and then transforming.

(b) Check the result obtained (a) by using the operational transform given by Eq. (8.33).

8.14. **(a)** Find $\mathcal{L}\left\{\displaystyle\int_{0-}^{t} e^{-ax}\,dx\right\}$.

(b) Find $\mathcal{L}\left\{\displaystyle\int_{0-}^{t} y\,dy\right\}$.

(c) Check the results of (a) and (b) by first integrating and then transforming.

8.15. Show that

$$\mathcal{L}\{e^{-at} f(t)\} = F(s + a).$$

8.16. Find the Laplace transform of each of the following functions:

(a) $f(t) = -20e^{-5(t-2)}u(t-2)$.

(b) $f(t) = (8t-8)[u(t-1)-u(t-2)]+(24-8t)[u(t-2) - u(t-4)] + (8t - 40)[u(t-4) - u(t-5)]$.

8.17. (a) Find the Laplace transform of the function illustrated in Fig. P8.17.

(b) Find the Laplace transform of the first derivative of the function illustrated in Fig. P8.17.

(c) Find the Laplace transform of the second derivative of the function illustrated in Fig. P8.17.

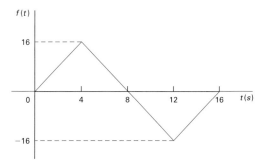

Figure P8.17

8.18. Find the Laplace transform for (a) and (b).

(a) $f(t) = \dfrac{d}{dt}(e^{-at} \sin \omega t u(t))$.

(b) $f(t) = \displaystyle\int_{0^-}^{t} e^{-ax} \cos \omega x \, dx$.

(c) Verify the results obtained in (a) and (b) by first carrying out the indicated mathematical operation and then finding the Laplace transform.

8.19. (a) Show that

$$\mathcal{L}\{f(at)\} = \frac{1}{a} F\left(\frac{s}{a}\right).$$

(b) Use the result of (a) along with the answer derived in Problem 8.8(d) to find

$$\mathcal{L}\{\cosh \beta t\}.$$

8.20. **(a)** Given that $F(s) = \mathcal{L}\{f(t)\}$, show that

$$-\frac{dF(s)}{ds} = \mathcal{L}\{tf(t)\}.$$

(b) Show that

$$(-1)^n \frac{d^n F(s)}{ds^n} = \mathcal{L}\{t^n f(t)\}.$$

(c) Use the result of (b) to find $\mathcal{L}\{t^5\}$, $\mathcal{L}\{t \sin \beta t\}$, and $\mathcal{L}\{te^{-t} \cosh t\}$.

8.21. **(a)** Show that if $F(s) = \mathcal{L}\{f(t)\}$ and $\{f(t)/t\}$ is Laplace-transformable, then

$$\int_s^\infty F(u)du = \mathcal{L}\left\{\frac{f(t)}{t}\right\}.$$

(*Hint:* Use the defining integral to write

$$\int_s^\infty F(u)du = \int_s^\infty \left(\int_{0-}^\infty f(t)e^{-ut}\, dt\right) du$$

and then reverse the order of integration.)

(b) Start with the result obtained in Problem 8.20(c) for $\mathcal{L}\{t \sin \beta t\}$ and use the operational transform given in (a) of this problem to find $\mathcal{L}\{\sin \beta t\}$.

8.22. There is no energy stored in the circuit shown in Fig. P8.22 at the time the switch is opened.

(a) Derive the integrodifferential equation that governs the behavior of the voltage v_o.

(b) Show that

$$V_o(s) = \frac{I_{dc}/C}{s^2 + (1/RC)s + (1/LC)}.$$

(c) Show that

$$I_o(s) = \frac{sI_{dc}}{s^2 + (1/RC)s + (1/LC)}.$$

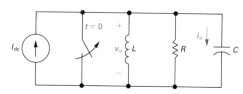

Figure P8.22

8.23. The circuit parameters in the circuit seen in Fig. P8.22 have the following values: $R = 4$ kΩ, $L = 2.5$ H, $C = 25$ nF, and $I_{dc} = 3$ mA.

(a) Find $v_o(t)$ for $t \geq 0$.

(b) Find $i_o(t)$ for $t \geq 0$.

(c) Does your solution for $i_o(t)$ make sense when $t = 0$? Explain.

8.24. The switch in the circuit in Fig. P8.24 has been open for a long time. At $t = 0$, the switch closes.

(a) Derive the integrodifferential equation that governs the behavior of the voltage v_o for $t \geq 0$.

(b) Show that

$$V_o(s) = \frac{V_{dc}/LC}{s[s^2 + (1/RC)s + (1/LC)]}.$$

(c) Show that

$$I_o(s) = \frac{V_{dc}[s + (1/RC)]}{sL[s^2 + (1/RC)s + (1/LC)]}.$$

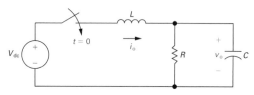

Figure P8.24

 8.25. The circuit parameters in the circuit in Fig. P8.24 are $R = 10\ k\Omega$; $L = 800\ mH$; and $C = 100\ nF$. If V_{dc} is 70 V, find

(a) $v_o(t)$ for $t \geq 0$

(b) $i_o(t)$ for $t \geq 0$

 8.26. Find $f(t)$ for each of the following functions:

(a) $F(s) = \dfrac{18s^2 + 66s + 54}{(s+1)(s+2)(s+3)}.$

(b) $F(s) = \dfrac{25s^2 + 86s + 40}{s(s+2)(s+4)}.$

(c) $F(s) = \dfrac{11s^2 + 172s + 700}{(s+2)(s^2 + 12s + 100)}.$

(d) $F(s) = \dfrac{56s^2 + 112s + 5000}{s(s^2 + 14s + 625)}.$

 8.27. Find $f(t)$ for each of the following functions.

(a) $F(s) = \dfrac{8(s^2 - 5s + 50)}{s^2(s + 10)}.$

(b) $F(s) = \dfrac{10(3s^2 + 4s + 4)}{s(s+2)^2}.$

(c) $F(s) = \dfrac{s^3 - 6s^2 + 15s + 50}{s^2(s^2 + 4s + 5)}.$

(d) $F(s) = \dfrac{s^2 + 6s + 5}{(s+2)^3}.$

(e) $F(s) = \dfrac{16s^3 + 72s^2 + 216s - 128}{(s^2 + 2s + 5)^2}.$

 8.28. Find $f(t)$ for each of the following functions.

(a) $F(s) = \dfrac{10s^2 + 85s + 95}{s^2 + 6s + 5}.$

(b) $F(s) = \dfrac{5(s^2 + 8s + 5)}{s^2 + 4s + 5}.$

(c) $F(s) = \dfrac{s^2 + 25s + 150}{s + 20}.$

8.29. Find $f(t)$ for each of the following functions.

(a) $F(s) = \dfrac{20s^2}{(s+1)^3}$.

(b) $F(s) = \dfrac{5(s+2)^2}{s^4(s+1)}$.

8.30. The switch in the circuit in Fig. P8.30 has been in position a for a long time. At $t = 0$, the switch moves instantaneously to position b.

(a) Derive the integrodifferential equation that governs the behavior of the voltage v_o for $t \geq 0^+$.

(b) Show that

$$V_o(s) = \frac{V_{dc}[s + (R/L)]}{[s^2 + (R/L)s + (1/LC)]}.$$

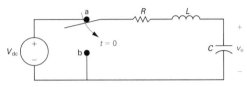

Figure P8.30

8.31. The circuit parameters in the circuit in Fig. P8.30 are $R = 5000\ \Omega$, $L = 1$ H, and $C = 0.25\ \mu\text{F}$. If $V_{dc} = 15$ V, find $v_o(t)$ for $t \geq 0$.

8.32. There is no energy stored in the circuit shown in Fig. P8.32 at the time the switch is opened.

(a) Derive the integrodifferential equations that govern the behavior of the node voltages v_1 and v_2.

(b) Show that

$$V_2(s) = \frac{sI_g}{C[s^2 + (R/L)s + (1/LC)]}.$$

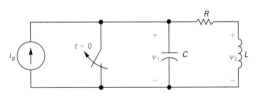

Figure P8.32

M
P

8.33. The circuit parameters in the circuit in Fig. P8.32 are $R = 1600 \ \Omega$; $L = 200$ mH; and $C = 0.2 \ \mu$F. If $i_g(t) = 6u(t)$ mA, find $v_2(t)$.

8.34. Derive the transform pair given by Eq. (8.64).

8.35. **(a)** Derive the transform pair given by Eq. (8.83).
(b) Derive the transform pair given by Eq. (8.84).

8.36. Apply the initial- and final-value theorems to each transform pair in Problem 8.26.

8.37. Apply the initial- and final-value theorems to each transform pair in Problem 8.27.

8.38. Apply the initial- and final-value theorems to each transform pair in Problem 8.28.

8.39. Apply the initial- and final-value theorems to each transform pair in Problem 8.29.

8.40. Use the initial- and final-value theorems to check the initial and final values of the current and voltage in Problem 8.22.

8.41. Use the initial- and final-value theorems to check the initial and final values of the current and voltage in Problem 8.24.

8.42. Use the initial- and final-value theorems to check the initial and final values of the voltage in Problem 8.30.

◆ **8.43.** Assume the parameter values in the two-stage RC-ladder network in Fig. 8.18 are $R_1 = 800 \ \Omega$; $R_2 = 1200 \ \Omega$; $C_1 = C_2 = (5/6)$ nF; and $V_{dc} = 5$ V.
(a) Find the time-domain expressions for v_1 and v_2.
(b) How long does it take v_1 to reach 90% of its final value?
(c) How long does it take v_2 to reach 90% of its final value?
(d) Using a computer program of your choice, make plots of v_1 and v_2 vs. t for $0 \le t \le 8 \ \mu$s.

◆ **8.44.** Assume that 4 μs after the switch in the circuit in Fig. 8.18 closes, the 5 V source suddenly drops to zero volts.
(a) Assuming the circuit parameter values are the same as in Problem 8.43, derive the time-domain expressions for v_1 and v_2 for $4 \ \mu$s $\le t \le \infty$.
(b) Using a computer program of your choice, plot on a single graph v_1 and v_2 vs. t for $4 \ \mu$s $\le t \le 12 \ \mu$s.
(c) Using the solutions for v_1 and v_2 from Problems 8.43 and 8.44, plot on a single graph v_1 and v_2 vs. t for $0 \le t \le 12 \ \mu$s.

Creation of a Voltage Surge

With the advent of personal computers, modems, fax machines, and other electronic equipment it is necessary to provide protection from voltage surges that can occur in a household circuit due to switching. A group of commercially available surge suppressors is shown in the accompanying figure.

At the end of this chapter we will illustrate how the Laplace transform is used to calculate a voltage surge created by switching off a load in a circuit operating in the sinusoid steady state.

Commercially available surge suppressors.

9 The Laplace Transform in Circuit Analysis

Chapter Contents

The Laplace transform has two characteristics that make it an attractive tool in circuit analysis. First, it transforms a set of linear constant-coefficient differential equations into a set of linear polynomial equations, which are easier to manipulate. Second, it automatically introduces into the polynomial equations the initial values of the current and voltage variables. Thus, initial conditions are an inherent part of the transform process. (This contrasts with the classical approach to the solution of differential equations, in which initial conditions are considered when the unknown coefficients are evaluated.)

We begin this chapter by showing how we can skip the step of writing time-domain integrodifferential equations and transforming them into the s domain. In Section 9.1, we'll develop the s-domain circuit models for resistors, inductors, and capacitors so that we can write s-domain equations for all circuits directly. Section 9.2 reviews Ohm's and Kirchhoff's laws in the context of the s domain. After establishing these fundamentals, we apply the Laplace transform method to a variety of circuit problems in Section 9.3.

Analytical and simplification techniques first introduced with resistive circuits—such as mesh-current and node-voltage methods and source transformations—can be used in the s domain as well. After solving for the circuit response in the s domain, we inverse-transform back to the time domain, using partial fraction expansion (as demonstrated in the preceding chapter). As before, checking the final time-domain equations in terms of the initial conditions and final values is an important step in the solution process.

The s-domain descriptions of circuit input and output lead us, in Section 9.4, to the concept of the transfer function. The transfer function for a particular circuit is the ratio of the Laplace transform of its output to the Laplace transform of its input. We look at the role of partial fraction expansion (Section 9.5) and the convolution integral (Section 9.6) in employing the transfer function in circuit analysis. We also look at the role the impulse function plays in circuit analysis.

Finally we illustrate how the Laplace transform can be used to predict the possible creation of a voltage surge as a result of switching off a load in a circuit operating in the sinusoidal steady state.

9.1 CIRCUIT ELEMENTS IN THE s DOMAIN

The procedure for developing an s-domain equivalent circuit for each circuit element is simple. First, we write the time-domain equation that relates the terminal voltage to the terminal current. Next, we take the Laplace transform of the time-domain equation. This step generates an algebraic relationship between the s-domain current and voltage. Note that the dimension of a transformed voltage is volt-seconds, and the dimension of a transformed current is ampere-seconds. A voltage-to-current ratio in the s domain carries the dimension of volts per ampere. An impedance in the s domain is measured in ohms, and an admittance is measured in siemens. Finally, we construct a circuit model that satisfies the relationship between the s-domain current and voltage. We use the passive sign convention in all the derivations.

A Resistor in the s Domain

We begin with the resistance element. From Ohm's law,

$$v = Ri. \tag{9.1}$$

Because R is a constant, the Laplace transform of Eq. (9.1) is

$$V = RI, \qquad (9.2)$$

where

$$V = \mathcal{L}\{v\} \quad \text{and} \quad I = \mathcal{L}\{i\}.$$

Equation (9.2) states that the s-domain equivalent circuit of a resistor is simply a resistance of R ohms that carries a current of I ampere-seconds and has a terminal voltage of V volt-seconds.

Figure 9.1 shows the time- and frequency-domain circuits of the resistor. Note that going from the time domain to the frequency domain does not change the resistance element.

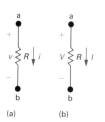

Figure 9.1 The resistance element. (a) Time domain. (b) Frequency domain.

An Inductor in the s Domain

Figure 9.2 shows an inductor carrying an initial current of I_0 amperes. The time-domain equation that relates the terminal voltage to the terminal current is

$$v = L\frac{di}{dt}. \qquad (9.3)$$

The Laplace transform of Eq. (9.3) gives

$$V = L[sI - i(0^-)] = sLI - LI_0. \qquad (9.4)$$

Figure 9.2 An inductor of L henrys carrying an initial current of I_0 amperes.

Two different circuit configurations satisfy Eq. (9.4). The first consists of an impedance of sL ohms in series with an independent voltage source of LI_0 volt-seconds, as shown in Fig. 9.3. Note that the polarity marks on the voltage source LI_0 agree with the minus sign in Eq. (9.4). Note also that LI_0 carries its own algebraic sign; that is, if the initial value of i is opposite to the reference direction for i, then I_0 has a negative value.

The second s-domain equivalent circuit that satisfies Eq. (9.4) consists of an impedance of sL ohms in parallel with an independent current source of I_0/s ampere-seconds, as shown in Fig. 9.4. We can derive the alternative equivalent circuit shown in Fig. 9.4 in several ways. One way is simply to solve Eq. (9.4) for the current I and then construct the circuit to satisfy the resulting equation. Thus

$$I = \frac{V + LI_0}{sL} = \frac{V}{sL} + \frac{I_0}{s}. \qquad (9.5)$$

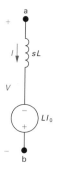

Figure 9.3 The series equivalent circuit for an inductor of L henrys carrying an initial current of I_0 amperes.

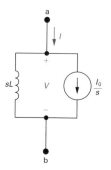

Figure 9.4 The parallel equivalent circuit for an inductor of L henrys carrying an initial current of I_0 amperes.

Figure 9.5 The s-domain circuit for an inductor when the initial current is zero.

Figure 9.6 A capacitor of C farads initially charged to V_0 volts.

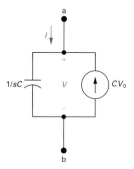

Figure 9.7 The parallel equivalent circuit for a capacitor initially charged to V_0 volts.

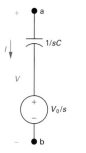

Figure 9.8 The series equivalent circuit for a capacitor initially charged to V_0 volts.

Figure 9.9 The s-domain circuit for a capacitor when the initial voltage is zero.

Two other ways are: (1) find the Norton equivalent of the circuit shown in Fig. 9.3 and (2) start with the inductor current as a function of the inductor voltage and then find the Laplace transform of the resulting integral equation. We leave these two approaches to Problems 9.1 and 9.2.

If the initial energy stored in the inductor is zero, that is, if $I_0 = 0$, the s-domain equivalent circuit of the inductor reduces to an inductor with an impedance of sL ohms. Figure 9.5 shows this circuit.

A Capacitor in the s Domain

An initially charged capacitor also has two s-domain equivalent circuits. Figure 9.6 shows a capacitor initially charged to V_0 volts. The terminal current is

$$i = C\frac{dv}{dt}. \qquad (9.6)$$

Transforming Eq. (9.6) yields

$$I = C[sV - v(0^-)]$$

or

$$I = sCV - CV_0, \qquad (9.7)$$

which indicates that the s-domain current I is the sum of two branch currents. One branch consists of an admittance of sC siemens, and the second branch consists of an independent current source of CV_0 ampere-seconds. Figure 9.7 shows this parallel equivalent circuit.

We derive the series equivalent circuit for the charged capacitor by solving Eq. (9.7) for V:

$$V = \left(\frac{1}{sC}\right)I + \frac{V_0}{s}. \qquad (9.8)$$

Figure 9.8 shows the circuit that satisfies Eq. (9.8).

In the equivalent circuits shown in Figs. 9.7 and 9.8, V_0 carries its own algebraic sign. In other words, if the polarity of V_0 is opposite to the reference polarity for v, V_0 is a negative quantity. If the initial voltage on the capacitor is zero, both equivalent circuits reduce to an impedance of $1/sC$ ohms, as shown in Fig. 9.9.

In this chapter, an important first problem-solving step will be to choose between the parallel or series equivalents when inductors and capacitors are present. With a little forethought and some experience, the correct choice will often be quite evident. The equivalent circuits are summarized in Table 9.1.

TABLE 9.1 **Summary of the s-Domain Equivalent Circuits**

TIME DOMAIN	FREQUENCY DOMAIN

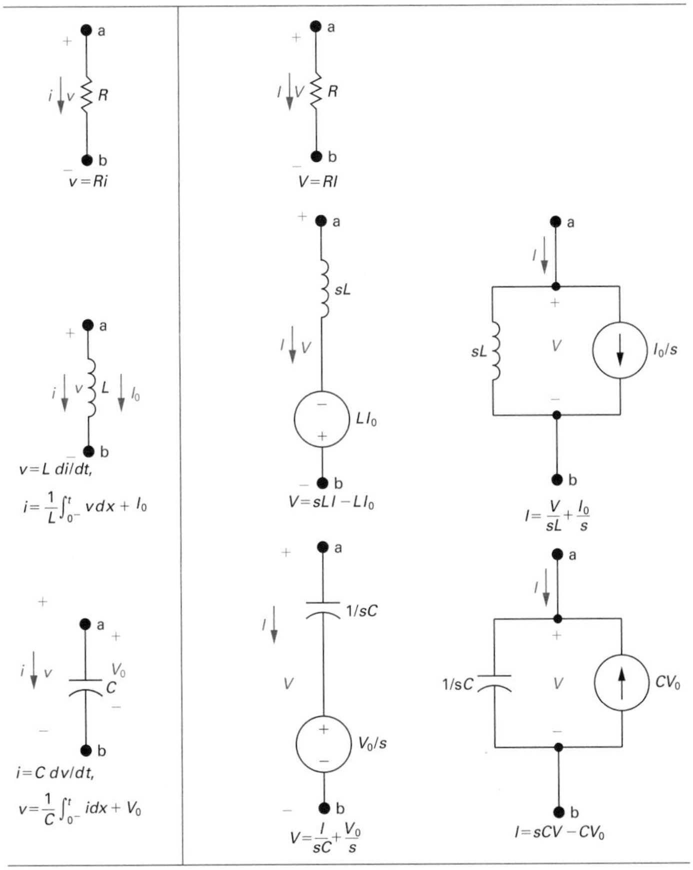

$v = Ri$

$V = RI$

$v = L\, di/dt,$

$i = \dfrac{1}{L}\displaystyle\int_{0^-}^{t} v\,dx + I_0$

$V = sLI - LI_0$

$I = \dfrac{V}{sL} + \dfrac{I_0}{s}$

$i = C\, dv/dt,$

$v = \dfrac{1}{C}\displaystyle\int_{0^-}^{t} i\,dx + V_0$

$V = \dfrac{I}{sC} + \dfrac{V_0}{s}$

$I = sCV - CV_0$

9.2 CIRCUIT ANALYSIS IN THE s DOMAIN

Before illustrating how to use the s-domain equivalent circuits in analysis, we need to lay some groundwork.

First, we know that if no energy is stored in the inductor or capacitor, the relationship between the terminal voltage and current for each passive element takes the form:

$$V = ZI, \tag{9.9}$$

where Z refers to the s-domain impedance of the element. Thus a resistor has an impedance of R ohms, an inductor has an impedance of sL ohms, and a capacitor has an impedance of $1/sC$ ohms. The relationship contained in Eq. (9.9) is also contained in Figs. 9.1(b), 9.5, and 9.9. Equation (9.9) is sometimes referred to as Ohm's law for the s domain.

The reciprocal of the impedance is admittance. Therefore, the s-domain admittance of a resistor is $1/R$ siemens, an inductor has an admittance of $1/sL$ siemens, and a capacitor has an admittance of sC siemens.

The rules for combining impedances and admittances in the s domain are the same as those for frequency-domain circuits. Thus series-parallel simplifications are applicable to s-domain analysis.

In addition, Kirchhoff's laws apply to s-domain currents and voltages. Their applicability stems from the operational transform stating that the Laplace transform of a sum of time-domain functions is the sum of the transforms of the individual functions (see Table 8.2). Because the algebraic sum of the currents at a node is zero in the time domain, the algebraic sum of the transformed currents is also zero. A similar statement holds for the algebraic sum of the transformed voltages around a closed path. The s-domain version of Kirchhoff's laws is

$$\text{alg} \sum I = 0, \tag{9.10}$$

$$\text{alg} \sum V = 0. \tag{9.11}$$

Because the voltage and current at the terminals of a passive element are related by an algebraic equation and because Kirchhoff's laws still hold, all the techniques of circuit analysis developed for pure resistive networks may be used in s-domain analysis. Thus node voltages, mesh currents, source transformations, and Thévenin-Norton equivalents are all valid techniques, even when energy is stored initially in the inductors and capacitors. Initially stored energy requires that we modify Eq. (9.9) by simply adding independent sources either in series or parallel with the element impedances. The addition of these sources is governed by Kirchhoff's laws.

DRILL EXERCISES

9.1 A 500 Ω resistor, a 16 mH inductor, and a 25 nF capacitor are connected in parallel.

(a) Express the admittance of this parallel combination of elements as a rational function of s.

(b) Compute the numerical values of the zeros and poles.

ANSWER: (a) $25 \times 10^{-9}(s^2 + 80{,}000s + 25 \times 10^8)/s$;
(b) $-z_1 = -40{,}000 - j30{,}000$ rad/s;
$-z_2 = -40{,}000 + j30{,}000$ rad/s; $p_1 = 0$ rad/s.

9.2 The parallel circuit in Drill Exercise 9.1 is placed in
series with a 2000 Ω resistor.

    (a) Express the impedance of this series combina-
tion as a rational function of s.

    (b) Compute the numerical values of the zeros and
poles.

ANSWER:
(a) $2000(s + 50,000)^2/(s^2 + 80,000s + 25 \times 10^8)$;
(b) $-z_1 = -z_2 = -50,000$ rad/s;
$-p_1 = -40,000 - j30,000$ rad/s,
$-p_2 = -40,000 + j30,000$ rad/s.

9.3 APPLICATIONS

We now illustrate how to use the Laplace transform to determine the transient behavior of several linear lumped-parameter
circuits. We start by analyzing familiar circuits from Chapters 5
and 6 because they represent a simple starting place and because
they show that the Laplace transform approach yields the same
results. In all the examples, the ease of manipulating algebraic
equations instead of differential equations should be apparent.

The Natural Response of an RC Circuit

We first revisit the natural response of an RC circuit (Fig. 9.10)
via Laplace transform techniques. (You may want to review the
classical analysis of this same circuit in Section 5.4.)

    The capacitor is initially charged to V_0 volts, and we are in-
terested in the time-domain expressions for i and v. We start
by finding i. In transferring the circuit in Fig. 9.10 to the s domain, we have a choice of two equivalent circuits for the charged
capacitor. Because we are interested in the current, the series-
equivalent circuit is more attractive; it results in a single-mesh
circuit in the frequency domain. Thus we construct the s-domain
circuit shown in Fig. 9.11.

    Summing the voltages around the mesh generates the ex-
pression

$$\frac{V_0}{s} = \frac{1}{sC}I + RI. \qquad (9.12)$$

Solving Eq. (9.12) for I yields

$$I = \frac{CV_0}{RCs + 1} = \frac{V_0/R}{s + (1/RC)}. \qquad (9.13)$$

Figure 9.10 The capacitor discharge circuit.

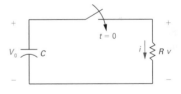

Figure 9.11 An s-domain equivalent circuit for
the circuit shown in Fig. 9.10.

Note that the expression for I is a proper rational function of s and can be inverse-transformed by inspection:

$$i = \frac{V_0}{R} e^{-t/RC} u(t), \qquad (9.14)$$

which is equivalent to the expression for the current derived by the classical methods discussed in Chapter 5. In that chapter, the current is given by Eq. (5.56), where τ is used in place of RC.

After we have found i, the easiest way to determine v is simply to apply Ohm's law; that is, from the circuit,

$$v = Ri = V_0 e^{-t/RC} u(t). \qquad (9.15)$$

We now illustrate a way to find v from the circuit without first finding i. In this alternative approach, we return to the original circuit of Fig. 9.10 and transfer it to the s domain using the parallel equivalent circuit for the charged capacitor. Using the parallel equivalent circuit is attractive now because we can describe the resulting circuit in terms of a single node voltage. Figure 9.12 shows the new s-domain equivalent circuit.

The node-voltage equation that describes the new circuit is

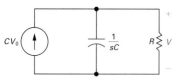

Figure 9.12 An s-domain equivalent circuit for the circuit shown in Fig. 9.10.

$$\frac{V}{R} + sCV = CV_0. \qquad (9.16)$$

Solving Eq. (9.16) for V gives

$$V = \frac{V_0}{s + (1/RC)}. \qquad (9.17)$$

Inverse-transforming Eq. (9.17) leads to the same expression for v given by Eq. (9.15), namely,

$$v = V_0 e^{-t/RC} = V_0 e^{-t/\tau} u(t). \qquad (9.18)$$

Our purpose in deriving v by direct use of the transform method is to show that the choice of which s-domain equivalent circuit to use is influenced by which response signal is of interest.

DRILL EXERCISE

9.3   The switch in the circuit shown has been in position a for a long time. At $t = 0$, the switch is thrown to position b.

(a) Find I, V_1, and V_2 as rational functions of s.

(b) Find the time-domain expressions for i, v_1, and v_2.

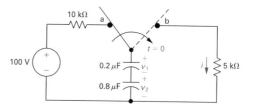

ANSWER: (a) $I = 0.02/(s + 1250)$,
$V_1 = 80/(s + 1250)$, $V_2 = 20/(s + 1250)$;
(b) $i = 20e^{-1250t} u(t)$ mA, $v_1 = 80e^{-1250t} u(t)$ V,
$v_2 = 20e^{-1250t} u(t)$ V.

The Step Response of a Parallel Circuit

Next we analyze the parallel RLC circuit, shown in Fig. 9.13, that we first analyzed in Example 6.7. The problem is to find the expression for i_L after the constant current source is switched across the parallel elements. The initial energy stored in the circuit is zero.

As before, we begin by constructing the s-domain equivalent circuit shown in Fig. 9.14. Note how easily an independent source can be transformed from the time domain to the frequency domain. We transform the source to the s domain simply by determining the Laplace transform of its time-domain function. Here, opening the switch results in a step change in the current applied to the circuit. Therefore the s-domain current source is $\mathcal{L}\{I_{dc}u(t)\}$, or I_{dc}/s. To find I_L, we first solve for V and then use

$$I_L = \frac{V}{sL} \qquad (9.19)$$

to establish the s-domain expression for I_L. Summing the currents away from the top node generates the expression

$$sCV + \frac{V}{R} + \frac{V}{sL} = \frac{I_{dc}}{s}. \qquad (9.20)$$

Solving Eq. (9.20) for V gives

$$V = \frac{I_{dc}/C}{s^2 + (1/RC)s + (1/LC)}. \qquad (9.21)$$

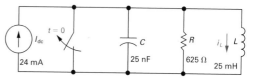

Figure 9.13 The step response of a parallel RLC circuit.

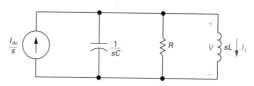

Figure 9.14 The s-domain equivalent circuit for the circuit shown in Fig. 9.13.

Substituting Eq. (9.21) into Eq. (9.19) gives

$$I_L = \frac{I_{\mathrm{dc}}/LC}{s[s^2 + (1/RC)s + (1/LC)]}. \tag{9.22}$$

Substituting the numerical values of R, L, C, and I_{dc} into Eq. (9.22) yields

$$I_L = \frac{384 \times 10^5}{s(s^2 + 64{,}000s + 16 \times 10^8)}. \tag{9.23}$$

Before expanding Eq. (9.23) into a sum of partial fractions, we factor the quadratic term in the denominator:

$$I_L = \frac{384 \times 10^5}{s(s + 32{,}000 - j24{,}000)(s + 32{,}000 + j24{,}000)}. \tag{9.24}$$

Now, we can test the s-domain expression for I_L by checking to see whether the final-value theorem predicts the correct value for i_L at $t = \infty$. All the poles of I_L, except for the first-order pole at the origin, lie in the left half of the s plane, so the theorem is applicable. We know from the behavior of the circuit that after the switch has been open for a long time, the inductor will short-circuit the current source. Therefore, the final value of i_L must be 24 mA. The limit of sI_L as $s \to 0$ is

$$\lim_{s \to 0} sI_L = \frac{384 \times 10^5}{16 \times 10^8} = 24 \text{ mA}. \tag{9.25}$$

(Currents in the s domain carry the dimension of ampere-seconds, so the dimension of sI_L will be amperes.) Thus our s-domain expression checks out.

We now proceed with the partial fraction expansion of Eq. (9.24):

$$I_L = \frac{K_1}{s} + \frac{K_2}{s + 32{,}000 - j24{,}000}$$

$$+ \frac{K_2^*}{s + 32{,}000 + j24{,}000}. \tag{9.26}$$

The partial fraction coefficients are

$$K_1 = \frac{384 \times 10^5}{16 \times 10^8} = 24 \times 10^{-3} \text{ A}, \tag{9.27}$$

$$K_2 = \frac{384 \times 10^5}{(-32{,}000 + j24{,}000)(j48{,}000)}$$

$$= 20 \times 10^{-3} \; \underline{/126.87°} \text{ A}. \tag{9.28}$$

Substituting the numerical values of K_1 and K_2 into Eq. (9.26) and inverse-transforming the resulting expression yields

$$i_L = [24 + 40e^{-32,000t} \cos(24,000t + 126.87°)]u(t) \text{ mA.} \quad (9.29)$$

The answer given by Eq. (9.29) is equivalent to the answer given for Example 6.7 because

$$40 \cos(24,000t + 126.87°) = -24 \cos 24,000t - 32 \sin 24,000t.$$

If we weren't using a previous solution as a check, we would test Eq. (9.29) to make sure that $i_L(0)$ satisfied the given initial conditions and $i_L(\infty)$ satisfied the known behavior of the circuit.

DRILL EXERCISE

9.4   The energy stored in the circuit shown is zero at the time when the switch is closed.

   (a) Find the s-domain expression for I.

   (b) Find the time-domain expression for i when $t > 0$.

   (c) Find the s-domain expression for V.

   (d) Find the time-domain expression for v when $t > 0$.

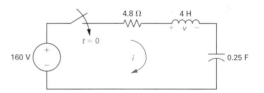

ANSWER: (a) $I = 40/(s^2 + 1.2s + 1)$;
(b) $i = (50e^{-0.6t} \sin 0.8t)u(t)$ A;
(c) $V = 160s/(s^2 + 1.2s + 1)$;
(d) $v = [200e^{-0.6t} \cos(0.8t + 36.87°)]u(t)$ V.

The Transient Response of a Parallel RLC Circuit

Another example of using the Laplace transform to find the transient behavior of a circuit arises from replacing the dc current source in the circuit shown in Fig. 9.13 with a sinusoidal current source. The new current source is

$$i_g = I_m \cos \omega t \text{ A,} \quad (9.30)$$

where $I_m = 24$ mA and $\omega = 40,000$ rad/s. As before, we assume that the initial energy stored in the circuit is zero.

The s-domain expression for the source current is

$$I_g = \frac{s I_m}{s^2 + \omega^2}. \tag{9.31}$$

The voltage across the parallel elements is

$$V = \frac{(I_g/C)s}{s^2 + (1/RC)s + (1/LC)}. \tag{9.32}$$

Substituting Eq. (9.31) into Eq. (9.32) results in

$$V = \frac{(I_m/C)s^2}{(s^2 + \omega^2)[s^2 + (1/RC)s + (1/LC)]}, \tag{9.33}$$

from which

$$I_L = \frac{V}{sL} = \frac{(I_m/LC)s}{(s^2 + \omega^2)[s^2 + (1/RC)s + (1/LC)]}. \tag{9.34}$$

Substituting the numerical values of I_m, ω, R, L, and C into Eq. (9.34) gives

$$I_L = \frac{384 \times 10^5 s}{(s^2 + 16 \times 10^8)(s^2 + 64{,}000s + 16 \times 10^8)}. \tag{9.35}$$

We now write the denominator in factored form:

$$I_L = \frac{384 \times 10^5 s}{(s - j\omega)(s + j\omega)(s + \alpha - j\beta)(s + \alpha + j\beta)}, \tag{9.36}$$

where $\omega = 40{,}000$, $\alpha = 32{,}000$, and $\beta = 24{,}000$.

We can't test the final value of i_L with the final-value theorem because I_L has a pair of poles on the imaginary axis; that is, poles at $\pm j4 \times 10^4$. Thus we must first find i_L and then check the validity of the expression from known circuit behavior.

When we expand Eq. (9.36) into a sum of partial fractions, we generate the equation

$$I_L = \frac{K_1}{s - j40{,}000} + \frac{K_1^*}{s + j40{,}000} + \frac{K_2}{s + 32{,}000 - j24{,}000}$$

$$+ \frac{K_2^*}{s + 32{,}000 + j24{,}000}. \tag{9.37}$$

The numerical values of the coefficients K_1 and K_2 are

$$K_1 = \frac{384 \times 10^5(j40,000)}{(j80,000)(32,000 + j16,000)(32,000 + j64,000)}$$

$$= 7.5 \times 10^{-3} \ \angle{-90°} \ \text{A}, \tag{9.38}$$

$$K_2 = \frac{384 \times 10^5(-32,000 + j24,000)}{(-32,000 - j16,000)(-32,000 + j64,000)(j48,000)}$$

$$= 12.5 \times 10^{-3} \ \angle{90°} \ \text{A}. \tag{9.39}$$

Substituting the numerical values from Eqs. (9.38) and (9.39) into Eq. (9.37) and inverse-transforming the resulting expression yields

$$i_L = [15\cos(40,000t - 90°)$$

$$+ \ 25e^{-32,000t}\cos(24,000t + 90°)] \ \text{mA},$$

$$= (15\sin 40,000t - 25e^{-32,000t}\sin 24,000t)u(t) \ \text{mA}. \tag{9.40}$$

We now test Eq. (9.40) to see whether it makes sense in terms of the given initial conditions and the known circuit behavior after the switch has been open for a long time. For $t = 0$, Eq. (9.40) predicts zero initial current, which agrees with the initial energy of zero in the circuit. Equation (9.40) also predicts a steady-state current of

$$i_{L_{ss}} = 15\sin 40,000t \ \text{mA}, \tag{9.41}$$

which can be verified by the phasor method (Chapter 7).

The Step Response of a Multiple Mesh Circuit

Until now, we avoided circuits that required two or more node-voltage or mesh-current equations, because the techniques for solving simultaneous differential equations are beyond the scope of this text. However, using Laplace techniques, we can solve a problem like the one posed by the multiple-mesh circuit in Fig. 9.15.

Here we want to find the branch currents i_1 and i_2 that arise when the 336 V dc voltage source is applied suddenly to the circuit. The initial energy stored in the circuit is zero. Figure 9.16 shows the s-domain equivalent circuit of Fig. 9.15. The two mesh-current equations are

$$\frac{336}{s} = (42 + 8.4s)I_1 - 42I_2, \tag{9.42}$$

$$0 = -42I_1 + (90 + 10s)I_2. \tag{9.43}$$

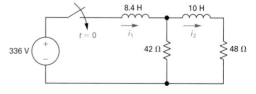

Figure 9.15 A multiple-mesh RL circuit.

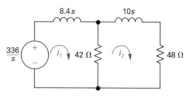

Figure 9.16 The s-domain equivalent circuit for the circuit shown in Fig. 9.15.

Using Cramer's method to solve for I_1 and I_2, we obtain

$$\Delta = \begin{vmatrix} 42 + 8.4s & -42 \\ -42 & 90 + 10s \end{vmatrix}$$

$$= 84(s^2 + 14s + 24)$$

$$= 84(s + 2)(s + 12), \tag{9.44}$$

$$N_1 = \begin{vmatrix} 336/s & -42 \\ 0 & 90 + 10s \end{vmatrix}$$

$$= \frac{3360(s + 9)}{s}, \tag{9.45}$$

$$N_2 = \begin{vmatrix} 42 + 8.4s & 336/s \\ -42 & 0 \end{vmatrix}$$

$$= \frac{14,112}{s}. \tag{9.46}$$

Based on Eqs. (9.44)–(9.46),

$$I_1 = \frac{N_1}{\Delta} = \frac{40(s + 9)}{s(s + 2)(s + 12)}, \tag{9.47}$$

$$I_2 = \frac{N_2}{\Delta} = \frac{168}{s(s + 2)(s + 12)}. \tag{9.48}$$

Expanding I_1 and I_2 into a sum of partial fractions gives

$$I_1 = \frac{15}{s} - \frac{14}{s + 2} - \frac{1}{s + 12}, \tag{9.49}$$

$$I_2 = \frac{7}{s} - \frac{8.4}{s + 2} + \frac{1.4}{s + 12}. \tag{9.50}$$

We obtain the expressions for i_1 and i_2 by inverse-transforming Eqs. (9.49) and (9.50), respectively:

$$i_1 = (15 - 14e^{-2t} - e^{-12t})u(t) \text{ A}, \tag{9.51}$$

$$i_2 = (7 - 8.4e^{-2t} + 1.4e^{-12t})u(t) \text{ A}. \tag{9.52}$$

Next we test the solutions to see whether they make sense in terms of the circuit. Because no energy is stored in the circuit at the instant the switch is closed, both $i_1(0^-)$ and $i_2(0^-)$ must be zero. The solutions agree with these initial values. After the switch has been closed for a long time, the two inductors appear as short circuits. Therefore, the final values of i_1 and i_2 are

$$i_1(\infty) = \frac{336(90)}{42(48)} = 15 \text{ A}, \tag{9.53}$$

$$i_2(\infty) = \frac{15(42)}{90} = 7 \text{ A}. \tag{9.54}$$

One final test involves the numerical values of the exponents and calculating the voltage drop across the 42 Ω resistor by three different methods. From the circuit, the voltage across the 42 Ω resistor (positive at the top) is

$$v = 42(i_1 - i_2) = 336 - 8.4\frac{di_1}{dt} = 48i_2 + 10\frac{di_2}{dt}. \tag{9.55}$$

You should verify that regardless of which form of Eq. (9.55) is used, the voltage is

$$v = (336 - 235.20e^{-2t} - 100.80e^{-12t})u(t) \text{ V.}$$

We are thus confident that the solutions for i_1 and i_2 are correct.

DRILL EXERCISE

9.5   The dc current and voltage sources are applied simultaneously to the circuit shown. No energy is stored in the circuit at the instant of application.

(a) Derive the s-domain expressions for V_1 and V_2.

(b) For $t > 0$, derive the time-domain expressions for v_1 and v_2.

(c) Calculate $v_1(0^+)$ and $v_2(0^+)$.

(d) Compute the steady-state values of v_1 and v_2.

ANSWER: (a) $V_1 = [5(s+3)]/[s(s+0.5)(s+2)]$, $V_2 = [2.5(s^2+6)]/[s(s+0.5)(s+2)]$;

(b) $v_1 = (15 - \frac{50}{3}e^{-0.5t} + \frac{5}{3}e^{-2t})u(t)$ V,

$v_2 = (15 - \frac{125}{6}e^{-0.5t} + \frac{25}{3}e^{-2t})u(t)$ V; (c) $v_1(0^+) = 0$, $v_2(0^+) = 2.5$ V; (d) $v_1 = v_2 = 15$ V.

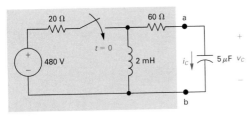

Figure 9.17 A circuit showing the use of Thévenin's equivalent in the s domain.

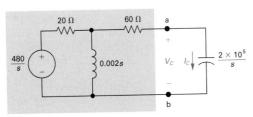

Figure 9.18 The s-domain model of the circuit shown in Fig. 9.17.

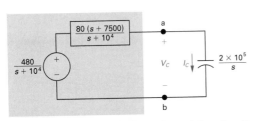

Figure 9.19 A simplified version of the circuit shown in Fig. 9.18, using a Thévenin equivalent.

The Use of Thévenin's Equivalent

In this section we show how to use Thévenin's equivalent in the s domain. Figure 9.17 shows the circuit to be analyzed. The problem is to find the capacitor current that results from closing the switch. The energy stored in the circuit prior to closing is zero.

To find i_C, we first construct the s-domain equivalent circuit and then find the Thévenin equivalent of this circuit with respect to the terminals of the capacitor. Figure 9.18 shows the s-domain circuit.

The Thévenin voltage is the open-circuit voltage across terminals a, b. Under open-circuit conditions, there is no voltage across the 60 Ω resistor. Hence

$$V_{\text{Th}} = \frac{(480/s)(0.002s)}{20 + 0.002s} = \frac{480}{s + 10^4}. \tag{9.56}$$

The Thévenin impedance seen from terminals a and b equals the 60 Ω resistor in series with the parallel combination of the 20 Ω resistor and the 2 mH inductor. Thus

$$Z_{\text{Th}} = 60 + \frac{0.002s(20)}{20 + 0.002s} = \frac{80(s + 7500)}{s + 10^4}. \tag{9.57}$$

Using the Thévenin equivalent, we reduce the circuit shown in Fig. 9.18 to the one shown in Fig. 9.19. It indicates that the capacitor current I_C equals the Thévenin voltage divided by the total series impedance. Thus,

$$I_C = \frac{480/(s + 10^4)}{[80(s + 7500)/(s + 10^4)] + [(2 \times 10^5)/s]}. \tag{9.58}$$

We simplify Eq. (9.58) to

$$I_C = \frac{6s}{s^2 + 10{,}000s + 25 \times 10^6}$$

$$= \frac{6s}{(s + 5000)^2}. \tag{9.59}$$

A partial fraction expansion of Eq. (9.59) generates

$$I_C = \frac{-30{,}000}{(s + 5000)^2} + \frac{6}{(s + 5000)}, \tag{9.60}$$

the inverse transform of which is

$$i_C = (-30,000te^{-5000t} + 6e^{-5000t})u(t) \text{ A}. \qquad (9.61)$$

We now test Eq. (9.61) to see whether it makes sense in terms of known circuit behavior. From Eq. (9.61),

$$i_C(0) = 6 \text{ A}. \qquad (9.62)$$

This result agrees with the initial current in the capacitor, as calculated from the circuit in Fig. 9.17. The initial inductor current is zero and the initial capacitor voltage is zero, so the initial capacitor current is 480/80, or 6 A. The final value of the current is zero, which also agrees with Eq. (9.61). Note also from this equation that the current reverses sign when t exceeds 6/30,000, or 200 μs. The fact that i_C reverses sign makes sense because, when the switch first closes, the capacitor begins to charge. Eventually this charge is reduced to zero because the inductor is a short circuit at $t = \infty$. The sign reversal of i_C reflects the charging and discharging of the capacitor.

Let's assume that the voltage drop across the capacitor v_C is also of interest. Once we know i_C, we find v_C by integration in the time domain; that is,

$$v_C = 2 \times 10^5 \int_{0^-}^{t} (6 - 30,000x)e^{-5000x} dx. \qquad (9.63)$$

Although the integration called for in Eq. (9.63) is not difficult, we may avoid it altogether by first finding the s-domain expression for V_C and then finding v_C by an inverse transform. Thus

$$V_C = \frac{1}{sC}I_C = \frac{2 \times 10^5}{s}\frac{6s}{(s+5000)^2}$$

$$= \frac{12 \times 10^5}{(s+5000)^2}, \qquad (9.64)$$

from which

$$v_C = 12 \times 10^5 te^{-5000t}u(t) \text{ V}. \qquad (9.65)$$

You should verify that Eq. (9.65) is consistent with Eq. (9.63) and that it also supports the observations made with regard to the behavior of i_C (see Problem 9.33).

DRILL EXERCISE

9.6   The initial charge on the capacitor in the circuit shown is zero.

 (a) Find the s-domain Thévenin equivalent circuit with respect to terminals a and b.

 (b) Find the s-domain expression for the current that the circuit delivers to a load consisting of a 1 H inductor in series with a 2 Ω resistor.

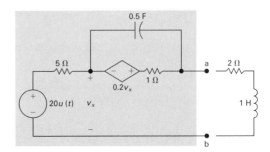

ANSWER: (a) $V_{Th} = V_{ab} = [20(s+2.4)]/[s(s+2)]$, $Z_{Th} = 5(s+2.8)/(s+2)$;
(b) $I_{ab} = [20(s+2.4)]/[s(s^2 + 9s + 18)]$.

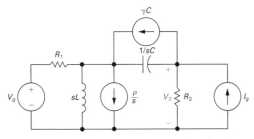

Figure 9.20 A circuit showing the use of superposition in s-domain analysis.

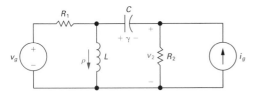

Figure 9.21 The s-domain equivalent for the circuit of Fig. 9.20.

The Use of Superposition

Because we are analyzing linear lumped-parameter circuits, we can use superposition to divide the response into components that can be identified with particular sources and initial conditions. Distinguishing these components is critical to being able to use the transfer function, which we introduce in the next section.

Figure 9.20 shows our illustrative circuit. We assume that at the instant when the two sources are applied to the circuit, the inductor is carrying an initial current of ρ amperes and the capacitor is carrying an initial voltage of γ volts. The desired response of the circuit is the voltage across the resistor R_2, labeled v_2.

Figure 9.21 shows the s-domain equivalent circuit. We opted for the parallel equivalents for L and C because we anticipated solving for V_2 using the node-voltage method.

To find V_2 by superposition, we calculate the component of V_2 resulting from each source acting alone, and then we sum the components. We begin with V_g acting alone. Opening each of the three current sources deactivates them. Figure 9.22 shows the resulting circuit. We added the node voltage V_1' to aid the analysis. The primes on V_1 and V_2 indicate that they are the components of V_1 and V_2 attributable to V_g acting alone. The two equations that describe the circuit in Fig. 9.22 are

$$\left(\frac{1}{R_1} + \frac{1}{sL} + sC\right) V_1' - sCV_2' = \frac{V_g}{R_1}, \qquad (9.66)$$

$$-sCV_1' + \left(\frac{1}{R_2} + sC\right) V_2' = 0. \qquad (9.67)$$

For convenience, we introduce the notation

$$Y_{11} = \frac{1}{R_1} + \frac{1}{sL} + sC,$$ (9.68)

$$Y_{12} = -sC;$$ (9.69)

$$Y_{22} = \frac{1}{R_2} + sC.$$ (9.70)

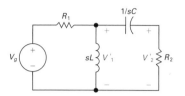

Figure 9.22 The circuit shown in Fig. 9.21 with V_g acting alone.

Substituting Eqs. (9.68)–(9.70) into Eqs. (9.66) and (9.67) gives

$$Y_{11} V_1' + Y_{12} V_2' = V_g / R_1,$$ (9.71)

$$Y_{12} V_1' + Y_{22} V_2' = 0.$$ (9.72)

Solving Eqs. (9.71) and (9.72) for V_2' gives

$$V_2' = \frac{-Y_{12}/R_1}{Y_{11} Y_{22} - Y_{12}^2} V_g.$$ (9.73)

With the current source I_g acting alone, the circuit shown in Fig. 9.21 reduces to the one shown in Fig. 9.23. Here, V_1'' and V_2'' are the components of V_1 and V_2 resulting from I_g. If we use the notation introduced in Eqs. (9.68)–(9.70), the two node-voltage equations that describe the circuit in Fig. 9.23 are

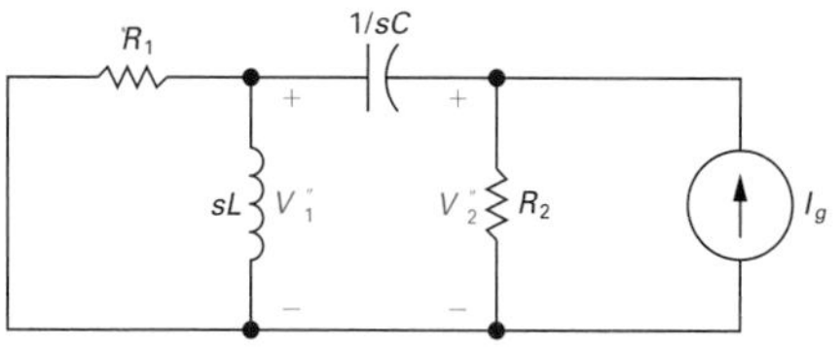

$$Y_{11} V_1'' + Y_{12} V_2'' = 0$$ (9.74)

$$Y_{12} V_1'' + Y_{22} V_2'' = I_g.$$ (9.75)

Figure 9.23 The circuit shown in Fig. 9.21, with I_g acting alone.

Solving Eqs. (9.74) and (9.75) for V_2'' yields

$$V_2'' = \frac{Y_{11}}{Y_{11} Y_{22} - Y_{12}^2} I_g.$$ (9.76)

To find the component of V_2 resulting from the initial energy stored in the inductor (V_2'''), we must solve the circuit shown in Fig. 9.24, where

$$Y_{11} V_1''' + Y_{12} V_2''' = -\rho/s,$$ (9.77)

$$Y_{12} V_1''' + Y_{22} V_2''' = 0.$$ (9.78)

Thus

$$V_2''' = \frac{Y_{12}/s}{Y_{11} Y_{22} - Y_{12}^2} \rho.$$ (9.79)

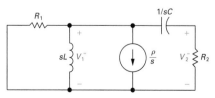

Figure 9.24 The circuit shown in Fig. 9.21, with the energized inductor acting alone.

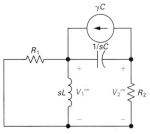

Figure 9.25 The circuit shown in Fig. 9.21, with the energized capacitor acting alone.

From the circuit shown in Fig. 9.25, we find the component of V_2 (V_2'''') resulting from the initial energy stored in the capacitor. The node-voltage equations describing this circuit are

$$Y_{11}V_1'''' + Y_{12}V_2'''' = \gamma C, \tag{9.80}$$

$$Y_{12}V_1'''' + Y_{22}V_2'''' = -\gamma C. \tag{9.81}$$

Solving for V_2'''' yields

$$V_2'''' = \frac{-(Y_{11} + Y_{12})C}{Y_{11}Y_{22} - Y_{12}^2}\gamma. \tag{9.82}$$

The expression for V_2 is

$$V_2 = V_2' + V_2'' + V_2''' + V_2''''$$

$$= \frac{-(Y_{12}/R_1)}{Y_{11}Y_{22} - Y_{12}^2}V_g + \frac{Y_{11}}{Y_{11}Y_{22} - Y_{12}^2}I_g$$

$$+ \frac{Y_{12}/s}{Y_{11}Y_{22} - Y_{12}^2}\rho + \frac{-C(Y_{11} + Y_{12})}{Y_{11}Y_{22} - Y_{12}^2}\gamma. \tag{9.83}$$

We can find V_2 without using superposition by solving the two node-voltage equations that describe the circuit shown in Fig. 9.21. Thus

$$Y_{11}V_1 + Y_{12}V_2 = \frac{V_g}{R_1} + \gamma C - \frac{\rho}{s}, \tag{9.84}$$

$$Y_{12}V_1 + Y_{22}V_2 = I_g - \gamma C. \tag{9.85}$$

You should verify in Problem 9.41 that the solution of Eqs. (9.84) and (9.85) for V_2 gives the same result as Eq. (9.83).

DRILL EXERCISE

9.7 The energy stored in the circuit shown is zero at the instant the two sources are turned on.

  (a) Find the component of v for $t > 0$ owing to the voltage source.

  (b) Find the component of v for $t > 0$ owing to the current source.

  (c) Find the expression for v when $t > 0$.

ANSWER: (a) $[(100/3)e^{-2t} - (100/3)e^{-8t}]u(t)$ V;
(b) $[(50/3)e^{-2t} - (50/3)e^{-8t}]u(t)$ V;
(c) $[50e^{-2t} - 50e^{-8t}]u(t)$ V.

9.4 THE TRANSFER FUNCTION

The **transfer function** is defined as the s-domain ratio of the Laplace transform of the output (response) to the Laplace transform of the input (source). In computing the transfer function, we restrict our attention to circuits where all initial conditions are zero. If a circuit has multiple independent sources, we can find the transfer function for each source and use superposition to find the response to all sources.

The transfer function is

$$H(s) = \frac{Y(s)}{X(s)}, \tag{9.86}$$

where $Y(s)$ is the Laplace transform of the output signal, and $X(s)$ is the Laplace transform of the input signal. Note that the transfer function depends on what is defined as the output signal. Consider, for example, the series circuit shown in Fig. 9.26. If the current is defined as the response signal of the circuit,

$$H(s) = \frac{I}{V_g} = \frac{1}{R + sL + 1/sC} = \frac{sC}{s^2 LC + RCs + 1}. \tag{9.87}$$

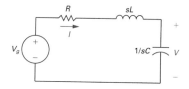

Figure 9.26 A series RLC circuit.

In deriving Eq. (9.87), we recognized that I corresponds to the output $Y(s)$ and V_g corresponds to the input $X(s)$.

If the voltage across the capacitor is defined as the output signal of the circuit shown in Fig. 9.26, the transfer function is

$$H(s) = \frac{V}{V_g} = \frac{1/sC}{R + sL + 1/sC} = \frac{1}{s^2 LC + RCs + 1}. \tag{9.88}$$

Thus, because circuits may have multiple sources and because the definition of the output signal of interest can vary, a single circuit can generate many transfer functions. Remember that when multiple sources are involved, no single transfer function can represent the total output—transfer functions associated with each source must be combined using superposition to yield the total response. Example 9.1 illustrates the computation of a transfer function for known numerical values of R, L, and C.

EXAMPLE 9.1

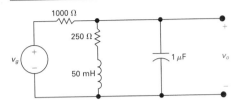

Figure 9.27 The circuit for Example 9.1.

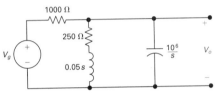

Figure 9.28 The s-domain equivalent circuit for the circuit shown in Fig. 9.27.

The voltage source v_g drives the circuit shown in Fig. 9.27. The response signal is the voltage across the capacitor, v_o.

(a) Calculate the numerical expression for the transfer function.

(b) Calculate the numerical values for the poles and zeros of the transfer function.

SOLUTION

(a) The first step in finding the transfer function is to construct the s-domain equivalent circuit, as shown in Fig. 9.28. By definition, the transfer function is the ratio of V_o/V_g, which can be computed from a single node-voltage equation. Summing the currents away from the upper node generates

$$\frac{V_o - V_g}{1000} + \frac{V_o}{250 + 0.05s} + \frac{V_o s}{10^6} = 0.$$

Solving for V_o yields

$$V_o = \frac{1000(s + 5000) V_g}{s^2 + 6000s + 25 \times 10^6}.$$

Hence the transfer function is

$$H(s) = \frac{V_o}{V_g} = \frac{1000(s + 5000)}{s^2 + 6000s + 25 \times 10^6}.$$

(b) The poles of $H(s)$ are the roots of the denominator polynomial. Therefore

$$-p_1 = -3000 - j4000 \text{ rad/s},$$

$$-p_2 = -3000 + j4000 \text{ rad/s}.$$

The zeros of $H(s)$ are the roots of the numerator polynomial; thus $H(s)$ has a zero at

$$-z_1 = -5000 \text{ rad/s}.$$

DRILL EXERCISE

9.8   **(a)** Derive the numerical expression for the transfer function V_o/I_g for the circuit shown.

   (b) Give the numerical value of each pole and zero of $H(s)$.

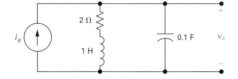

ANSWER: (a) $H(s) = 10(s + 2)/(s^2 + 2s + 10)$;
(b) $-p_1 = -1 + j3$ rad/s, $-p_2 = -1 - j3$ rad/s,
$-z = -2$ rad/s.

The Location of Poles and Zeros of H(s)

For linear lumped-parameter circuits, $H(s)$ is always a rational function of s. Complex poles and zeros always appear in conjugate pairs. The poles of $H(s)$ must lie in the left half of the s plane if the response to a bounded source (one whose values lie within some finite bounds) is to be bounded. The zeros of $H(s)$ may lie in either the right half or the left half of the s plane.

   With these general characteristics in mind, we next discuss the role that $H(s)$ plays in determining the response function. We begin with the partial fraction expansion technique for finding $y(t)$.

9.5 THE TRANSFER FUNCTION IN PARTIAL FRACTION EXPANSIONS

From Eq. 9.86 we can write the circuit output as the product of the transfer function and the driving function:

$$Y(s) = H(s)X(s). \qquad (9.89)$$

We have already noted that $H(s)$ is a rational function of s. Reference to Table 8.1 shows that $X(s)$ also is a rational function of s for the excitation functions of most interest in circuit analysis.

   Expanding the right-hand side of Eq. (9.89) into a sum of partial fractions produces a term for each pole of $H(s)$ and $X(s)$. Remember from Chapter 8 that poles are the roots of the denominator polynomial; zeros are the roots of the numerator polynomial. The terms generated by the poles of $H(s)$ give rise to the transient component of the total response, whereas the terms generated by the poles of $X(s)$ give rise to the steady-state component of the response. By steady-state response, we mean the response that exists after the transient components have become negligible. Example 9.2 illustrates these general observations.

EXAMPLE 9.2

The circuit in Example 9.1 (Fig. 9.27) is driven by a voltage source whose voltage increases linearly with time, namely, $v_g = 50tu(t)$.

(a) Use the transfer function to find v_o.

(b) Identify the transient component of the response.

(c) Identify the steady-state component of the response.

(d) Sketch v_o versus t for $0 \le t \le 1.5$ ms.

SOLUTION

(a) From Example 9.1,

$$H(s) = \frac{1000(s + 5000)}{s^2 + 6000s + 25 \times 10^6}.$$

The transform of the driving voltage is $50/s^2$; therefore, the s-domain expression for the output voltage is

$$V_o = \frac{1000(s + 5000)}{(s^2 + 6000s + 25 \times 10^6)} \frac{50}{s^2}.$$

The partial fraction expansion of V_o is

$$V_o = \frac{K_1}{s + 3000 - j4000} + \frac{K_1^*}{s + 3000 + j4000} + \frac{K_2}{s^2} + \frac{K_3}{s}.$$

We evaluate the coefficients K_1, K_2, and K_3 by using the techniques described in Section 8.7:

$$K_1 = 5\sqrt{5} \times 10^{-4} \ \underline{/79.70^\circ} \ \text{V};$$

$$K_1^* = 5\sqrt{5} \times 10^{-4} \ \underline{/-79.70^\circ} \ \text{V},$$

$$K_2 = 10 \ \text{V},$$

$$K_3 = -4 \times 10^{-4} \ \text{V}.$$

The time-domain expression for v_o is

$$v_o = [10\sqrt{5} \times 10^{-4}e^{-3000t}\cos(4000t + 79.70°)$$
$$+ 10t - 4 \times 10^{-4}]u(t) \text{ V}.$$

(b) The transient component of v_o is

$$10\sqrt{5} \times 10^{-4}e^{-3000t}\cos(4000t + 79.70°).$$

Note that this term is generated by the poles $(-3000+j4000)$ and $(-3000 - j4000)$ of the transfer function.

(c) The steady-state component of the response is

$$(10t - 4 \times 10^{-4})u(t).$$

These two terms are generated by the second-order pole (K/s^2) of the driving voltage.

(d) Figure 9.29 shows a sketch of v_o versus t. Note that the deviation from the steady-state solution $10{,}000t - 0.4$ mV is imperceptible after approximately 1 ms.

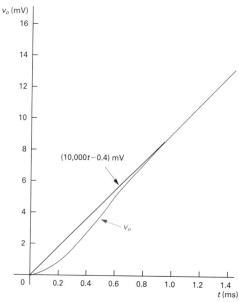

Figure 9.29 The graph of v_o versus t for Example 9.2.

DRILL EXERCISES

9.9 Find (a) the unit step and (b) the unit impulse response of the circuit shown in Drill Exercise 9.8.

ANSWER: (a) $[2+(10/3)e^{-t}\cos(3t-126.87°)]u(t)$ V; (b) $10.54e^{-t}\cos(3t - 18.43°)u(t)$ V.

9.10 The unit impulse response of a circuit is

$$v_o(t) = 10{,}000e^{-70t}\cos(240t + \theta)u(t) \text{ V},$$

where $\tan\theta = \frac{7}{24}$.

(a) Find the transfer function of the circuit.

(b) Find the unit step response of the circuit.

ANSWER: (a) $9600s/(s^2 + 140s + 62{,}500)$; (b) $40e^{-70t}\sin 240t \, u(t)$ V.

Observations on the Use of H(s) in Circuit Analysis

Example 9.2 clearly shows how the transfer function $H(s)$ relates to the response of a circuit through a partial fraction expansion. However, the example raises questions about the practicality of driving a circuit with an increasing ramp voltage that generates an increasing ramp response. Eventually the circuit components will fail under the stress of excessive voltage, and when that happens our linear model is no longer valid. The ramp response is of interest in practical applications where the ramp function increases to a maximum value over a finite time interval. If the time taken to reach this maximum value is long compared with the time constants of the circuit, the solution assuming an unbounded ramp is valid for this finite time interval.

We make two additional observations regarding Eq. (9.89). First, let's look at the response of the circuit due to a delayed input. If the input is delayed by a seconds,

$$\mathcal{L}\{x(t-a)u(t-a)\} = e^{-as}X(s),$$

and, from Eq. (9.89), the response becomes

$$Y(s) = H(s)X(s)e^{-as}. \tag{9.90}$$

If $y(t) = \mathcal{L}^{-1}\{H(s)X(s)\}$, then, from Eq. (9.90),

$$y(t-a)u(t-a) = \mathcal{L}^{-1}\{H(s)X(s)e^{-as}\}. \tag{9.91}$$

Therefore, delaying the input by a seconds simply delays the response function by a seconds. A circuit that exhibits this characteristic is said to be **time invariant**.

Second, if a unit impulse source drives the circuit, the response of the circuit equals the inverse transform of the transfer function. Thus if

$$x(t) = \delta(t), \quad \text{then } X(s) = 1$$

and

$$Y(s) = H(s). \tag{9.92}$$

Hence, from Eq. (9.92),

$$y(t) = h(t), \tag{9.93}$$

where the inverse transform of the transfer function equals the unit impulse response of the circuit. Note that this is also the natural response of the circuit because the application of an impulsive source is equivalent to instantaneously storing energy in the circuit (see Section 9.8). The subsequent release of this stored energy gives rise to the natural response (see Problem 9.74).

Actually, the unit impulse response of a circuit, $h(t)$, contains enough information to compute the response to any source that drives the circuit. The convolution integral is used to extract the response of a circuit to an arbitrary source as demonstrated in the next section.

9.6 THE TRANSFER FUNCTION AND THE CONVOLUTION INTEGRAL

The convolution integral relates the output $y(t)$ of a linear time-invariant circuit to the input $x(t)$ of the circuit and the circuit's impulse response $h(t)$. The integral relationship can be expressed in two ways:

$$y(t) = \int_{-\infty}^{\infty} h(\lambda)x(t - \lambda)\, d\lambda = \int_{-\infty}^{\infty} h(t - \lambda)x(\lambda)\, d\lambda. \quad (9.94)$$

We are interested in the convolution integral for several reasons. First, it allows us to work entirely in the time domain. Doing so may be beneficial in situations where $x(t)$ and $h(t)$ are known only through experimental data. In such cases, the transform method may be awkward or even impossible, as it would require us to compute the Laplace transform of experimental data. Second, the convolution integral introduces the concepts of memory and the weighting function into analysis. We will show how the concept of memory enables us to look at the impulse response (or the weighting function) $h(t)$ and predict, to some degree, how closely the output waveform replicates the input waveform. Finally, the convolution integral provides a formal procedure for finding the inverse transform of products of Laplace transforms.

We based the derivation of Eq. (9.94) on the assumption that the circuit is linear and time invariant. Because the circuit is linear, the principle of superposition is valid, and because it is time invariant, the amount of the response delay is exactly the same as that of the input delay. Now consider Fig. 9.30, in which the block containing $h(t)$ represents any linear time-invariant circuit whose impulse response is known, $x(t)$ represents the excitation signal and $y(t)$ represents the desired output signal.

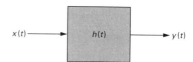

Figure 9.30 A block diagram of a general circuit.

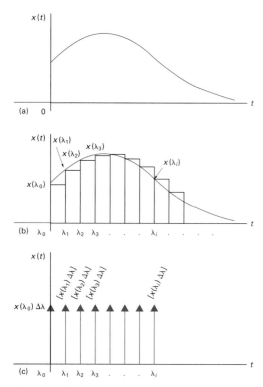

Figure 9.31 The excitation signal of $x(t)$. (a) A general excitation signal. (b) Approximating $x(t)$ with a series of pulses. (c) Approximating $x(t)$ with a series of impulses.

We assume that $x(t)$ is the general excitation signal shown in Fig. 9.31(a).

For convenience we also assume that $x(t) = 0$ for $t < 0^-$. Once you see the derivation of the convolution integral assuming $x(t) = 0$ for $t < 0^-$, the extension of the integral to include excitation functions that exist over all time becomes apparent. Note also that we permit a discontinuity in $x(t)$ at the origin, that is, a jump between 0^- and 0^+.

Now we approximate $x(t)$ by a series of rectangular pulses of uniform width $\Delta\lambda$, as shown in Fig. 9.31(b). Thus

$$x(t) = x_0(t) + x_1(t) + \cdots + x_i(t) + \cdots, \qquad (9.95)$$

where $x_i(t)$ is a rectangular pulse that equals $x(\lambda_i)$ between λ_i and λ_{i+1} and is zero elsewhere. Note that the ith pulse can be expressed in terms of step functions; that is,

$$x_i(t) = x(\lambda_i)\{u(t - \lambda_i) - u[t - (\lambda_i + \Delta\lambda)]\}.$$

The next step in the approximation of $x(t)$ is to make $\Delta\lambda$ small enough that the ith component can be approximated by an impulse function of strength $x(\lambda_i)\Delta\lambda$. Figure 9.31(c) shows the impulse representation, with the strength of each impulse shown in brackets beside each arrow. The impulse representation of $x(t)$ is

$$x(t) = x(\lambda_0)\Delta\lambda\delta(t - \lambda_0) + x(\lambda_1)\Delta\lambda\delta(t - \lambda_1)$$
$$+ \cdots + x(\lambda_i)\Delta\lambda\delta(t - \lambda_i) + \cdots. \qquad (9.96)$$

Now when $x(t)$ is represented by a series of impulse functions (which occur at equally spaced intervals of time, that is, at $\lambda_0, \lambda_1, \lambda_2, \ldots$), the response function $y(t)$ consists of the sum of a series of uniformly delayed impulse responses. The strength of each response depends on the strength of the impulse driving the circuit. For example, let's assume that the unit impulse response of the circuit contained in the box in Fig. 9.30 is the exponential decay function shown in Fig. 9.32(a).

Then the approximation of $y(t)$ is the sum of the impulse responses shown in Fig. 9.32(b).

Analytically, the expression for $y(t)$ is

$$y(t) = x(\lambda_0)\Delta\lambda h(t - \lambda_0) + x(\lambda_1)\Delta\lambda h(t - \lambda_1)$$
$$+ x(\lambda_2)\Delta\lambda h(t - \lambda_2) + \cdots$$
$$+ x(\lambda_i)\Delta\lambda h(t - \lambda_i) + \cdots. \qquad (9.97)$$

As $\Delta\lambda \to 0$, the summation in Eq. (9.97) approaches a continuous integration, or

$$\sum_{i=0}^{\infty} x(\lambda_i)h(t - \lambda_i)\Delta\lambda \to \int_{0}^{\infty} x(\lambda)h(t - \lambda)\,d\lambda. \qquad (9.98)$$

Therefore,

$$y(t) = \int_{0}^{\infty} x(\lambda)h(t - \lambda)\,d\lambda. \qquad (9.99)$$

If $x(t)$ exists over all time, then the lower limit on Eq. (9.99) becomes $-\infty$; thus, in general,

$$y(t) = \int_{-\infty}^{\infty} x(\lambda)h(t - \lambda)\,d\lambda, \qquad (9.100)$$

which is the second form of the convolution integral given in Eq. (9.94). We derive the first form of the integral from Eq. (9.100) by making a change in the variable of integration. We let $u = t - \lambda$, and then we note that $du = -d\lambda$, $u = -\infty$ when $\lambda = \infty$, and $u = +\infty$ when $\lambda = -\infty$. Now we can write Eq. (9.100) as

$$y(t) = \int_{\infty}^{-\infty} x(t - u)h(u)(-du),$$

or

$$y(t) = \int_{-\infty}^{\infty} x(t - u)h(u)(du). \qquad (9.101)$$

But because u is just a symbol of integration, Eq. (9.101) is equivalent to the first form of the convolution integral, Eq. (9.94).

The integral relationship between $y(t), h(t)$, and $x(t)$, expressed in Eq. (9.94), often is written in a shorthand notation:

$$y(t) = h(t) * x(t) = x(t) * h(t), \qquad (9.102)$$

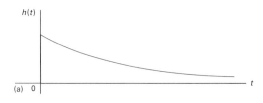
(a)

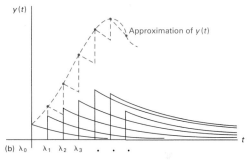
(b)

Figure 9.32 The approximation of $y(t)$. (a) The impulse response of the box shown in Fig. 9.30. (b) Summing the impulse responses.

where the asterisk signifies the integral relationship between $h(t)$ and $x(t)$. Thus $h(t) * x(t)$ is read as "$h(t)$ is convolved with $x(t)$" and implies that

$$h(t) * x(t) = \int_{-\infty}^{\infty} h(\lambda)x(t - \lambda)\,d\lambda,$$

whereas $x(t) * h(t)$ is read as "$x(t)$ is convolved with $h(t)$" and implies that

$$x(t) * h(t) = \int_{-\infty}^{\infty} x(\lambda)h(t - \lambda)\,d\lambda.$$

The integrals in Eq. (9.94) give the most general relationship for the convolution of two functions. However, in our applications of the convolution integral, we can change the lower limit to zero and the upper limit to t. Then we can write Eq. (9.94) as

$$y(t) = \int_{0}^{t} h(\lambda)x(t - \lambda)\,d\lambda = \int_{0}^{t} x(\lambda)h(t - \lambda)\,d\lambda. \qquad (9.103)$$

We change the limits for two reasons. First, for physically realizable circuits, $h(t)$ is zero for $t < 0$. In other words, there can be no impulse response before an impulse is applied. Second, we start measuring time at the instant the excitation $x(t)$ is turned on; therefore $x(t) = 0$ for $t < 0^-$.

A graphic interpretation of the convolution integrals contained in Eq. (9.103) is important in the use of the integral as a computational tool. We begin with the first integral. For purposes of discussion, we assume that the impulse response of our circuit is the exponential decay function shown in Fig. 9.33(a) and that the excitation function has the waveform shown in Fig. 9.33(b). In each of these plots, we replaced t with λ, the symbol of integration. Replacing λ with $-\lambda$ simply folds the excitation function over the vertical axis, and replacing $-\lambda$ with $t-\lambda$ slides the folded function to the right. See Figures 9.33(c) and (d). This folding operation gives rise to the term **convolution.** At any specified value of t, the response function $y(t)$ is the area under the product function $h(\lambda)x(t - \lambda)$, as shown in Fig. 9.33(e). It should be apparent from this plot why the lower limit on the convolution integral is zero and the upper limit is t. For $\lambda < 0$, the product $h(\lambda)x(t - \lambda)$ is zero because $h(\lambda)$ is zero. For $\lambda > t$, the product $h(\lambda)x(t - \lambda)$ is zero because $x(t - \lambda)$ is zero.

Figure 9.34 shows the second form of the convolution integral. Note that the product function in Fig. 9.34(e) confirms the use of zero for the lower limit and t for the upper limit.

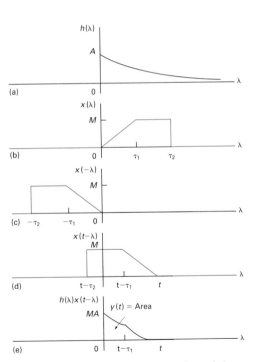

Figure 9.33 A graphic interpretation of the convolution integral $\int_{0}^{t} h(\lambda)x(t - \lambda)\,d\lambda$. (a) The impulse response. (b) The excitation function. (c) The folded excitation function. (d) The folded excitation function displaced t units. (e) The product $h(\lambda)x(t - \lambda)$.

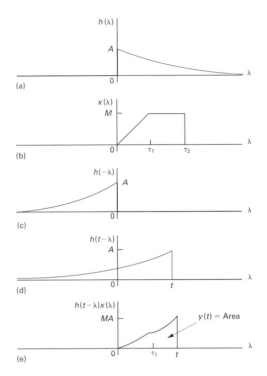

Figure 9.34 A graphic interpretation of the convolution integral $\int_0^t h(t - \lambda)x(\lambda) \, d\lambda$. (a) The impulse response. (b) The excitation function. (c) The folded excitation function. (d) The folded excitation function displaced t units. (e) the product $h(t - \lambda)x(\lambda)$.

Example 9.3 illustrates how to use the convolution integral, in conjunction with the unit impulse response, to find the response of a circuit.

EXAMPLE 9.3

The excitation voltage v_i for the circuit shown in Fig. 9.35(a) is shown in Fig. 9.35(b).

(a) Use the convolution integral to find v_o.

(b) Plot v_o over the range of $0 \le t \le 15$ s.

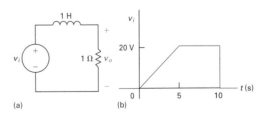

(a)   (b)

Figure 9.35 The circuit and excitation voltage for Example 9.3. (a) The circuit. (b) The excitation voltage.

SOLUTION

(a) The first step in using the convolution integral is to find the unit impulse response of the circuit. We obtain the expression for V_o from the s-domain equivalent of the circuit in Fig. 9.35(a):

$$V_o = \frac{V_i}{s + 1}(1).$$

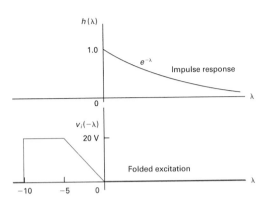

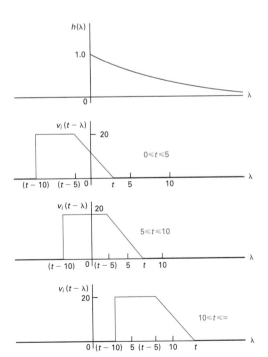

Figure 9.36 The impulse response and the folded excitation function for Example 9.3.

Figure 9.37 The displacement of $v_i(t - \lambda)$ for three different time intervals.

When v_i is a unit impulse function $\delta(t)$,

$$v_o = h(t)$$
$$= e^{-t}u(t),$$

from which

$$h(\lambda) = e^{-\lambda}u(\lambda).$$

Using the first form of the convolution integral in Eq. (9.103), we construct the impulse response and folded excitation function shown in Fig. 9.36, which are helpful in selecting the limits on the convolution integral. Sliding the folded excitation function to the right requires breaking the integration into three intervals: $0 \leq t \leq 5$; $5 \leq t \leq 10$; and $10 \leq t \leq \infty$. The breaks in the excitation function at 0, 5, and 10 s dictate these break points. Figure 9.37 shows the positioning of the folded excitation for each of these intervals. The analytical expression for v_i in the time interval $0 \leq t \leq 5$ is

$$v_i = 4t, \quad 0 \leq t \leq 5 \text{ s}.$$

Hence, the analytical expression for the folded excitation function in the interval $t - 5 \leq \lambda \leq t$ is

$$v_i(t - \lambda) = 4(t - \lambda), \quad t - 5 \leq \lambda \leq t.$$

We can now set up the three integral expressions for v_o. For $0 \leq t \leq 5$ s:

$$v_o = \int_0^t 4(t - \lambda)e^{-\lambda} \, d\lambda$$

$$= 4(e^{-t} + t - 1) \text{ V}.$$

For $5 \leq t \leq 10$ s,

$$v_o = \int_0^{t-5} 20e^{-\lambda} \, d\lambda + \int_{t-5}^t 4(t - \lambda)e^{-\lambda} \, d\lambda$$

$$= 4(5 + e^{-t} - e^{-(t-5)}) \text{ V}.$$

And for $10 \leq t \leq \infty$ s,

$$v_o = \int_{t-10}^{t-5} 20e^{-\lambda} \, d\lambda + \int_{t-5}^t 4(t - \lambda)e^{-\lambda} \, d\lambda$$

$$= 4(e^{-t} - e^{-(t-5)} + 5e^{-(t-10)}) \text{ V}.$$

(b) We have computed v_o for 1 s intervals of time, using the appropriate equation. The results are tabulated in Table 9.2 and shown graphically in Fig. 9.38.

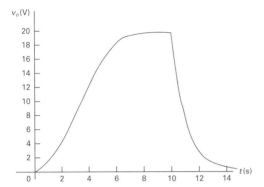

Figure 9.38 The voltage response versus time for Example 9.3.

TABLE 9.2   Numerical Values of $v_o(t)$

t	v_o	t	v_o	t	v_o
1	1.47	6	18.54	11	7.35
2	4.54	7	19.56	12	2.70
3	8.20	8	19.80	13	0.99
4	12.07	9	19.93	14	0.37
5	16.03	10	19.97	15	0.13

The Concepts of Memory and the Weighting Function

We mentioned at the beginning of this section that the convolution integral introduces the concepts of memory and the weighting function into circuit analysis. The graphic interpretation of the convolution integral is the easiest way to begin to grasp these concepts. We can view the folding and sliding of the excitation function on a timescale characterized as past, present, and future. The vertical axis, over which the excitation function $x(t)$ is folded, represents the present value; past values of $x(t)$ lie to the right of the vertical axis, and future values lie to the left. Figure 9.39 shows this description of $x(t)$. For illustrative purposes, we used the excitation function from Example 9.3.

When we combine the past, present, and future views of $x(t - \lambda)$ with the impulse response of the circuit, we see that the impulse response weights $x(t)$ according to present and past values. For example, Fig. 9.37 shows that the impulse response in Example 9.3 gives less weight to past values of $x(t)$ than to the present value of $x(t)$. In other words, the circuit retains less and less about past input values. Therefore, in Fig. 9.38, v_o quickly approaches zero when the present value of the input is zero (that is, when $t > 10$ s). In other words, because the present value of the input receives more weight than the past values, the output quickly approaches the present value of the input.

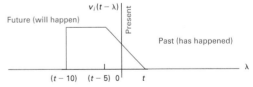

Figure 9.39 The past, present, and future values of the excitation function.

(a)

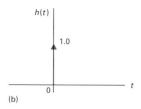

(b)

Figure 9.40 Weighting functions. (a) Perfect memory. (b) No memory.

The multiplication of $x(t - \lambda)$ by $h(\lambda)$ gives rise to the practice of referring to the impulse response as the circuit **weighting function**. The weighting function, in turn, determines how much memory the circuit has. **Memory** is the extent to which the circuit's response matches its input. For example, if the impulse response, or weighting function, is flat, as shown in Fig. 9.40(a), it gives equal weight to all values of $x(t)$, past and present. Such a circuit has a perfect memory. However, if the impulse response is an impulse function, as shown in Fig. 9.40(b), it gives no weight to past values of $x(t)$. Such a circuit has no memory. Thus the more memory a circuit has, the more distortion there is between the waveform of the excitation function and the waveform of the response function. We can show this relationship by assuming that the circuit has no memory, that is, $h(t) = A\delta(t)$, and then noting from the convolution integral that

$$y(t) = \int_0^t h(\lambda)x(t - \lambda)\,d\lambda$$

$$= \int_0^t A\delta(\lambda)x(t - \lambda)\,d\lambda$$

$$= Ax(t). \tag{9.104}$$

Equation (9.104) shows that, if the circuit has no memory, the output is a scaled replica of the input.

The circuit shown in Example 9.3 illustrates the distortion between input and output for a circuit that has some memory. This distortion is clear when we plot the input and output waveforms on the same graph, as in Fig. 9.41.

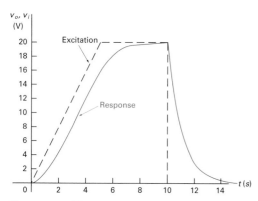

Figure 9.41 The input and output waveforms for Example 9.3.

DRILL EXERCISES

9.11 A rectangular voltage pulse $v_i = [u(t) - u(t - 1)]$ V is applied to the circuit shown. Use the convolution integral to find v_o.

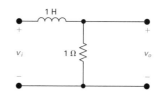

ANSWER: $v_o = 1 - e^{-t}$ V, $0 \le t \le 1$;
$v_o = (e - 1)e^{-t}$ V, $1 \le t \le \infty$.

9.12 Interchange the inductor and resistor in Drill Exercise 9.11 and again use the convolution integral to find v_o.

ANSWER: $v_o = e^{-t}$ V, $0 < t < 1$;
$v_o = (1 - e)e^{-t}$ V, $1 < t \leq \infty$.

9.7 THE TRANSFER FUNCTION AND THE STEADY-STATE SINUSOIDAL RESPONSE

Once we have computed a circuit's transfer function, we no longer need to perform a separate phasor analysis of the circuit to determine its steady-state response. Instead, we use the transfer function to relate the steady-state response to the excitation source. First we assume that

$$x(t) = A \cos(\omega t + \phi), \tag{9.105}$$

and then we use Eq. (9.89) to find the steady-state solution of $y(t)$. To find the Laplace transform of $x(t)$, we first write $x(t)$ as

$$x(t) = A \cos \omega t \cos \phi - A \sin \omega t \sin \phi, \tag{9.106}$$

from which

$$X(s) = \frac{(A \cos \phi)s}{s^2 + \omega^2} - \frac{(A \sin \phi)\omega}{s^2 + \omega^2}$$

$$= \frac{A(s \cos \phi - \omega \sin \phi)}{s^2 + \omega^2}. \tag{9.107}$$

Substituting Eq. (9.107) into Eq. (9.89) gives the s-domain expression for the response:

$$Y(s) = H(s)\frac{A(s \cos \phi - \omega \sin \phi)}{s^2 + \omega^2}. \tag{9.108}$$

We now visualize the partial fraction expansion of Eq. (9.108). The number of terms in the expansion depends on the number of poles of $H(s)$. Because $H(s)$ is not specified beyond being the transfer function of a physically realizable circuit, the expansion of Eq. (9.108) is

$$Y(s) = \frac{K_1}{s - j\omega} + \frac{K_1^*}{s + j\omega}$$

$$+ \sum \text{ terms generated by the poles of } H(s). \quad \text{(9.109)}$$

In Eq. (9.109), the first two terms result from the complex conjugate poles of the driving source; that is, $s^2 + \omega^2 = (s - j\omega)(s + j\omega)$. However, the terms generated by the poles of $H(s)$ do not contribute to the steady-state response of $y(t)$, because all these poles lie in the left half of the s plane; consequently, the corresponding time-domain terms approach zero as t increases. Thus the first two terms on the right-hand side of Eq. (9.109) determine the steady-state response. The problem is reduced to finding the partial fraction coefficient K_1:

$$K_1 = \left. \frac{H(s)A(s\cos\phi - \omega\sin\phi)}{s + j\omega} \right|_{s=j\omega}$$

$$= \frac{H(j\omega)A(j\omega\cos\phi - \omega\sin\phi)}{2j\omega}$$

$$= \frac{H(j\omega)A(\cos\phi + j\sin\phi)}{2} = \frac{1}{2}H(j\omega)Ae^{j\phi}. \quad \text{(9.110)}$$

In general, $H(j\omega)$ is a complex quantity, which we recognize by writing it in polar form; thus

$$H(j\omega) = |H(j\omega)|e^{j\theta(\omega)}. \quad \text{(9.111)}$$

Note from Eq. (9.111) that both the magnitude, $|H(j\omega)|$, and phase angle, $\theta(\omega)$, of the transfer function vary with the frequency ω. When we substitute Eq. (9.111) into Eq. (9.110), the expression for K_1 becomes

$$K_1 = \frac{A}{2}|H(j\omega)|e^{j[\theta(\omega)+\phi]}. \quad \text{(9.112)}$$

We obtain the steady-state solution for $y(t)$ by inverse-transforming Eq. (9.109) and, in the process, ignoring the terms generated by the poles of $H(s)$. Thus

$$y_{ss}(t) = A|H(j\omega)|\cos[\omega t + \phi + \theta(\omega)], \quad \text{(9.113)}$$

which indicates how to use the transfer function to find the steady-state sinusoidal response of a circuit. The amplitude of the response equals the amplitude of the source, A, times the magnitude of the transfer function, $|H(j\omega)|$. The phase angle of the response, $\phi + \theta(\omega)$, equals the phase angle of the source, ϕ, plus the phase angle of the transfer function, $\theta(\omega)$. We evaluate both $|H(j\omega)|$ and $\theta(\omega)$ at the frequency of the source, ω.

Example 9.4 illustrates how to use the transfer function to find the steady-state sinusoidal response of a circuit.

EXAMPLE 9.4

The circuit from Example 9.1 is shown in Fig. 9.42.

The sinusoidal source voltage is $120\cos(5000t + 30°)$ V. Find the steady-state expression for v_o.

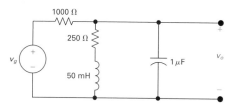

Figure 9.42 The circuit for Example 9.4.

SOLUTION

From Example 9.1,

$$H(s) = \frac{1000(s + 5000)}{s^2 + 6000s + 25 \times 10^6}.$$

The frequency of the voltage source is 5000 rad/s; hence we evaluate $H(s)$ at $H(j5000)$:

$$H(j5000) = \frac{1000(5000 + j5000)}{-25 \times 10^6 + j5000(6000) + 25 \times 10^6}$$

$$= \frac{1 + j1}{j6} = \frac{1 - j1}{6} = \frac{\sqrt{2}}{6} \; \underline{/-45°}.$$

Then, from Eq. (9.113),

$$v_{oss} = \frac{(120)\sqrt{2}}{6} \cos(5000t + 30° - 45°)$$

$$= 20\sqrt{2}\cos(5000t - 15°) \text{ V}.$$

The ability to use the transfer function to calculate the steady-state sinusoidal response of a circuit is important. Note that if we know $H(j\omega)$, we also know $H(s)$, at least theoretically. In other words, we can reverse the process; instead of using $H(s)$ to find $H(j\omega)$, we use $H(j\omega)$ to find $H(s)$. Once we know $H(s)$, we can find the response to other excitation sources. In this application, we determine $H(j\omega)$ experimentally and then construct

$H(s)$ from the data. Practically, this experimental approach is not always possible; however, in some cases it does provide a useful method for deriving $H(s)$. In theory, the relationship between $H(s)$ and $H(j\omega)$ provides a link between the time domain and the frequency domain. The transfer function is also a very useful tool in problems concerning the frequency response of a circuit.

DRILL EXERCISES

9.13 The current source in the circuit shown is delivering $10\cos 4t$ A. Use the transfer function to compute the steady-state expression for v_o.

ANSWER: $44.7\cos(4t - 63.43°)$ V.

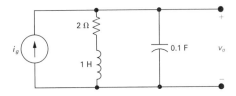

9.14 **(a)** For the circuit shown, find the steady-state expression for v_o when $v_g = 10\cos 50{,}000t$ V.

(b) Replace the 50 kΩ resistor with a variable resistor and compute the value of resistance necessary to cause v_o to lead v_g by 120°.

ANSWER: (a) $10\cos(50{,}000t + 90°)$ V;
(b) $28{,}867.51$ Ω.

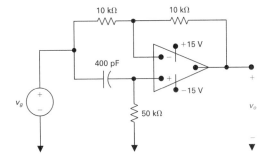

9.8 THE IMPULSE FUNCTION IN CIRCUIT ANALYSIS

Impulse functions occur in circuit analysis either because of a switching operation or because a circuit is excited by an impulsive source. The Laplace transform can be used to predict the impulsive currents and voltages created during switching and the response of a circuit to an impulsive source. We begin our discussion by showing how to create an impulse function with a switching operation.

Switching Operations

We use two different circuits to illustrate how an impulse function can be created with a switching operation: a capacitor circuit, and a series inductor circuit.

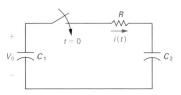

Figure 9.43 A circuit showing the creation of an impulsive current.

Capacitor Circuit

In the circuit shown in Fig. 9.43, the capacitor C_1 is charged to an initial voltage of V_0 at the time the switch is closed. The initial charge on C_2 is zero. The problem is to find the expression for $i(t)$ as $R \to 0$. Figure 9.44 shows the s-domain equivalent circuit. From Fig. 9.44,

$$I = \frac{V_0/s}{R + (1/sC_1) + (1/sC_2)}$$

$$= \frac{V_0/R}{s + (1/RC_e)}, \tag{9.114}$$

where the equivalent capacitance $C_1C_2/(C_1 + C_2)$ is replaced by C_e.

We inverse-transform Eq. (9.114) by inspection to obtain

$$i = \left(\frac{V_0}{R} e^{-t/RC_e} \right) u(t), \tag{9.115}$$

which indicates that as R decreases, the initial current (V_0/R) increases and the time constant (RC_e) decreases. Thus, as R gets smaller, the current starts from a larger initial value and then drops off more rapidly. Figure 9.45 shows these characteristics of i.

Apparently i is approaching an impulse function as R approaches zero because the initial value of i is approaching infinity and the duration of i is approaching zero. We still have to determine whether the area under the current function is independent of R. Physically, the total area under the i versus t curve represents the total charge transferred to C_2 after the switch is closed. Thus

$$\text{Area} = q = \int_{0^-}^{\infty} \frac{V_0}{R} e^{-t/RC_e} dt = V_0 C_e, \tag{9.116}$$

which says that the total charge transferred to C_2 is independent of R and equals $V_0 C_e$ coulombs. Thus, as R approaches zero, the current approaches an impulse strength $V_0 C_e$; that is,

$$i \to V_0 C_e \delta(t). \tag{9.117}$$

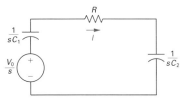

Figure 9.44 The s-domain equivalent circuit for the circuit shown in Fig. 9.43.

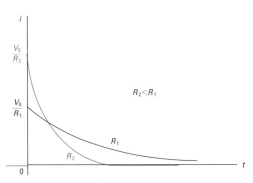

Figure 9.45 The plot of $i(t)$ versus t for two different values of R.

The physical interpretation of Eq. (9.117) is that when $R = 0$, a finite amount of charge is transferred to C_2 instantaneously. Making R zero in the circuit shown in Fig. 9.43 shows why we get an instantaneous transfer of charge. With $R = 0$, we create a contradiction when we close the switch; that is, we apply a voltage across a capacitor that has a zero initial voltage. The only way to have an instantaneous change in capacitor voltage is to have an instantaneous transfer of charge. When the switch is closed, the voltage across C_2 does not jump to V_0 but to its final value of

$$v_2 = \frac{C_1 V_0}{C_1 + C_2}. \tag{9.118}$$

We leave the derivation of Eq. (9.118) to you (see Problem 9.69).

If we set R equal to zero at the outset, the Laplace transform analysis will predict the impulsive current response. Thus,

$$I = \frac{V_0/s}{(1/sC_1) + (1/sC_2)} = \frac{C_1 C_2 V_0}{C_1 + C_2} = C_e V_0. \tag{9.119}$$

In writing Eq. (9.119), we use the capacitor voltages at $t = 0^-$. The inverse transform of a constant is the constant times the impulse function; therefore, from Eq. (9.119),

$$i = C_e V_0 \delta(t). \tag{9.120}$$

The ability of the Laplace transform to predict correctly the occurrence of an impulsive response is one reason why the transform is widely used to analyze the transient behavior of linear lumped-parameter time-invariant circuits.

Series Inductor Circuit

The circuit shown in Fig. 9.46 illustrates a second switching operation that produces an impulsive response. The problem is to find the time-domain expression for v_o after the switch has been opened. Note that opening the switch forces an instantaneous change in the current of L_2, which causes v_o to contain an impulsive component.

Figure 9.47 shows the s-domain equivalent with the switch open. In deriving this circuit, we recognized that the current in the 3 H inductor at $t = 0^-$ is 10 A, and the current in the 2 H inductor at $t = 0^-$ is zero. Using the initial conditions at $t = 0^-$ is a direct consequence of our using 0^- as the lower limit on the defining integral of the Laplace transform.

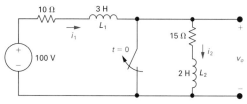

Figure 9.46 A circuit showing the creation of an impulsive voltage.

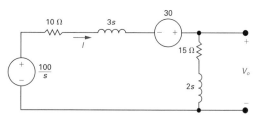

Figure 9.47 The s-domain equivalent circuit for the circuit shown in Fig. 9.46.

We derive the expression for V_0 from a single node-voltage equation. Summing the currents away from the node between the 15 Ω resistor and the 30 V source gives

$$\frac{V_0}{2s + 15} + \frac{V_0 - [(100/s) + 30]}{3s + 10} = 0. \qquad (9.121)$$

Solving for V_0 yields

$$V_0 = \frac{40(s + 7.5)}{s(s + 5)} + \frac{12(s + 7.5)}{s + 5}. \qquad (9.122)$$

We anticipate that v_o will contain an impulse term because the second term on the right-hand side of Eq. (9.122) is an improper rational function. We can express this improper fraction as a constant plus a rational function by simply dividing the denominator into the numerator; that is,

$$\frac{12(s + 7.5)}{s + 5} = 12 + \frac{30}{s + 5}. \qquad (9.123)$$

Combining Eq. (9.123) with the partial fraction expansion of the first term on the right-hand side of Eq. (9.122) gives

$$V_0 = \frac{60}{s} - \frac{20}{s + 5} + 12 + \frac{30}{s + 5}$$

$$= 12 + \frac{60}{s} + \frac{10}{s + 5}, \qquad (9.124)$$

from which

$$v_o = 12\delta(t) + (60 + 10e^{-5t})u(t) \text{ V}. \qquad (9.125)$$

Does this solution make sense? Before answering that question, let's first derive the expression for the current when $t > 0^-$. After the switch has been opened, the current in L_1 is the same as the current in L_2. If we reference the current clockwise around the mesh, the s-domain expression is

$$I = \frac{(100/s) + 30}{5s + 25} = \frac{20}{s(s + 5)} + \frac{6}{s + 5}$$

$$= \frac{4}{s} - \frac{4}{s + 5} + \frac{6}{s + 5}$$

$$= \frac{4}{s} + \frac{2}{s + 5}. \qquad (9.126)$$

Inverse-transforming Eq. (9.126) gives

$$i = (4 + 2e^{-5t})u(t) \text{ A.} \qquad (9.127)$$

Before the switch is opened, the current in L_1 is 10 A, and the current in L_2 is 0 A; from Eq. (9.127) we know that at $t = 0^+$, the current in L_1 and in L_2 is 6 A. Then, the current in L_1 changes instantaneously from 10 to 6 A, while the current in L_2 changes instantaneously from 0 to 6 A. From this value of 6 A, the current decreases exponentially to a final value of 4 A. This final value is easily verified from the circuit; that is, it should equal 100/25, or 4 A. Figure 9.48 shows these characteristics of i_1 and i_2.

How can we verify that these instantaneous jumps in the inductor current make sense in terms of the physical behavior of the circuit? First, we note that the switching operation places the two inductors in series. Any impulsive voltage appearing across the 3 H inductor must be exactly balanced by an impulsive voltage across the 2 H inductor, because the sum of the impulsive voltages around a closed path must equal zero. Faraday's law states that the induced voltage is proportional to the change in flux linkage ($v = d\lambda/dt$). Therefore, the change in flux linkage must sum to zero. In other words, the total flux linkage immediately after switching is the same as that before switching. For the circuit here, the flux linkage before switching is

$$\lambda = L_1 i_1 + L_2 i_2 = 3(10) + 2(0) = 30 \text{ Wb-turns.} \qquad (9.128)$$

Immediately after switching, it is

$$\lambda = (L_1 + L_2)i(0^+) = 5i(0^+). \qquad (9.129)$$

Combining Eqs. (9.128) and (9.129) gives

$$i(0^+) = 30/5 = 6 \text{ A.} \qquad (9.130)$$

Thus the solution for i [Eq. (9.127)] agrees with the principle of the conservation of flux linkage.

We now test the validity of Eq. (9.125). First we check the impulsive term $12\delta(t)$. The instantaneous jump of i_2 from 0 to 6 A at $t = 0$ gives rise to an impulse of strength $6\delta(t)$ in the derivative of i_2. This impulse gives rise to the $12\delta(t)$ in the voltage across the 2 H inductor. For $t > 0^+$, di_2/dt is $-10e^{-5t}$ A/s; therefore, the voltage v_o is

$$v_o = 15(4 + 2e^{-5t}) + 2(-10e^{-5t})$$

$$= (60 + 10e^{-5t})u(t) \text{ V.} \qquad (9.131)$$

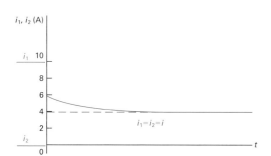

Figure 9.48 The inductor currents versus t for the circuit shown in Fig. 9.46.

Equation (9.131) agrees with the last two terms on the right-hand side of Eq. (9.125); thus we have confirmed that Eq. (9.125) does make sense in terms of known circuit behavior.

We can also check the instantaneous drop from 10 to 6 A in the current i_1. This drop gives rise to an impulse of $-4\delta(t)$ in the derivative of i_1. Therefore the voltage across L_1 contains an impulse of $-12\delta(t)$ at the origin. This impulse exactly balances the impulse across L_2; that is, the sum of the impulsive voltages around a closed path equals zero.

Impulsive Sources

Impulse functions can occur in sources as well as responses; such sources are called **impulsive sources**. An impulsive source driving a circuit imparts a finite amount of energy into the system instantaneously. A mechanical analogy is striking a bell with an impulsive clapper blow. After the energy has been transferred to the bell, the natural response of the bell determines the tone emitted (that is, the frequency of the resulting sound waves) and the tone's duration.

In the circuit shown in Fig. 9.49, an impulsive voltage source having a strength of V_0 volt-seconds is applied to a series connection of a resistor and an inductor. When the voltage source is applied, the initial energy in the inductor is zero; therefore the initial current is zero. There is no voltage drop across R, so the impulsive voltage source appears directly across L. An impulsive voltage at the terminals of an inductor establishes an instantaneous current. The current is

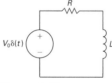

Figure 9.49 An RL circuit excited by an impulsive voltage source.

$$i = \frac{1}{L}\int_{0^-}^{t} V_0\delta(x)\,dx. \qquad (9.132)$$

Given that the integral of $\delta(t)$ over any interval that includes zero is 1, we find that Eq. (9.132) yields

$$i(0^+) = \frac{V_0}{L}\ \text{A}. \qquad (9.133)$$

Thus, in an infinitesimal moment, the impulsive voltage source has stored

$$w = \frac{1}{2}L\left(\frac{V_0}{L}\right)^2 = \frac{1}{2}\frac{V_0^2}{L}\ \text{J} \qquad (9.134)$$

in the inductor.

The current V_0/L now decays to zero in accordance with the natural response of the circuit; that is,

$$i = \frac{V_0}{L}e^{-t/\tau}u(t), \qquad (9.135)$$

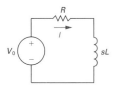

Figure 9.50 The s-domain equivalent circuit for the circuit shown in Fig. 9.49.

where $\tau = L/R$. Remember from Chapter 5 that the natural response is attributable only to passive elements releasing or storing energy, and not to the effects of sources. When a circuit is driven by only an impulsive source, the total response is completely defined by the natural response; the duration of the impulsive source is so infinitesimal that it does not contribute to any forced response.

We may also obtain Eq. (9.135) by direct application of the Laplace transform method. Figure 9.50 shows the s-domain equivalent of the circuit in Fig. 9.49.

Hence

$$I = \frac{V_0}{R + sL} = \frac{V_0/L}{s + (R/L)}, \tag{9.136}$$

$$i = \frac{V_0}{L} e^{-(R/L)t} = \frac{V_0}{L} e^{-t/\tau} u(t). \tag{9.137}$$

Thus the Laplace transform method gives the correct solution for $i \ge 0^+$.

Finally, we consider the case in which internally generated impulses and externally applied impulses occur simultaneously. The Laplace transform approach automatically ensures the correct solution for $t > 0^+$ if inductor currents and capacitor voltages at $t = 0^-$ are used in constructing the s-domain equivalent circuit and if externally applied impulses are represented by their transforms. To illustrate, we add an impulsive voltage source of $50\delta(t)$ in series with the 100 V source to the circuit shown in Fig. 9.46. Figure 9.51 shows the new arrangement.

At $t = 0^-$, $i_1(0^-) = 10$ A and $i_2(0^-) = 0$ A. The Laplace transform of $50\delta(t) = 50$. If we use these values, the s-domain equivalent circuit is as shown in Fig. 9.52.

The expression for I is

$$I = \frac{50 + (100/s) + 30}{25 + 5s}$$

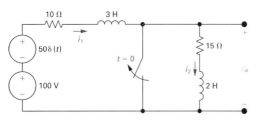

Figure 9.51 The circuit shown in Fig. 9.46 with an impulsive voltage source added in series with the 100 V source.

$$= \frac{16}{s+5} + \frac{20}{s(s+5)}$$

$$= \frac{16}{s+5} + \frac{4}{s} - \frac{4}{s+5}$$

$$= \frac{12}{s+5} + \frac{4}{s}, \tag{9.138}$$

from which

$$i(t) = (12e^{-5t} + 4)u(t) \text{ A}. \tag{9.139}$$

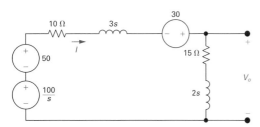

Figure 9.52 The s-domain equivalent circuit for the circuit shown in Fig. 9.51.

The expression for V_0 is

$$V_0 = (15 + 2s)I = \frac{32(s + 7.5)}{s + 5} + \frac{40(s + 7.5)}{s(s + 5)}$$

$$= 32\left(1 + \frac{2.5}{s + 5}\right) + \frac{60}{s} - \frac{20}{s + 5}$$

$$= 32 + \frac{60}{s + 5} + \frac{60}{s}, \tag{9.140}$$

from which

$$v_0 = 32\delta(t) + (60e^{-5t} + 60)u(t)\ \text{V}. \tag{9.141}$$

Now we test the results to see whether they make sense. From Eq. (9.139), we see that the current in L_1 and L_2 is 16 A at $t = 0^+$. As in the previous case, the switch operation causes i_1 to decrease instantaneously from 10 to 6 A and, at the same time, causes i_2 to increase from 0 to 6 A. Superimposed on these changes is the establishment of 10 A in L_1 and L_2 by the impulsive voltage source; that is,

$$i = \frac{1}{3 + 2}\int_{0^-}^{t} 50\delta(x)\,dx = 10\ \text{A}. \tag{9.142}$$

Therefore i_1 increases suddenly from 10 to 16 A, while i_2 increases suddenly from 0 to 16 A. The final value of i is 4 A. Figure 9.53 shows i_1, i_2, and i graphically.

We may also find the abrupt changes in i_1 and i_2 without using superposition. The sum of the impulsive voltages across L_1 (3 H) and L_2 (2 H) equals $50\delta(t)$. Thus the change in flux linkage must sum to 50; that is,

$$\Delta\lambda_1 + \Delta\lambda_2 = 50. \tag{9.143}$$

Because $\lambda = Li$, we express Eq. (9.143) as

$$3\Delta i_1 + 2\Delta i_2 = 50. \tag{9.144}$$

But because i_1 and i_2 must be equal after the switching takes place,

$$i_1(0^-) + \Delta i_1 = i_2(0^-) + \Delta i_2. \tag{9.145}$$

Then,

$$10 + \Delta i_1 = 0 + \Delta i_2. \tag{9.146}$$

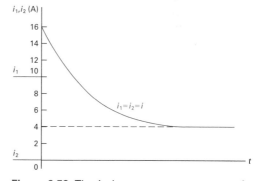

Figure 9.53 The inductor currents versus t for the circuit shown in Fig. 9.51.

Solving Eqs. (9.144) and (9.146) for Δi_1 and Δi_2 yields

$$\Delta i_1 = 6 \text{ A}, \qquad (9.147)$$

$$\Delta i_2 = 16 \text{ A}. \qquad (9.148)$$

These expressions agree with the previous check.

Figure 9.53 also indicates that the derivatives of i_1 and i_2 will contain an impulse at $t = 0$. Specifically, the derivative of i_1 will have an impulse of $6\delta(t)$, and the derivative of i_2 will have an impulse of $16\delta(t)$. The left- and right-hand sides, respectively, of Figure 9.54 illustrate the derivatives of i_1 and i_2.

Now let's turn to Eq. (9.141). The impulsive component $32\delta(t)$ agrees with the impulse of $16\delta(t)$ of di_2/dt at the origin. The terms $60e^{-5t} + 60$ agree with the fact that for $t > 0^+$,

$$v_o = 15i + 2\frac{di}{dt}.$$

We test the impulsive component of di_1/dt by noting that it produces an impulsive voltage of $(3)6\delta(t)$, or $18\delta(t)$, across L_1. This voltage, along with $32\delta(t)$ across L_2, adds to $50\delta(t)$. Thus the algebraic sum of the impulsive voltages around the mesh adds to zero.

To summarize, the Laplace transform will correctly predict the creation of impulsive currents and voltages that arise from switching. However, the s-domain equivalent circuits must be based on initial conditions at $t = 0^-$, that is, on the initial conditions that exist prior to the disturbance caused by the switching. The Laplace transform will correctly predict the response to impulsive driving sources by simply representing these sources in the s domain by their correct transforms.

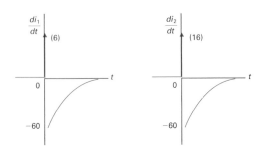

Figure 9.54 The derivatives of i_1 and i_2.

DRILL EXERCISES

9.15 The switch in the circuit shown has been in position a for a long time. The energy stored in the 1.6 μF capacitor is zero at $t = 0$. At $t = 0$, the switch moves to position b. Compute (a) $v_1(0^-)$; (b) $v_2(0^-)$; (c) $v_3(0^-)$; (d) $i(t)$; (e) $v_1(0^+)$; (f) $v_2(0^+)$; and (g) $v_3(0^+)$.

ANSWER: (a) 80 V; (b) 20 V; (c) 0 V; (d) $32\delta(t)$ μA; (e) 16 V; (f) 4 V; (g) 20 V.

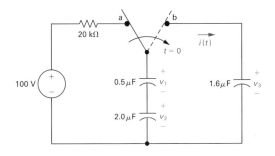

9.16 The switch in the circuit shown has been closed for a long time.
The switch opens at $t = 0$. Compute (a) $i_1(0^-)$; (b) $i_1(0^+)$; (c) $i_2(0^-)$; (d) $i_2(0^+)$; (e) $i_1(t)$; (f) $i_2(t)$; and (g) $v(t)$.

ANSWER: (a) 0.8 A; (b) 0.6 A; (c) 0.2 A; (d) −0.6 A; (e) $0.6e^{-2 \times 10^6 t}u(t)$ A; (f) $-0.6e^{-2 \times 10^6 t}u(t)$ A; (g) $-1.6 \times 10^{-3}\delta(t) - 7200e^{-2 \times 10^6 t}u(t)$ V.

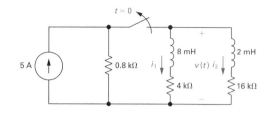

Practical Perspective

Creation of a Voltage Surge

As mentioned at the beginning of this chapter, voltage surges can occur in a circuit that is operating in the sinusoidal steady state. Our purpose is to show how the Laplace transform is used to determine the creation of a surge in voltage between the line and neutral conductors of a household circuit when a load is switched off during sinusoidal steady-state operation.

EXAMPLE 9.5

Assume the line-to-neutral voltage $\mathbf{V_0}$ in the 60 Hz circuit of Fig. 9.55 is 120 $\angle 0°$ V rms. Load R_a is absorbing 1200 W; load R_b is absorbing 1800 W; and load X_a is absorbing 350 magnetizing VAR. The inductive reactance of the line (X_ℓ) is 1 Ω. Assume $\mathbf{V_g}$ does not change after the switch opens.

(a) Calculate the initial value of $i_2(t)$ and $i_1(t)$.

(b) Construct the s-domain equivalent circuit for $t > 0$.

(c) Find $\mathbf{V_0}$, $v_0(t)$, and $v_0(0^+)$.

(d) Test the steady-state component of v_0 using phasor domain analysis.

(e) Using a computer program of your choice plot v_0 vs. t for $0 \le t \le 20$ ms.

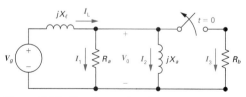

Figure 9.55 Circuit used to introduce a switching surge voltage.

SOLUTION

(a) The circuit parameters are

$$R_a = \frac{120^2}{1200} = 12\ \Omega,$$

$$R_b = \frac{(120)^2}{1800} = 8\ \Omega,$$

$$X_a = \frac{(120)^2}{350} = \frac{1440}{35}\ \Omega.$$

The branch currents are

$$\mathbf{I}_1 = \frac{120\ \angle 0°}{12} = 10\ \angle 0°\ \text{A(rms)},$$

$$\mathbf{I}_2 = \frac{120\ \angle 0°}{j\frac{1440}{35}} = -j\frac{35}{12} = \frac{35}{12}\ \angle -90°\ \text{A(rms)},$$

$$\mathbf{I}_3 = \frac{120\ \angle 0°}{8} = 15\ \angle 0°\ \text{A(rms)},$$

$$\mathbf{I}_L = \mathbf{I}_1 + \mathbf{I}_2 + \mathbf{I}_3 = 25 - j\frac{35}{12}$$

$$= 25.17\ \angle -6.65°\ \text{(rms)}.$$

Therefore

$$i_2 = \left(\frac{35}{12}\right)\sqrt{2}\cos(\omega t - 90°)\ \text{A},$$

and

$$i_L = 25.17\sqrt{2}\cos(\omega t - 6.65°)\ \text{A}.$$

It follows that

$$i_2 = (0^-) = i_2(0^+) = 0\ \text{A}$$

and

$$i_L = (0^-) = i_L(0^+) = 25\sqrt{2}\ \text{A}$$

(b) The source voltage $\mathbf{V}_g$ is

$$\mathbf{V}_g = 120\ \angle 0° + (25.17\ \angle -6.65°)(1\ \angle 90°)$$

$$= 122.92 + j25$$

$$= 125.43\ \angle 11.50°\ \text{V(rms)}.$$

Therefore

$$v_g = 125.43\sqrt{2}\cos(\omega t + 11.50°) \text{ V}$$

$$= (122.92\sqrt{2}\cos\omega t - 25\sqrt{2}\sin\omega t) \text{ V}.$$

Knowing that $\omega = 120\pi$ rad/s, we can write the Laplace transform of v_g as

$$V_g = \frac{122.92\sqrt{2}s - 25\sqrt{2}(120\pi)}{s^2 + (120\pi)^2}.$$

The next step in the construction of the s-domain circuit is to calculate the line and load inductances. Hence

$$L_\ell = \frac{1}{120\pi} = 2.65 \text{ mH},$$

$$L_a = \frac{1440}{35(120\pi)} = 109.13 \text{ mH},$$

Therefore the s domain equivalent circuit is as shown in Fig. 9.56.

(c) We begin by first solving for V_0 using circuit symbols in place of numerical values. The circuit in Fig. 9.56 is redrawn in Fig. 9.57 using circuit symbols to represent the components.

The single node-voltage equation that describes the circuit in Fig. 9.57 is

$$\frac{V_0 - (V_g + L_\ell I_o)}{sL_\ell} + \frac{V_o}{R_a} + \frac{V_o}{sL_a} = 0.$$

Solving for V_0 yields

$$V_0 = \frac{(R_a/L_\ell)V_g + I_oR_a}{s + [R_a(L_a + L_\ell)]/L_aL_\ell},$$

where $L_\ell = 1/120\pi$ H, $L_a = 12/35\pi$ H, $R_a = 12\ \Omega$, and $I_oR_a = 300\sqrt{2}$ V.

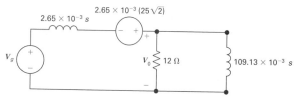

Figure 9.56 s-Domain equivalent circuit.

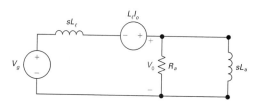

Figure 9.57 Symbolic s-domain circuit.

The numerical expression for V_0 becomes

$$V_0 = \frac{1440\pi[122.92\sqrt{2}s - 3000\pi\sqrt{2}]}{(s + 1475\pi)(s^2 + 14,440\pi^2)} + \frac{300\sqrt{2}}{s + 1475\pi}.$$

The partial fraction expansion of V_0 is

$$V_0 = \frac{K_1}{s + 1475\pi} + \frac{K_2}{s - j120\pi} + \frac{K_2^*}{s + j120\pi} + \frac{300\sqrt{2}}{s + 1475\pi}.$$

The partial fraction coefficients are

$$K_1 = -121.18\sqrt{2}\ \text{V},$$

$$K_2 = 61.03\sqrt{2}\ \underline{/6.85°}\ \text{V},$$

and

$$K_2^* = 61.03\sqrt{2}\ \underline{/-6.85°}\ \text{V}.$$

After observing that $K_1 + 300\sqrt{2}$ equals $178.82\sqrt{2}$ V, we can inverse-transform V_0 to the time domain. Thus

$$v_0 = 178.82\sqrt{2}e^{-1475\pi t} + 122.06\sqrt{2}\cos(120\pi t + 6.85°)\ \text{V}.$$

If follows directly that

$$v_0(0^+) = 178.82\sqrt{2} + 122.06\sqrt{2}\cos\ 6.85°$$

$$= 300\sqrt{2}\ \text{V}.$$

We can test this initial value of V_0 by noting at the $t = 0^+$ the initial value of i_L, i.e., $25\sqrt{2}$ A, exists in $R_a(12\ \Omega)$. Hence the initial value of V_o is $(25\sqrt{2})(12)$ or $300\sqrt{2}$ V.

(d) We now test the steady-state component of v_o using phasor-domain analysis. The phasor-domain equivalent circuit is shown in Fig. 9.58. From the solution to part (b) we have $\mathbf{V}_g = 125.43\sqrt{2}\ \underline{/11.50°}$ V(rms).

Therefore the phasor-domain node-voltage equation is

$$\frac{V_0 - 125.43\ \underline{/11.50°}}{j1} + \frac{V_0}{12} + \frac{V_0(35)}{j1440} = 0.$$

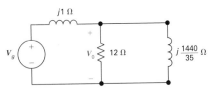

Figure 9.58 Phasor-domain equivalent circuit.

Solving for $\mathbf{V}_0$ gives

$$\mathbf{V}_0 = 122.06\ \underline{/6.85°}\ \text{V(rms)}.$$

Therefore

$$v_0 = 122.06\sqrt{2}\cos(120\pi t + 6.85°)\ \text{V}$$

which is in agreement with the steady-state component of v_0.

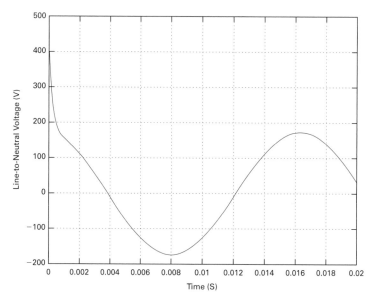

Figure 9.59 A voltage surge created by a switching operation.

(e) A MATLAB program for plotting v_0 vs. t is

```
t = 0:0.2e - 3: 20e - 3
v₀ = 178.82* sqrt(2)* exp(-1475*pi*t)
+122.06*sqrt(2)*cos(120*pi*t + 0.1196)
plot(t, V₀, 'r')
grid
ylabel('Line-to-Neutral Voltage')
xlabel('Time')
print
```

The plot is shown in Fig. 9.59.

SUMMARY

- We can represent each of the circuit elements as an s-domain equivalent circuit by Laplace-transforming the voltage-current equation for each element:

  - Resistor: $V = RI$
  - Inductor: $V = sLI - LI_0$
  - Capacitor: $V = (1/sC)I + V_0/s$

In these equations, $V = \mathcal{L}\{v\}$, $I = \mathcal{L}\{i\}$, I_0 is the initial current through the inductor, and V_0 is the initial voltage across the capacitor.

• We can perform circuit analysis in the s domain by replacing each circuit element with its s-domain equivalent circuit. The resulting equivalent circuit is solved by writing algebraic equations using the circuit analysis techniques from resistive circuits. Table 9.1 summarizes the equivalent circuits for resistors, inductors, and capacitors in the s domain.

• Circuit analysis in the s domain is particularly advantageous for solving transient response problems in linear lumped-parameter circuits when initial conditions are known. It is also useful for problems involving multiple simultaneous mesh-current or node-voltage equations, because it reduces problems to algebraic rather than differential equations.

• The **transfer function** is the s-domain ratio of a circuit's output to its input. It is represented as

$$H(s) = \frac{Y(s)}{X(s)},$$

where $Y(s)$ is the Laplace transform of the output signal, and $X(s)$ is the Laplace transform of the input signal.

• The partial fraction expansion of the product $H(s)X(s)$ yields a term for each pole of $H(s)$ and $X(s)$. The $H(s)$ terms correspond to the transient component of the total response; the $X(s)$ terms correspond to the steady-state component.

• If a circuit is driven by a unit impulse, $x(t) = \delta(t)$, then the response of the circuit equals the inverse Laplace transform of the transfer function, $y(t) = \mathcal{L}^{-1}\{H(s)\} = h(t)$.

• A **time-invariant** circuit is one for which, if the input is delayed by a seconds, the response function is also delayed by a seconds.

• The output of a circuit, $y(t)$, can be computed by convolving the input, $x(t)$, with the impulse response of the circuit, $h(t)$:

$$y(t) = h(t) * x(t) = \int_0^t h(\lambda)x(t - \lambda)\,d\lambda$$

$$= x(t) * h(t) = \int_0^t x(\lambda)h(t - \lambda)\,d\lambda.$$

A graphical interpretation of the convolution integral often provides an easier computational method to generate $y(t)$.

- We can use the transfer function of a circuit to compute its steady-state response to a sinusoidal source. To do so, make the substitution $s = j\omega$ in $H(s)$ and represent the resulting complex number as a magnitude and phase angle. If

$$x(t) = A\cos(\omega t + \phi),$$

$$H(j\omega) = |H(j\omega)|e^{j\theta(\omega)},$$

then

$$y_{ss}(t) = A|H(j\omega)|\cos[\omega t + \phi + \theta(\omega)].$$

- Laplace transform analysis correctly predicts impulsive currents and voltages arising from switching and impulsive sources. You must ensure that the s-domain equivalent circuits are based on initial conditions at $t = 0^-$, that is, prior to the switching.

PROBLEMS

9.1. Find the Norton equivalent of the circuit shown in Fig. 9.3.

9.2. Derive the s-domain equivalent circuit shown in Fig. 9.4 by expressing the inductor current i as a function of the terminal voltage v and then finding the Laplace transform of this time-domain integral equation.

9.3. Find the Thévenin equivalent of the circuit shown in Fig. 9.7.

9.4. A 1 kΩ resistor is in series with a 500 mH inductor. This series combination is in parallel with a 0.4 μF capacitor.

    (a) Express the equivalent s-domain impedance of these parallel branches as a rational function.

    (b) Determine the numerical values of the poles and zeros.

9.5. A 10 kΩ resistor, a 5 H inductor, and a 20 nF capacitor are in series.

    (a) Express the s-domain impedance of this series combination as a rational function.

    (b) Give the numerical value of the poles and zeros of the impedance.

9.6. A 5 kΩ resistor, a 6.25 H inductor, and a 40 nF capacitor are in parallel.

    (a) Express the s-domain impedance of this parallel combination as a rational function.

    (b) Give the numerical values of the poles and zeros of the impedance.

9.7. Find the poles and zeros of the impedance seen looking into the terminals a,b of the circuit shown in Fig. P9.7.

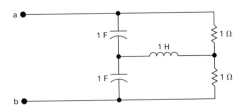

Figure P9.7

9.8. Find the poles and zeros of the impedance seen looking into the terminals a,b of the circuit shown in Fig. P9.8.

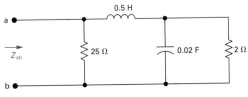

Figure P9.8

9.9. The switch in the circuit in Fig. P9.9 has been in position a for a long time. At $t = 0$, the switch moves instantaneously to position b.

(a) Construct the s-domain circuit for $t > 0$.

(b) Find V_o.

(c) Find I_L.

(d) Find v_o for $t > 0$.

(e) Find i_L for $t > 0$.

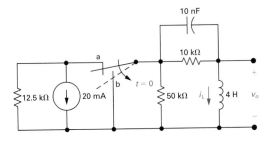

Figure P9.9

9.10. The switch in the circuit shown in Fig. P9.10 has been in position x for a long time. At $t = 0$, the switch moves instantaneously to position y.

(a) Construct an s-domain circuit for $t > 0$.

(b) Find V_o.

(c) Find v_o.

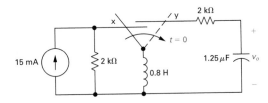

Figure P9.10

9.11. Find v_o in the circuit shown in Fig. P9.11 if $i_g = 15u(t)$ A. There is no energy stored in the circuit at $t = 0$.

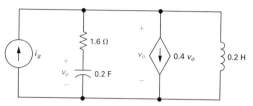

Figure P9.11

 9.12. The switch in the circuit in Fig. P9.12 has been in position a for a long time. At $t = 0$, it moves instantaneously from a to b.

(a) Construct the s-domain circuit for $t > 0$.

(b) Find $I_o(s)$.

(c) Find $i_o(t)$ for $t \geq 0$.

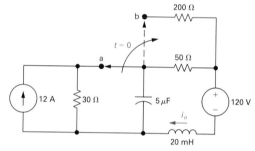

Figure P9.12

 9.13. There is no energy stored in the circuit in Fig. P9.13 at the time the switch is closed.

(a) Find v_o for $t \geq 0$.

(b) Does your solution make sense in terms of known circuit behavior? Explain.

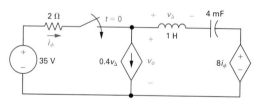

Figure P9.13

 9.14. The make-before-break switch in the circuit in Fig. P9.14 has been in position a for a long time. At $t = 0$, it moves instantaneously to position b. Find v_o for $t \geq 0$.

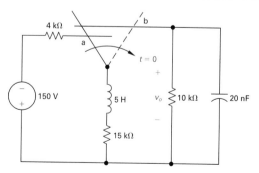

Figure P9.14

9.15. The make-before-break switch in the circuit seen in Fig. P9.15 has been in position a for a long time before moving instantaneously to position b at $t = 0$.

    (a) Construct the s-domain equivalent circuit for $t > 0$.

    (b) Find I_o and i_o.

    (c) Find V_o and v_o.

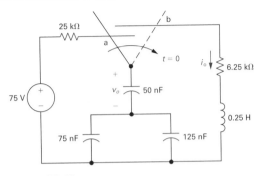

Figure P9.15

9.16. The switch in the circuit in Fig. P9.16 has been closed for a long time before opening at $t = 0$. Find v_o for $t \geq 0$.

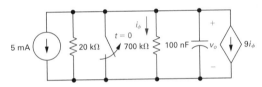

Figure P9.16

9.17. There is no energy stored in the capacitors in the circuit in Fig. P9.17 at the time the switch is closed.

    (a) Construct the s-domain circuit for $t > 0$.

    (b) Find I_1, V_1, and V_2.

    (c) Find i_1, v_1, and v_2.

    (d) Do your answers for i_1, v_1, and v_2 make sense in terms of known circuit behavior? Explain.

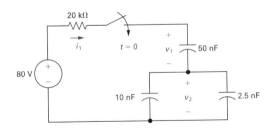

Figure P9.17

9.18. (a) Find the s-domain expression for V_o in the circuit in Fig. P9.18.

   (b) Use the s-domain expression derived in (a) to predict the initial and final values of v_o.

   (c) Find the time-domain expression for v_o.

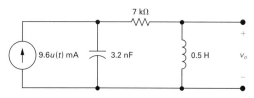

Figure P9.18

9.19. Find the time-domain expression for the current in the capacitor in Fig. P9.18. Assume the reference direction for i_C is down.

9.20. There is no energy stored in the circuit in Fig. P9.20 at the time the voltage source is energized.

   (a) Find I_o and V_o.

   (b) Find i_o and v_o for $t \geq 0$.

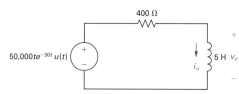

Figure P9.20

9.21. The switch in the circuit in Fig. P9.21 has been closed for a long time before opening at $t = 0$.

   (a) Construct the s-domain equivalent circuit for $t > 0$.

   (b) Find V_o.

   (c) Find v_o for $t \geq 0$.

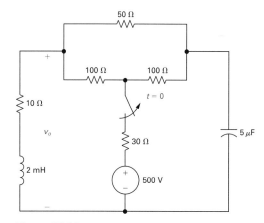

Figure P9.21

9.22. The switch in the circuit in Fig. P9.22 has been closed for a long time. At $t = 0$, the switch is opened.

    (a) Find $v_o(t)$ for $t \geq 0$.

    (b) Find $i_o(t)$ for $t \geq 0$.

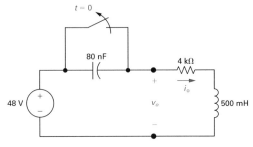

Figure P9.22

9.23. The switch in the circuit seen in Fig. P9.23 has been in position a for a long time. At $t = 0$, it moves instantaneously to position b.

    (a) Find V_o.

    (b) Find v_o.

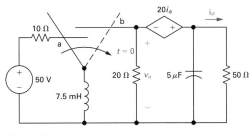

Figure P9.23

9.24. Find V_o and v_o in the circuit shown in Fig. P9.24 if the initial energy is zero and the switch is closed at $t = 0$.

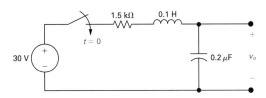

Figure P9.24

9.25. Repeat Problem 9.24 if the initial voltage on the capacitor is 20 V positive at the lower terminal.

9.26. The initial energy in the circuit in Fig. P9.26 is zero. The ideal voltage source is $600u(t)$ V.

(a) Find $V_o(s)$.

(b) Use the initial- and final-value theorems to find $v_o(0^+)$ and $v_o(\infty)$.

(c) Do the values obtained in (b) agree with known circuit behavior? Explain.

(d) Find $v_o(t)$.

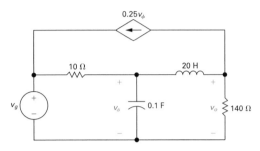

Figure P9.26

9.27. There is no energy stored in the circuit in Fig. P9.27 at the time the voltage source is turned on, and $v_g = 75u(t)$ V.

(a) Find V_o and I_o.

(b) Find v_o and i_o.

(c) Do the solutions for v_o and i_o make sense in terms of known circuit behavior? Explain.

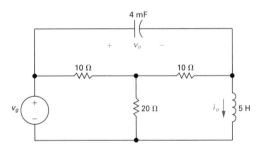

Figure P9.27

9.28. There is no energy stored in the circuit in Fig. P9.28 at $t = 0^-$.

(a) Find V_o.

(b) Find v_o.

(c) Does your solution for v_o make sense in terms of known circuit behavior? Explain.

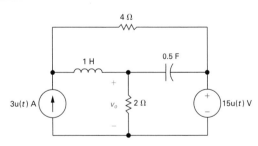

Figure P9.28

9.29. There is no energy stored in the circuit in Fig. P9.29 at the time the current source is energized.

(a) Find I_a and I_b.

(b) Find i_a and i_b.

(c) Find V_a, V_b, and V_c.

(d) Find v_a, v_b, and v_c.

(e) Assume a capacitor will break down whenever its terminal voltage is 1000 V. How long after the current source turns on will one of the capacitors break down?

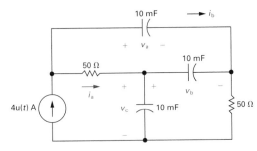

Figure P9.29

9.30. There is no energy stored in the circuit in Fig. P9.30 at the time the sources are energized.

(a) Find $I_1(s)$ and $I_2(s)$.

(b) Use the initial- and final-value theorems to check the initial and final values of $i_1(t)$ and $i_2(t)$.

(c) Find $i_1(t)$ and $i_2(t)$ for $t \geq 0$.

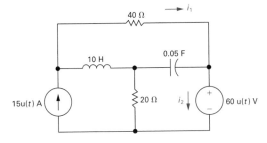

Figure P9.30

9.31. There is no energy stored in the circuit in Fig. P9.31 at the time the current source turns on. Given that $i_g = 100u(t)$ A:

(a) Find $I_o(s)$,

(b) Use the initial- and final-value theorems to find $i_o(0^+)$ and $i_o(\infty)$.

(c) Determine if the results obtained in (b) agree with known circuit behavior.

(d) Find $i_o(t)$.

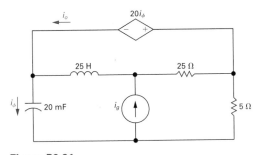

Figure P9.31

9.32. There is no energy stored in the circuit in Fig. P9.32 at $t = 0^-$.

   (a) Use the node-voltage method to find v_o.

   (b) Find the time-domain expression for i_o.

   (c) Do your answers in (a) and (b) make sense in terms of known circuit behavior? Explain.

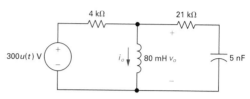

Figure P9.32

9.33. Beginning with Eq. (9.65), show that the capacitor current in the circuit in Fig. 9.17 is positive for $0 < t < 200$ μs and negative for $t > 200$ μs. Also show that at 200 μs the current is zero and that this corresponds to when dv_C/dt is zero.

9.34. The switch in the circuit shown in Fig. P9.34 has been open for a long time. The voltage of the sinusoidal source is $v_g = V_m \sin(\omega t + \phi)$. The switch closes at $t = 0$. Note that the angle ϕ in the voltage expression determines the value of the voltage at the moment when the switch closes, that is, $v_g(0) = V_m \sin \phi$.

   (a) Use the Laplace transform method to find i for $t > 0$.

   (b) Using the expression derived in (a), write the expression for the current after the switch has been closed for a long time.

   (c) Using the expression derived in (a), write the expression for the transient component of i.

   (d) Find the steady-state expression for i using the phasor method. Verify that your expression is equivalent to that obtained in (b).

   (e) Specify the value of ϕ so that the circuit passes directly into steady-state operation when the switch is closed.

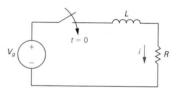

Figure P9.34

9.35. The two switches in the circuit shown in Fig. P9.35 operate simultaneously. There is no energy stored in the circuit at the instant the switches close. Find $i(t)$ for $t \geq 0^+$ by first finding the s-domain Thévenin equivalent of the circuit to the left of the terminals a,b.

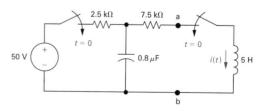

Figure P9.35

9.36. The op amp in the circuit shown in Fig. P9.36 is ideal. There is no energy stored in the capacitors at the instant the circuit is energized.

(a) Find v_o if $v_{g1} = 16u(t)$ V and $v_{g2} = 8u(t)$ V.

(b) How many milliseconds after the two voltage sources are turned on does the op amp saturate?

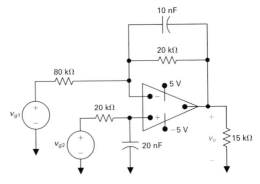

Figure P9.36

9.37. The op amp in the circuit shown in Fig. P9.37 is ideal. There is no energy stored in the circuit at the time it is energized. If $v_g = 20,000tu(t)$ V, find (a) V_o, (b) v_o, (c) how long it takes to saturate the operational amplifier, and (d) how small the rate of increase in v_g must be to prevent saturation.

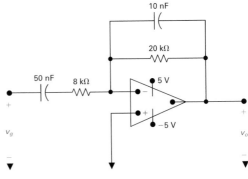

Figure P9.37

9.38. Find $v_o(t)$ in the circuit shown in Fig. P9.38 if the ideal op amp operates within its linear range and $v_g = 400u(t)$ mV. There is no energy stored in the circuit at the time it is energized.

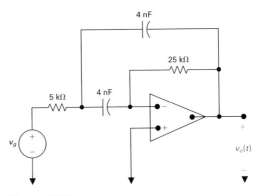

Figure P9.38

9.39. The op amp in the circuit shown in Fig. P9.39 are ideal. There is no energy stored in the capacitors at $t = 0^-$. If $v_g = 180u(t)$ mV, how many milliseconds elapse before an op amp saturates?

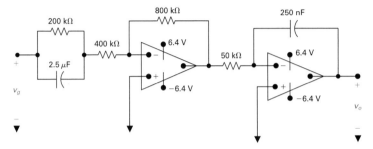

Figure P9.39

9.40. The op amp in the circuit seen in Fig. P9.40 is ideal. There is no energy stored in the capacitors at the time the circuit is energized. Determine (a) V_o, (b) v_o, and (c) how long it takes to saturate the operational amplifier.

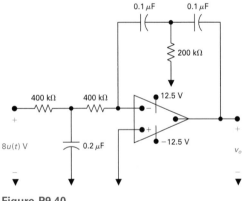

Figure P9.40

9.41. Verify that the solution of Eqs. (9.84) and (9.85) for V_2 yields the same expression as that given by Eq. (9.83).

9.42. There is no energy stored in the circuit seen in Fig. P9.42 at the time the two sources are energized.

(a) Use the principle of superposition to find V_o.

(b) Find v_o for $t > 0$.

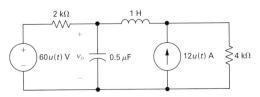

Figure P9.42

9.43. Find the numerical expression for the transfer function (V_o/V_i) of each circuit in Fig. P9.43 and give the numerical value of the poles and zeros of each transfer function.

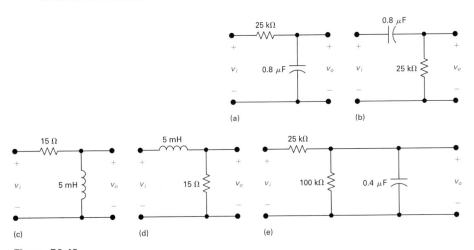

Figure P9.43

9.44. The operational amplifier in the circuit in Fig. P9.44 is ideal.

(a) Find the numerical expression for the transfer function $H(s) = V_o/V_g$.

(b) Give the numerical value of each zero and pole of $H(s)$.

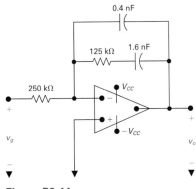

Figure P9.44

9.45. The operation amplifier in the circuit in Fig. P9.45 is ideal.

(a) Derive the numerical expression of the transfer function $H(s) = V_o/V_g$ for the circuit in Fig. P9.45.

(b) Give the numerical value of each pole and zero of $H(s)$.

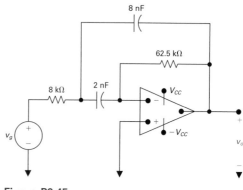

Figure P9.45

9.46. There is no energy stored in the circuit in Fig. P9.46 at the time the switch is opened. The sinusoidal current source is generating the signal $60\cos 4000t$ mA. The response signal is the current i_o.

(a) Find the transfer function I_o/I_g.

(b) Find $I_o(s)$.

(c) Describe the nature of the transient component of $i_o(t)$ without solving for $i_o(t)$.

(d) Describe the nature of the steady-state component of $i_o(t)$ without solving for $i_o(t)$.

(e) Verify the observations made in (c) and (d) by finding $i_o(t)$.

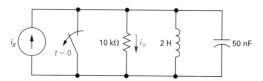

Figure P9.46

9.47. (a) Find the transfer function I_o/I_g as a function of μ for the circuit seen in Fig. P9.47.

(b) Find the largest value of μ that will produce a bounded output signal for a bounded input signal.

(c) Find i_o for $\mu = -0.5$, 0, 1, 1.5, and 2 if $i_g = 10u(t)$ A.

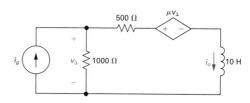

Figure P9.47

9.48. **(a)** Find $h(t) * x(t)$ when $h(t)$ and $x(t)$ are the rectangular pulses shown in Fig. P9.48(a).

(b) Repeat (a) when $x(t)$ changes to the rectangular pulse shown in Fig. P9.48(b).

(c) Repeat (a) when $h(t)$ changes to the rectangular pulse shown in Fig. P9.48(c).

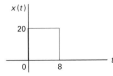

(a)

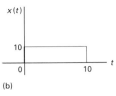

(b)

(c)

Figure P9.48

9.49. **(a)** Given $y(t) = h(t) * x(t)$, find $y(t)$ when $h(t)$ and $x(t)$ are the rectangular pulses shown in Fig. P9.49(a).

(b) Repeat (a) when $h(t)$ changes to the rectangular pulse shown in Fig. P9.49(b).

(c) Repeat (a) when $h(t)$ changes to the rectangular pulse shown in Fig. P9.49(c).

(d) Sketch $y(t)$ versus t for (a)–(c) on a single graph.

(e) Do the sketches in (d) make sense? Explain.

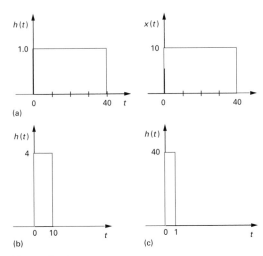

(a)

(b)

(c)

Figure P9.49

9.50. The voltage impulse response of a circuit is shown in Fig. P9.50(a). The input signal to the circuit is the rectangular voltage pulse shown in Fig. P9.50(b).

(a) Derive the equations for the output voltage. Note the range of time for which each equation is applicable.

(b) Sketch v_o for $0 \le t \le 27$ s.

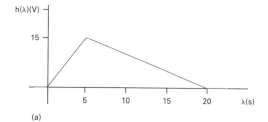

(a)

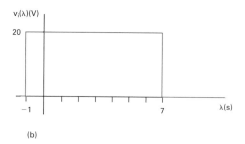

(b)

Figure P9.50

9.51. Assume the voltage impulse response of a circuit can be modeled by the triangular waveform shown in Fig. P9.51. The voltage input signal to this circuit is the step function $4u(t)$ V.

(a) Use the convolution integral to derive the expressions for the output voltage.

(b) Sketch the output voltage over the interval 0 to 25 s.

(c) Repeat parts (a) and (b) if the area under the voltage impulse response stays the same but the width of the impulse response narrows to 5 s.

(d) Which output waveform is closer to replicating the input waveform: (b) or (c)? Explain.

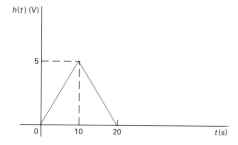

Figure P9.51

9.52. **(a)** Assume the voltage impulse response of a circuit is

$$h(t) = \begin{cases} 0, & t < 0 \\ 10e^{-4t}, & t \geq 0. \end{cases}$$

Use the convolution integral to find the output voltage if the input signal is $10u(t)$ V.

(b) Repeat (a) if the voltage impulse response is

$$h(t) = \begin{cases} 0, & t < 0 \\ 10(1 - 2t), & 0 \leq t \leq 0.5 \text{ s} \\ 0, & t \geq 0.5 \text{ s}. \end{cases}$$

(c) Plot the output voltage versus time for (a) and (b) for $0 \leq t \leq 1$ s.

9.53. **(a)** Use the convolution integral to find the output voltage of the circuit in Fig. P9.43(d) if the input voltage is the rectangular pulse shown in Fig. P9.53.

(b) Sketch $v_o(t)$ versus t for the time interval $0 \leq t \leq 100$ μs.

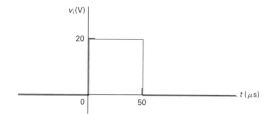

Figure P9.53

9.54. **(a)** Repeat Problem 9.53, given that the resistor in the circuit in Fig. P9.43(d) is increased to 400 Ω.

(b) Does increasing the resistor increase or decrease the memory of the circuit?

(c) Which circuit comes closest to transmitting a replica of the input voltage?

9.55. **(a)** Use the convolution integral to find i_o in the circuit in Fig. P9.55(a) if i_g is the pulse shown in Fig. P9.55(b).

(b) Use the convolution integral to find v_o.

(c) Show that your solutions for v_o and i_o are consistent by calculating i_o at 1^- ms, 1^+ms, 4^- ms, and 4^+ ms.

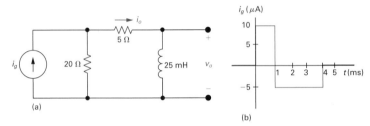

Figure P9.55

9.56. The sinusoidal voltage pulse shown in Fig. P9.56(a) is applied to the circuit shown in Fig. P9.56(b). Use the convolution integral to find the value of v_o at $t = 2.2$ s.

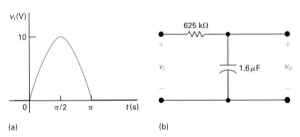

Figure P9.56

9.57. **(a)** Find the impulse response of the circuit shown in Fig. P9.57(a) if v_g is the input signal and i_o is the output signal.

(b) Given that v_g has the waveform shown in Fig. P9.57(b), use the convolution integral to find i_o.

(c) Does i_o have the same waveform as v_g? Why?

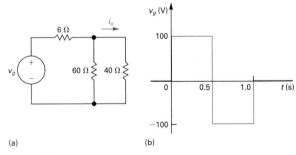

Figure P9.57

9.58. **(a)** Find the impulse response of the circuit seen in Fig. P9.58 if v_g is the input signal and v_o is the output signal.

(b) Assume that the voltage source has the waveform shown in Fig. P9.57(b). Use the convolution integral to find v_o.

(c) Sketch v_o for $0 \le t \le 1.5$ s.

(d) Does v_o have the same waveform as v_g? Why?

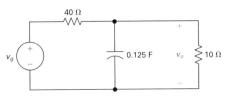

Figure P9.58

9.59. The current source in the circuit shown in Fig. P9.59(a) is generating the waveform shown in Fig. P9.59(b). Use the convolution integral to find v_o at $t = 7$ ms.

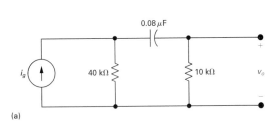

(a)

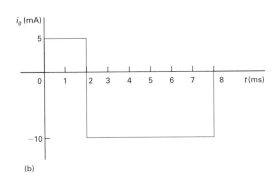

(b)

Figure P9.59

9.60. The input voltage in the circuit seen in Fig. P9.60 is

$$v_i = 10[u(t) - u(t - 0.1)] \text{ V}.$$

(a) Use the convolution integral to find v_o.

(b) Sketch v_o for $0 \le t \le 1$ s.

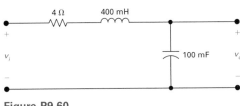

Figure P9.60

9.61. Use the convolution integral to find v_o in the circuit seen in Fig. P9.61 if $v_i = 75u(t)$ V.

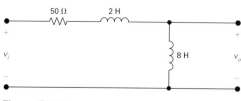

Figure P9.61

9.62. **(a)** Show that if $y(t) = h(t) * x(t)$, then $Y(s) = H(s)X(s)$.

(b) Use the result given in (a) to find $f(t)$ if

$$F(s) = \frac{a}{s(s+a)^2}.$$

 9.63. The transfer function for a linear time-invariant circuit is

$$H(s) = \frac{V_o}{V_g} = \frac{10^4(s + 6000)}{s^2 + 875s + 88 \times 10^6}.$$

If $v_g = 12.5\cos(8000t)$ V, what is the steady-state expression for v_o?

9.64. The inductor L_1 in the circuit shown in Fig. P9.64 is carrying an initial current of ρ A at the instant the switch opens and the initial current in L_2 is zero. Find (a) $v(t)$; (b) $i_1(t)$; (c) $i_2(t)$; and (d) $\lambda(t)$, where $\lambda(t)$ is the total flux linkage in the circuit.

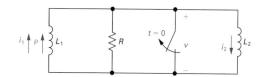

Figure P9.64

9.65. **(a)** Let $R \rightarrow \infty$ in the circuit shown in Fig. P9.64, and use the solutions derived in Problem 9.64 to find $v(t)$, $i_1(t)$, and $i_2(t)$.

(b) Let $R = \infty$ in the circuit shown in Fig. P9.64 and use the Laplace transform method to find $v(t)$, $i_1(t)$, and $i_2(t)$.

9.66. The op amp in the circuit seen in Fig. P9.66 is ideal.

    (a) Find the transfer function V_o/V_g.

    (b) Find v_o if $v_g = 10u(t)$ V.

    (c) Find the steady-state expression for v_o if $v_g = 8\cos 2000t$ V.

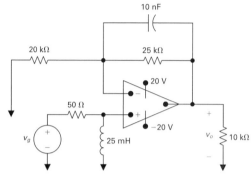

Figure P9.66

9.67. The operational amplifier in the circuit seen in Fig. P9.67 is ideal and is operating within its linear region.

    (a) Calculate the transfer function V_o/V_g.

    (b) If $v_g = 200\sqrt{10}\cos 8000t$ mV, what is the steady-state expression for v_o?

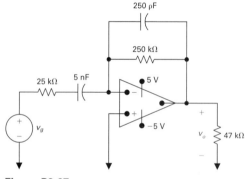

Figure P9.67

9.68. When an input voltage of $240u(t)$ V is applied to a circuit, the response is known to be

$$v_o = (75 - 100e^{-800t} + 25e^{-3200t})u(t) \text{ V.}$$

What will the steady-state response be if $v_g = 40\cos 1600t$ V?

9.69. Show that after V_0C_e coulombs are transferred from C_1 to C_2 in the circuit shown in Fig. 9.43, the voltage across each capacitor is $C_1V_0/(C_1 + C_2)$. (*Hint:* Use the conservation-of-charge principle.)

9.70. The parallel combination of R_2 and C_2 in the circuit shown in Fig. P9.70 represents the input circuit to a cathode-ray oscilloscope. The parallel combination of R_1 and C_1 is a circuit model of a compensating lead that is used to connect the CRO to the source. There is no energy stored in C_1 or C_2 at the time when the 10 V source is connected to the CRO via the compensating lead. The circuit values are $C_1 = 5$ pF, $C_2 = 20$ pF, $R_1 = 1$ MΩ, and $R_2 = 4$ MΩ.

(a) Find v_o.

(b) Find i_o.

(c) Repeat (a) and (b) given C_1 is changed to 80 pF.

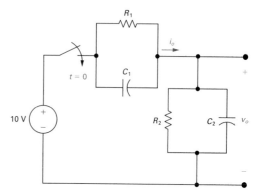

Figure P9.70

9.71. Show that if $R_1 C_1 = R_2 C_2$ in the circuit shown in Fig. P9.70, v_o will be a scaled replica of the source voltage.

9.72. The switch in the circuit shown in Fig. P9.72 has been open for a long time before closing at $t = 0$.

(a) Find v_o and i_o for $t \geq 0$.

(b) Test your solutions and make sure they are in agreement with known circuit behavior.

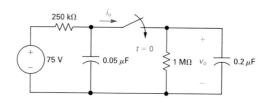

Figure P9.72

9.73. There is no energy stored in the circuit in Fig. P9.73 at the time the impulsive voltage is applied.

(a) Find $v_o(t)$ for $t \geq 0$.

(b) Does your solution make sense in terms of known circuit behavior? Explain.

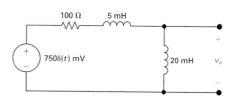

Figure P9.73

9.74. The voltage source in the circuit in Example 9.1 is changed to a unit impulse; that is, $v_g = \delta(t)$.

  (a) How much energy does the impulsive voltage source store in the capacitor?

  (b) How much energy does it store in the inductor?

  (c) Use the transfer function to find $v_o(t)$.

  (d) Show that the response found in (c) is identical to the response generated by first charging the capacitor to 1000 V and then releasing the charge to the circuit, as shown in Fig. P9.74.

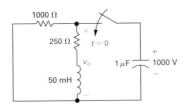

Figure P9.74

9.75. There is no energy stored in the circuit in Fig. P9.75 at the time the impulsive current is applied.

  (a) Find v_o for $t \geq 0^+$.

  (b) Does your solution make sense in terms of known circuit behavior? Explain.

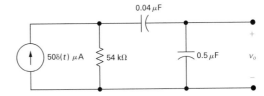

Figure P9.75

9.76. Assume the switch in the circuit in Fig. 9.55 opens at the instant the sinusoidal steady-state voltage v_o is zero and going positive, i.e., $v_o = 120\sqrt{2}\sin 120\pi t$ V.

  (a) Find $v_o(t)$ for $t \geq 0$.

  (b) Using a computer program of your choice, plot $v_o(t)$ vs. t for $0 \leq t \leq 20$ ms.

  (c) Compare the disturbance in the voltage in part (a) with that obtained in the Practical Perspective.

9.77. The purpose of this problem is to show that the line-to-neutral voltage in the circuit in Fig. 9.55 can go directly into steady state if the load R_b is disconnected from the circuit at precisely the right time. Let $v_o = V_m \cos(120\pi t - \theta°)$ V, where $V_m = 120\sqrt{2}$. Assume v_g does not change after R_b is disconnected.

  (a) Find the value of θ (in degrees) so that v_o goes directly into steady-state operation when the load R_b is disconnected from the circuit.

  (b) For the value of θ found in part (a) find $v_o(t)$ for $t \geq 0$.

  (c) Using a computer program of your choice, plot, on a single graph, for -10 ms $\leq t \leq 10$ ms, $v_o(t)$ before and after the load R_b is disconnected.

APPENDIX A

The International System of Units

Engineers compare theoretical results to experimental results and compare competing engineering designs using quantitative measures. Modern engineering is a multidisciplinary profession in which teams of engineers work together on projects, and they can communicate their results in a meaningful way only if they all use the same units of measure. The International System of Units (abbreviated SI) is used by all the major engineering societies and most engineers throughout the world; hence we use it in this book.

The SI units are based on six *defined* quantities:

- length
- mass
- time
- electric current
- thermodynamic temperature
- luminous intensity

These quantities, along with the basic unit and symbol for each, are listed in Table A.1. Although not strictly SI units, the familiar time units of minute (60 s), hour (3600 s), and so on are often used in engineering calculations. In addition, defined quantities are combined to form **derived** units. Some, such as force, energy,

TABLE A.1   **The International System of Units (SI)**

QUANTITY	BASIC UNIT	SYMBOL
Length	meter	m
Mass	kilogram	kg
Time	second	s
Electric current	ampere	A
Thermodynamic temperature	degree kelvin	K
Luminous intensity	candela	cd

TABLE A.2 Derived Units in SI

QUANTITY	UNIT NAME (SYMBOL)	FORMULA
Frequency	hertz(Hz)	s^{-1}
Force	newton (N)	$kg \cdot m/s^2$
Energy or work	joule (J)	$N \cdot m$
Power	watt (W)	J/s
Electric charge	coulomb (C)	$A \cdot s$
Electric potential	volt (V)	W/A
Electric resistance	ohm (Ω)	V/A
Electric conductance	siemens (S)	A/V
Electric capacitance	farad (F)	C/V
Magnetic flux	weber (Wb)	$V \cdot s$
Inductance	henry (H)	Wb/A

power, and electric charge, you already know through previous physics courses. Table A.2 lists the derived units used in this book.

In many cases, the SI unit is either too small or too large to use conveniently. Standard prefixes corresponding to powers of 10, as listed in Table A.3, are then applied to the basic unit. All of

TABLE A.3 Standardized Prefixes to Signify Powers of 10

PREFIX	SYMBOL	POWER
atto	a	10^{-18}
femto	f	10^{-15}
pico	p	10^{-12}
nano	n	10^{-9}
micro	μ	10^{-6}
milli	m	10^{-3}
centi	c	10^{-2}
deci	d	10^{-1}
deka	da	10
hecto	h	10^2
kilo	k	10^3
mega	M	10^6
giga	G	10^9
tera	T	10^{12}

these prefixes are correct, but engineers often use only the ones for powers divisible by 3; thus centi, deci, deka, and hecto are used rarely. Also, engineers often select the prefix that places the base number in the range between 1 and 1000. Suppose that a time calculation yields a result of 10^{-5} s, that is, 0.00001 s. Most engineers would describe this quantity as 10 μs, that is, $10^{-5} = 10 \times 10^{-6}$ s, rather than as 0.01 ms or 10,000,000 ps.

APPENDIX B

Complex Numbers

Complex numbers were invented to permit the extraction of the square roots of negative numbers. Complex numbers simplify the solution of problems that would otherwise be very difficult. The equation $x^2 + 8x + 41 = 0$, for example, has no solution in a number system that excludes complex numbers. These numbers, and the ability to manipulate them algebraically, are extremely useful in circuit analysis.

B.1 NOTATION

There are two ways to designate a complex number: with the cartesian, or rectangular, form or with the polar, or trigonometric, form. In the **rectangular form**, a complex number is written in terms of its real and imaginary components; hence

$$n = a + jb, \tag{B.1}$$

where a is the real component, b is the imaginary component, and j is by definition $\sqrt{-1}$.[1]

In the **polar form**, a complex number is written in terms of its magnitude (or modulus) and angle (or argument); hence

$$n = ce^{j\theta}, \tag{B.2}$$

where c is the magnitude, θ is the angle, e is the base of the natural logarithm, and, as before, $j = \sqrt{-1}$. In the literature, the symbol $\underline{/\theta^\circ}$ is frequently used in place of $e^{j\theta}$; that is, the polar form is written

$$n = c \; \underline{/\theta^\circ}. \tag{B.3}$$

---

[1] You may be more familiar with the notation $i = \sqrt{-1}$. In electrical and computer engineering, i is used as the symbol for current, and hence in electrical and computer engineering literature, j is used to denote $\sqrt{-1}$.

Although Eq. (B.3) is more convenient in printing text material, Eq. (B.2) is of primary importance in mathematical operations because the rules for manipulating an exponential quantity are well known. For example, because $(y^x)^n = y^{xn}$, then $(e^{j\theta})^n = e^{jn\theta}$; because $y^{-x} = 1/y^x$, then $e^{-j\theta} = 1/e^{j\theta}$; and so forth.

Because there are two ways of expressing the same complex number, we need to relate one form to the other. The transition from the polar to the rectangular form makes use of Euler's identity:

$$e^{\pm j\theta} = \cos\theta \pm j\sin\theta. \tag{B.4}$$

A complex number in polar form can be put in rectangular form by writing

$$ce^{j\theta} = c(\cos\theta + j\sin\theta)$$

$$= c\cos\theta + jc\sin\theta$$

$$= a + jb. \tag{B.5}$$

The transition from rectangular to polar form makes use of the geometry of the right triangle, namely,

$$a + jb = (\sqrt{a^2 + b^2})e^{j\theta}$$

$$= ce^{j\theta}, \tag{B.6}$$

where

$$\tan\theta = b/a. \tag{B.7}$$

It is not obvious from Eq. (B.7) in which quadrant the angle θ lies. The ambiguity can be resolved by a graphical representation of the complex number.

B.2 THE GRAPHICAL REPRESENTATION OF A COMPLEX NUMBER

A complex number is represented graphically on a complex-number plane, which uses the horizontal axis for plotting the real component and the vertical axis for plotting the imaginary component. The angle of the complex number is measured counterclockwise from the positive real axis. The graphical plot of the complex number $n = a + jb = c \angle\theta°$, if we assume that a and b are both positive, is shown in Fig. B.1.

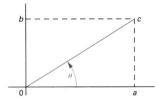

Figure B.1 The graphical representation of $a + jb$ when a and b are both positive.

This plot makes very clear the relationship between the rectangular and polar forms. Any point in the complex-number plane is uniquely defined by giving either its distance from each axis (that is, a and b) or its radial distance from the origin (c) and the angle of the radial measurement θ.

It follows from Fig. B.1 that θ is in the first quadrant when a and b are both positive, in the second quadrant when a is negative and b is positive, in the third quadrant when a and b are both negative, and in the fourth quadrant when a is positive and b is negative. These observations are illustrated in Fig. B.2, where we have plotted $4 + j3$, $-4 + j3$, $-4 - j3$, and $4 - j3$.

Note that we can also specify θ as a clockwise angle from the positive real axis. Thus in Fig. B.2(c) we could also designate $-4 - j3$ as $5\ \angle{-143.13°}$. In Fig. B.2(d) we observe that $5\ \angle{323.13°} = 5\ \angle{-36.87°}$. It is customary to express θ in terms of negative values when θ lies in the third or fourth quadrant.

The graphical interpretation of a complex number also shows the relationship between a complex number and its conjugate. The **conjugate of a complex number** is formed by reversing the sign of its imaginary component. Thus the conjugate of $a + jb$ is $a - jb$, and the conjugate of $-a + jb$ is $-a - jb$. When we write a complex number in polar form, we form its conjugate simply by reversing the sign of the angle θ. Therefore the conjugate of $c\ \angle{\theta°}$ is $c\ \angle{-\theta°}$. The conjugate of a complex number is designated with an asterisk. In other words, n^* is understood to be the conjugate of n. Figure B.3 shows two complex numbers and their conjugates plotted on the complex-number plane.

Note that conjugation simply reflects the complex numbers about the real axis.

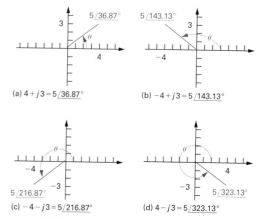

(a) $4 + j3 = 5\ \angle{36.87°}$

(b) $-4 + j3 = 5\ \angle{143.13°}$

(c) $-4 - j3 = 5\ \angle{216.87°}$

(d) $4 - j3 = 5\ \angle{323.13°}$

Figure B.2 The graphical representation of four complex numbers.

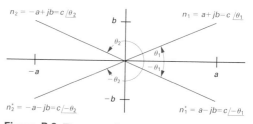

Figure B.3 The complex numbers n_1 and n_2 and their conjugates n_1^* and n_2^*

B.3   ARITHMETIC OPERATIONS

Addition(Subtraction)

To add or subtract complex numbers, we must express the numbers in rectangular form. Addition involves adding the real parts of the complex numbers to form the real part of the sum, and the imaginary parts to form the imaginary part of the sum. Thus, if we are given

$$n_1 = 8 + j16$$

and

$$n_2 = 12 - j3,$$

then

$$n_1 + n_2 = (8 + 12) + j(16 - 3) = 20 + j13.$$

Subtraction follows the same rule. Thus

$$n_2 - n_1 = (12 - 8) + j(-3 - 16) = 4 - j19.$$

If the numbers to be added or subtracted are given in polar form, they are first converted to rectangular form. For example, if

$$n_1 = 10 \; \underline{/53.13^\circ}$$

and

$$n_2 = 5 \; \underline{/-135^\circ},$$

then

$$n_1 + n_2 = 6 + j8 - 3.535 - j3.535$$

$$= (6 - 3.535) + j(8 - 3.535)$$

$$= 2.465 + j4.465 = 5.10 \; \underline{/61.10^\circ},$$

and

$$n_1 - n_2 = 6 + j8 - (-3.535 - j3.535)$$

$$= 9.535 + j11.535$$

$$= 14.97 \; \underline{/50.42^\circ}.$$

Multiplication (Division)

Multiplication or division of complex numbers can be carried out with the numbers written in either rectangular or polar form. However, in most cases, the polar form is more convenient. As an example, let's find the product $n_1 n_2$ when $n_1 = 8 + j10$ and $n_2 = 5 - j4$. Using the rectangular form, we have

$$n_1 n_2 = (8 + j10)(5 - j4) = 40 - j32 + j50 + 40$$

$$= 80 + j18$$

$$= 82 \; \underline{/12.68^\circ}.$$

If we use the polar form, the multiplication $n_1 n_2$ becomes

$$n_1 n_2 = (12.81 \; \underline{/51.34^\circ})(6.40 \; \underline{/-38.66^\circ})$$

$$= 82 \; \underline{/12.68^\circ}$$

$$= 80 + j18.$$

The first step in dividing two complex numbers in rectangular form is to multiply the numerator and denominator by the conjugate of the denominator. This reduces the denominator to a real number. We then divide the real number into the new numerator. As an example, let's find the value of n_1/n_2, where $n_1 = 6 + j3$ and $n_2 = 3 - j1$. We have

$$\frac{n_1}{n_2} = \frac{6 + j3}{3 - j1} = \frac{(6 + j3)(3 + j1)}{(3 - j1)(3 + j1)}$$

$$= \frac{18 + j6 + j9 - 3}{9 + 1}$$

$$= \frac{15 + j15}{10} = 1.5 + j1.5$$

$$= 2.12 \ \angle 45°.$$

In polar form, the division of n_1 by n_2 is

$$\frac{n_1}{n_2} = \frac{6.71 \ \angle 26.57°}{3.16 \ \angle -18.43°} = 2.12 \ \angle 45°$$

$$= 1.5 + j1.5.$$

B.4   USEFUL IDENTITIES

In working with complex numbers and quantities, the following identities are very useful:

$$\pm j^2 = \mp 1, \tag{B.8}$$

$$(-j)(j) = 1, \tag{B.9}$$

$$j = \frac{1}{-j}, \tag{B.10}$$

$$e^{\pm j\pi} = -1, \tag{B.11}$$

$$e^{\pm j\pi/2} = \pm j. \tag{B.12}$$

Given that $n = a + jb = c \ \angle \theta°$, it follows that

$$nn^* = a^2 + b^2 = c^2, \tag{B.13}$$

$$n + n^* = 2a, \tag{B.14}$$

$$n - n^* = j2b, \tag{B.15}$$

$$n/n^* = 1 \ \angle 2\theta°. \tag{B.16}$$

B.5 THE INTEGER POWER OF A COMPLEX NUMBER

To raise a complex number to an integer power k, it is easier to first write the complex number in polar form. Thus

$$n^k = (a + jb)^k$$
$$= (ce^{j\theta})^k = c^k e^{jk\theta}$$
$$= c^k (\cos k\theta + j \sin k\theta).$$

For example,

$$(2e^{j12°})^5 = 2^5 e^{j60°} = 32 e^{j60°}$$
$$= 16 + j27.71,$$

and

$$(3 + j4)^4 = (5e^{j53.13°})^4 = 5^4 e^{j212.52°}$$
$$= 625 e^{j212.52°}$$
$$= -527 - j336.$$

B.6 THE ROOTS OF A COMPLEX NUMBER

To find the kth root of a complex number, we must recognize that we are solving the equation

$$x^k - ce^{j\theta} = 0, \tag{B.17}$$

which is an equation of the kth degree and therefore has k roots. To find the k roots, we first note that

$$ce^{j\theta} = ce^{j(\theta+2\pi)} = ce^{j(\theta+4\pi)} = \cdots . \tag{B.18}$$

It follows from Eqs. (B.17) and (B.18) that

$$x_1 = (ce^{j\theta})^{1/k} = c^{1/k} e^{j\theta/k}, \tag{B.19}$$

$$x_2 = [ce^{j(\theta+2\pi)}]^{1/k} = c^{1/k} e^{j(\theta+2\pi)/k}, \tag{B.20}$$

$$x_3 = [ce^{j(\theta+4\pi)}]^{1/k} = c^{1/k} e^{j(\theta+4\pi)/k}, \tag{B.21}$$

$$\vdots$$

We continue the process outlined by Eqs. (B.19), (B.20), and (B.21) until the roots start repeating. This will happen when the multiple of π is equal to $2k$. For example, let's find the four roots of $81e^{j60°}$. We have

$$x_1 = 81^{1/4}e^{j60/4} = 3e^{j15°},$$

$$x_2 = 81^{1/4}e^{j(60+360)/4} = 3e^{j105°},$$

$$x_3 = 81^{1/4}e^{j(60+720)/4} = 3e^{j195°},$$

$$x_4 = 81^{1/4}e^{j(60+1080)/4} = 3e^{j285°},$$

$$x_5 = 81^{1/4}e^{j(60+1440)/4} = 3e^{j375°} = 3e^{j15°}.$$

Here, x_5 is the same as x_1, so the roots have started to repeat. Therefore we know the four roots of $81e^{j60°}$ are the values given by x_1, x_2, x_3, and x_4.

It is worth noting that the roots of a complex number lie on a circle in the complex-number plane. The radius of the circle is $c^{1/k}$. The roots are uniformly distributed around the circle, the angle between adjacent roots being equal to $2\pi/k$ radians, or $360/k$ degrees. The four roots of $81e^{j60°}$ are shown plotted in Fig. B.4.

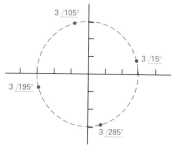

Figure B.4 The four roots of $81e^{j60°}$.

An Abbreviated Table of Trigonometric Identities

1. $\sin(\alpha \pm \beta) = \sin\alpha\cos\beta \pm \cos\alpha\sin\beta$

2. $\cos(\alpha \pm \beta) = \cos\alpha\cos\beta \mp \sin\alpha\sin\beta$

3. $\sin\alpha + \sin\beta = 2\sin\dfrac{\alpha+\beta}{2}\cos\dfrac{\alpha-\beta}{2}$

4. $\sin\alpha - \sin\beta = 2\cos\left(\dfrac{\alpha+\beta}{2}\right)\sin\left(\dfrac{\alpha-\beta}{2}\right)$

5. $\cos\alpha + \cos\beta = 2\cos\left(\dfrac{\alpha+\beta}{2}\right)\cos\left(\dfrac{\alpha-\beta}{2}\right)$

6. $\cos\alpha - \cos\beta = -2\sin\left(\dfrac{\alpha+\beta}{2}\right)\sin\left(\dfrac{\alpha-\beta}{2}\right)$

7. $2\sin\alpha\sin\beta = \cos(\alpha-\beta) - \cos(\alpha+\beta)$

8. $2\cos\alpha\cos\beta = \cos(\alpha-\beta) + \cos(\alpha+\beta)$

9. $2\sin\alpha\cos\beta = \sin(\alpha+\beta) + \sin(\alpha-\beta)$

10. $\sin 2\alpha = 2\sin\alpha\cos\alpha$

11. $\cos 2\alpha = 2\cos^2\alpha - 1 = 1 - 2\sin^2\alpha$

12. $\cos^2\alpha = \dfrac{1}{2} + \dfrac{1}{2}\cos 2\alpha$

13. $\sin^2\alpha = \dfrac{1}{2} - \dfrac{1}{2}\cos 2\alpha$

14. $\tan(\alpha \pm \beta) = \dfrac{\tan\alpha \pm \tan\beta}{1 \mp \tan\alpha\tan\beta}$

15. $\tan 2\alpha = \dfrac{2\tan\alpha}{1 - \tan^2\alpha}$

An Abbreviated Table of Integrals

1. $\displaystyle\int xe^{ax}\,dx = \frac{e^{ax}}{a^2}(ax - 1)$

2. $\displaystyle\int x^2e^{ax}\,dx = \frac{e^{ax}}{a^3}(a^2x^2 - 2ax + 2)$

3. $\displaystyle\int x\sin ax\,dx = \frac{1}{a^2}\sin ax - \frac{x}{a}\cos ax$

4. $\displaystyle\int x\cos ax\,dx = \frac{1}{a^2}\cos ax + \frac{x}{a}\sin ax$

5. $\displaystyle\int e^{ax}\sin bx\,dx = \frac{e^{ax}}{a^2 + b^2}(a\sin bx - b\cos bx)$

6. $\displaystyle\int e^{ax}\cos bx\,dx = \frac{e^{ax}}{a^2 + b^2}(a\cos bx + b\sin bx)$

7. $\displaystyle\int \frac{dx}{x^2 + a^2} = \frac{1}{a}\tan^{-1}\frac{x}{a}$

8. $\displaystyle\int \frac{dx}{(x^2 + a^2)^2} = \frac{1}{2a^2}\left(\frac{x}{x^2 + a^2} + \frac{1}{a}\tan^{-1}\frac{x}{a}\right)$

9. $\displaystyle\int \sin ax\sin bx\,dx = \frac{\sin(a - b)x}{2(a - b)} - \frac{\sin(a + b)x}{2(a + b)}, \quad a^2 \neq b^2$

10. $\displaystyle\int \cos ax\cos bx\,dx = \frac{\sin(a - b)x}{2(a - b)} + \frac{\sin(a + b)x}{2(a + b)}, \quad a^2 \neq b^2$

11. $\displaystyle\int \sin ax\cos bx\,dx = -\frac{\cos(a - b)x}{2(a - b)} - \frac{\cos(a + b)x}{2(a + b)}, \quad a^2 \neq b^2$

12. $\displaystyle\int \sin^2 ax\,dx = \frac{x}{2} - \frac{\sin 2ax}{4a}$

13. $\displaystyle\int \cos^2 ax\,dx = \frac{x}{2} + \frac{\sin 2ax}{4a}$

14. $\displaystyle\int_0^\infty \frac{a\,dx}{a^2 + x^2} = \begin{cases} \frac{\pi}{2}, & a > 0; \\ 0, & a = 0; \\ \frac{-\pi}{2}, & a < 0 \end{cases}$

15. $\displaystyle\int_0^\infty \frac{\sin ax}{x}\, dx = \begin{cases} \frac{\pi}{2}, & a > 0; \\ \frac{-\pi}{2}, & a < 0 \end{cases}$

16. $\displaystyle\int x^2 \sin ax\, dx = \frac{2x}{a^2}\sin ax - \frac{a^2x^2 - 2}{a^3}\cos ax$

17. $\displaystyle\int x^2 \cos ax\, dx = \frac{2x}{a^2}\cos ax + \frac{a^2x^2 - 2}{a^3}\sin ax$

18. $\displaystyle\int e^{ax} \sin^2 bx\, dx = \frac{e^{ax}}{a^2 + 4b^2}\left[(a\sin bx - 2b\cos bx)\sin bx + \frac{2b^2}{a}\right]$

19. $\displaystyle\int e^{ax} \cos^2 bx\, dx = \frac{e^{ax}}{a^2 + 4b^2}\left[(a\cos bx + 2b\sin bx)\cos bx + \frac{2b^2}{a}\right]$

APPENDIX E

Answers to Selected Problems

Chapter 1

1.1 $6 \sin 4000t$ mC

1.3   (a) 600 W from A to B;

     (b) -2000 W from B to A;

     (c) -2400 W from B to A;

     (d) 4800 W from A to B

1.7   16.2 kJ

1.9   (a) 5.39 W;

     (b) 9 J

1.11   (a) 0.5 W;

     (b) 2 mJ

1.15   (a) 0.634 s;

     (b) 5.196 mW;

     (c) 2.366 s;

     (d) 5.196 mW;

     (e) 0 mJ, 4 mJ, 4 mJ, 0 mJ

1.19   (a) a, c, e, and f;

     (b) 150 W, 150 W

1.21   Valid, 720 W

1.25   Not valid, it violates Kirchhoff's current law

1.29   (a) $i_g = 4$ A, $i_a = 2.4$ A;

     (b) 144 V;

     (c) 768 W

1.31   (a) 2 A;

     (b) 0.5 A;

     (c) 40 V;

     (d) $p_{4\Omega} = 25$ W, $p_{20\Omega} = 80$ W, $p_{80\Omega} = 20$ W;

     (e) 125 W

1.33   (a) -9 A;

     (b) $p_{8\Omega} = 32$ W, $p_{12\Omega} = 48$ W, $p_{4\Omega} = 16$ W, $p_{24\Omega} = 216$ W, $p_{4\Omega} = 144$ W, $p_{6\Omega} = 150$ W $p_{10\Omega} = 250$ W, $p_{12\Omega} = 192$ W;

     (c) 152 V;

     (d) 1368 W

1.39   (a) 60 Ω;

     (b) 2000 W

1.41   (a) 4.5 V;

     (b) 741 mW

Chapter 2

2.3   80 W

2.7   (a) 10 Ω, 27 Ω, 24 Ω;

     (b) 40 W, 768 W, 864 W

2.11   (a) 1 V;

     (b) $v_x = v_s/30$

2.13 $i_o = 2$ A, $i_g = 12.5$ A

2.17   (a) 20 V;

     (b) 15 V;

     (c) 0.80;

     (d) 0.60

2.21   (a) 3 mA;

     (b) 105 mW

2.23   (a) 120 V;

     (b) 3.75 kW;

     (c) 1300 W;

     (d) 5050 W

2.27 $i_{o\ 15\ A} = 4$ A, $i_{o\ 50\ V} = -2$ A, $i_o = 2$ A

Chapter 3

3.3 $v_1 = 20$ V, 180 W

3.5   10 V

3.9 $v_1 = 100$ V, $v_2 = 20$ V

3.11   1.5 V

3.15   (a) −30 W;

(b) 300 W;

(c) 300 W

3.19   26 V

3.21   (a) 50 V;

(b) 31.875 W;

(c) 270 W

3.23   (a) 165 W;

(b) 165 W

3.25   602.5 W

3.29   (a) $i_a = 9.8$ A, $i_b = −0.2$ A, $i_c = −10$ A;

(b) $i_a = −1.72$ A, $i_b = 1.08$ A, $i_c = 2.8$ A

3.33   1650 W

3.35   (a) $i_a = 5.7$ A, $i_b = 4.6$ A, $i_c = 0.97$ A, $i_d = −1.1$ A, $i_e = 3.63$ A;

(b) 1319.685 W

3.39   1484 W

3.41   (a) 800 W;

(b) 40.4%

3.43   2 A

3.49   (a) The node-voltage method has the advantage of having to solve three simultaneous equations for one unknown voltage provided the connection at either the top or bottom of the circuit is used as the reference node.

(b) 5329.74 W

3.53

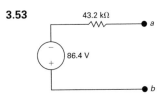

3.55

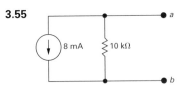

3.59

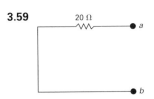

3.61   (a) 20 kΩ;

(b) 125 mW;

(c) 23.7%

3.63   22.5 Ω and 2.5 Ω

3.67   34.7%

3.71   (a) 35 Ω;

(b) 505.4 W;

(c) 21,364 W

3.75 $v_1 = 39.5833$ V, $v_2 = 102.50$ V; pspice solution: $v_1 = 39.583$ V, $v_2 = 102.50$ V

3.77 $v_1 = 52.0833$ V, $v_2 = 117.5$ V; pspice solution: $v_1 = 52.0830$ V, $v_2 = 117.5$ V

Chapter 4

4.1   8.25 V

4.3   (a) −12 V;

(b) −18 V;

(c) 10 V;

(d) −14 V;

(e) 18 V;

(f) $2.8125 \le v_a \le 7.3125$ V

4.5   −3.1 mA

4.9   (a) −2.4 V;

(b) $−61 \le v_c \le 19$ V

4.11   (a) −4 V;

(b) $−2.5 \le v_b \le −1.3$ V

4.17 $-84.6\ \mu A$

4.21   (a) 7.56 V;

(b) $-3.97 \le v_g \le 3.97$ V;

(c) 35 kΩ

4.23   (a) 4.44 V;

(b) $i_a = -4.86\ \mu A,\ i_b = 4.86\ \mu A$;

(c) 20/9 for v_a, 25/9 for v_b

4.29 $R_a = 20$ kΩ, $R_b = 200$ kΩ, $R_f = 47$ kΩ

4.31 $v_{o1} = 15.85$ V, $v_{o2} = 13.6$ V

Chapter 5

5.1 $W = 0.2$ J; this is the energy stored in the inductor at $t = \infty$

5.3   (a) $v = 100e^{-10t}(1 - 10t)$ mV;

(b) -18.32 mW;

(c) delivering;

(d) 1.83 mJ;

(e) 3.38 mJ, 100 ms

5.5   (a) $t < 0$: $i = 0$; $0 \le t \le 25$ ms: $i = 16t$ A; $25 \le t \le 50$ ms: $i = 0.8 - 16t$ A; 50 ms $\le t$: $i = 0$

(b) $t < 0$: $v = 0$, $p = 0$, $w = 0$; $0 < t < 25$ ms: $v = 6$ V, $p = 96t$ W, $w = 48t^2$ J; $25 < t < 50$ ms: $v = -6$ V, $p = 96t - 4.8$ W, $w = 48t^2 - 4.8t + 0.12$ J; 50 ms $< t$: $v = 0$, $p = 0$, $w = 0$

5.9   (a) 2.77 ms

(b) 64.27 V

5.11   1007 W, absorbing

5.15   (a) $250\ \mu J$;

(b) $19{,}307.24\ \mu J$

5.17   (a) $1.25\ \mu C$;

(b) 5 V;

(c) $2\ \mu J$

5.19   8.64 H

5.23

5.25   (a) 5 A, $t \ge 0$;

(b) $5e^{-25t}$ A, $t \ge 0$;

(c) $4e^{-25t} + 6$ A, $t \ge 0$;

(d) $e^{-25t} - 6$ A, $t \ge 0$;

(e) 845 J;

(f) 125 J;

(g) 720 J

5.31   (a) 25 Ω;

(b) 12.5 ms;

(c) 312.5 mH;

(d) 2.5 J;

(e) 10.06 ms

5.33   33.33%

5.37   48.64%

5.39   (a) 200 A;

(b) 220 A;

(c) 123.1 μs

5.43   (a) $i_o(0^+) = 2$ A, $i_o(\infty) = 4$ A;

(b) $i_o = 4 - 2e^{-1000t}$ A, $t \ge 0^+$;

(c) 2.30 ms

5.45 $v_o = -6e^{-250t}$ V, $t \ge 0^+$

5.47   693.15 Ω

5.51   (a) 20 kΩ;

(b) 0.05 μF;

(c) 1 ms;

(d) 250 μJ;

(e) 804.72 μs

5.55   (a) 125×10^{-6} A/V;

(b) $180e^{-1000t}$ V, $t \ge 0^+$

5.59   (a) $39.6e^{-2000t}$ mA, $t \ge 0^+$;

(b) 14.05%

5.61 $-625e^{-1000t}\ \mu A$

5.65   (a) 50 J;

(b) 100 J;

(c) 150 J

5.67   (a) $10\ \mu J$;

(b) $40\ \mu J$;

(c) $50\ \mu J$

5.73 $-6 \sin 5000t$ V

5.75 (a) $125 \text{ k}\Omega$;

(b) 100 ms

Chapter 6

6.3 (a) 800 mH;

(b) $2500 \ \Omega$;

(c) 125 V;

(d) 12.5 mA;

(e) $e^{-4000t}[12.5 \cos 3000t + 68.75 \sin 3000t]$ mA, $t \geq 0$

6.5 (a) $800 \ \Omega$, 200 mH, $\alpha = 12{,}500$ rads/s, $\omega = 10$ krad/s;

(b) $i_R = -6.25e^{-5000t} + 25e^{-20{,}000t}$ mA, $t \geq 0^+$, $i_L = 5e^{-5000t} - 5e^{-20{,}000t}$ mA, $t \geq 0^+$, $i_C = 1.25e^{-5000t} - 20e^{-20{,}000t}$ mA, $t \geq 0^+$

6.7 (a) $i_R(0) = 45$ mA, $i_L(0) = -30$ mA, $i_C(0) = -15$ mA;

(b) $70e^{-10{,}000t} + 20e^{-40{,}000t}$ V, $t \geq 0$;

(c) $-28e^{-10{,}000t} - 2e^{-40{,}000t}$ mA, $t \geq 0$

6.9 $(120 \times 10^4 t + 90)e^{-20{,}000t}$ V, $t \geq 0$

6.11 432 mA

6.15 $e^{-8000t}[45 \cos 6000t - 60 \sin 6000t]$ V, $t \geq 0$

6.19 $960{,}000te^{-40{,}000t}$ V, $t \geq 0$

6.21 $e^{-4000t}[60 \cos 3000t - 320 \sin 3000t]$ V, $t \geq 0$

6.23 (a) $25 \text{ k}\Omega$, 5 H;

(b) 0, 12 A/s;

(c) $(4e^{-1000t} - 4e^{-4000t})$ mA, $t \geq 0$;

(d) $462.10 \ \mu$s;

(e) 1.89 mA;

(f) $(-20e^{-1000t} + 80e^{-4000t})$ V, $t \geq 0^+$

6.25 $(100 + 5000te^{-50t} + 100e^{-50t})$ V, $t \geq 0$

6.29 $[40 - 40e^{-5000t}(\cos 5000t + \sin 5000t)]$ V, $t \geq 0$

6.35 $50e^{-10t}(7 \cos 70t + \sin 70t)$ mA, $t \geq 0$

6.37 (a) $(20e^{-1000t} - 5e^{-4000t})$ mA, $t \geq 0$;

(b) $(80e^{-1000t} - 5e^{-4000t})$ V, $t \geq 0$

6.39 $15 - e^{-80t}(45 \cos 60t + 10 \sin 60t)$ mA, $t \geq 0$

6.41 $e^{-5t}(70 \cos 5t - 30 \sin 5t)$ V, $t \geq 0$

6.45 (a) 8 mJ;

(b) 10 mJ;

(c) -2 mJ;

(d) 16 mJ;

(e) $w_{\text{del}} = w_{\text{abs}} = 18$ mJ

Chapter 7

7.1 (a) 2000π rad/s;

(b) $10 \cos(2000\pi t - 144°)$ A

7.3 (a) 170 V;

(b) 60 Hz;

(c) 376.99 rad/s;

(d) -1.05 rad;

(e) $-60°$;

(f) 16.67 ms;

(g) 2.78 ms;

(h) $-170 \sin 120\pi t$ V;

(i) (25/18) ms;

(j) (25/9) ms

7.11 (a) 50 Hz;

(b) $0°$;

(c) $-90°$;

(d) $40 \ \Omega$;

(e) 127.32 mH;

(f) $j40 \ \Omega$

7.13 $32 \cos(8000t + 90°)$ V

7.15 (a) $0.04 \ \mu$F, $0.16 \ \mu$F;

(b) 40 nF: $i_g = 25 \cos 1000t$ mA, 160 nF: $i_g = 100 \cos 1000t$ mA

7.17 5000 rad/s

7.19 $40 + j30$ mS $= 50 \ \underline{/36.87°}$ mS

7.23 $30 - j40 \ \Omega = 50 \ \underline{/-53.13°} \ \Omega$

7.29 (a) $5 \ \underline{/72°} \ \Omega$;

(b) $50 \ \mu$s

7.31 $50 \cos(5000t - 106.26°)$ V

7.35 $9.49 \ \underline{/71.57°}$ A, $7.5 - j2.5 \ \Omega = 7.91 \ \underline{/-18.43°} \ \Omega$

7.37   (a) 350 $\angle 0°$ V;

(b) $100 + j100\ \Omega$;

(c)

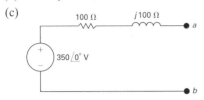

7.39

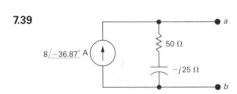

7.43

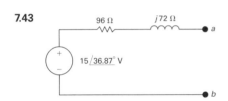

7.45   3 $\angle{-90°}$ A

7.47   15.81 $\angle{18.43°}$ V

7.49 $12\cos 5000t$ V

7.52 $\mathbf{I}_a = 30$ A, $\mathbf{I}_b = 30 - j20$ A, $\mathbf{I}_c = 30 + j10$ A,
$\mathbf{I}_d = -j30$ A

7.55   (a) $44.72\cos(8 \times 10^5 t - 10.30°)$ V;

(b) 224.82 ns

7.57 $72 + j96 = 120\ \angle{53.13°}$ V

7.63   (a) 60 V;

(b) 300 W

7.65   80 Ω

7.67   56.25 mW, −70.3125 mVAR, 90.044 mVA

7.69   (a) load 1: 0.96 lagging, 0.28

load 2: 0.80 leading, −0.60

load 3: 0.60 leading, −0.80

(b) 0.74 leading, −0.67

7.71   990 μW

7.75   (a) $72 - j24\ \Omega$;

(b) 0.9487 leading

7.77   2500 $\angle{16.26°}$ V(rms)

7.81   (a) $\mathbf{V}_{g1}$ is delivering 4200 W and 1200 magnetizing
VARS, $\mathbf{V}_{g2}$ is delivering 3600 W and is
absorbing 300 magnetizing VARS;

(b) $\sum P_{\text{gen}} = \sum P_{\text{abs}} = 7800$ W,
$\sum Q_{\text{del}} = \sum Q_{\text{abs}} = 2100$ VAR

Chapter 8

8.1   (a) $f(t) = 50\sin\frac{\pi}{2}t[u(t) - u(t-4)]$;

(b) $f(t) = (120 + 30t)[u(t+4) - u(t)]$
$+ (120 - 30t)[u(t) - u(t-8)]$
$+ (30t - 360)[u(t-8) - u(t-12)]$

8.5 $F(s) = \displaystyle\int_{-\epsilon}^{\epsilon} \frac{1}{2\epsilon}e^{-st}\,dt = \frac{e^{s\epsilon} - e^{-s\epsilon}}{2\epsilon s}$

$F(s) = \dfrac{1}{2s}\displaystyle\lim_{\epsilon\to 0}\left[\frac{se^{s\epsilon} + se^{-s\epsilon}}{1}\right] = 1$

8.8   (a) $1/(s+a)^2$;

(b) $\omega/(s^2 + \omega^2)$;

(c) $(\omega\cos\theta + s\sin\theta)/(s^2 + \omega^2)$;

(d) $s/(s^2 - 1)$;

(e) $(\sinh\theta + s\cosh\theta)/(s^2 - 1)$

8.16   (a) $-20e^{-2s}/(s+5)$;

(b) $(8/s^2)[e^{-s} - 2e^{-2s} + 2e^{-4s} - e^{-5s}]$

8.17   (a) $(4/s^2)[1 - 2e^{-4s} + 2e^{-12s} - e^{-16s}]$;

(b) $(4/s)[1 - 2e^{-4s} + 2e^{-12s} - e^{-16s}]$;

(c) $4[1 - 2e^{-4s} + 2e^{-12s} - e^{-16s}]$

8.23   (a) $(20e^{-2000t} - 20e^{-8000t})u(t)$ V;

(b) $(4e^{-8000t} - e^{-2000t})u(t)$ mA;

(c) $i_o(0) = 3$ mA. Yes, the initial inductor current is
zero by hypothesis; the initial resistor current is
zero because the initial capacitor voltage is zero
by hypothesis. Thus at $t = 0$ the source current
appears in the capacitor.

8.25   (a) $[70 + 10\sqrt{50}e^{-500t}\cos(3500t + 171.87°)]u(t)$ V;

(b) $[7 + 25e^{-500t}\cos(3500t - 106.26°)]u(t)$ mA

8.27   (a) $[40t - 8 + 16e^{-10t}]u(t)$;

(b) $[10 - 40te^{-2t} + 20e^{-2t}]u(t)$;

(c) $[10t - 5 + 10e^{-2t}\cos(t + 53.13°)]u(t)$;

(d) $[(2t - 1.5t^2 + 1)e^{-2t}]u(t)$;

(e) $[50te^{-t}\cos(2t - 16.26°)$
$+ 20e^{-t}\cos(2t + 36.87°)]u(t)$

8.31 $[20e^{-1000t} - 5e^{-4000t}]u(t)$ V

8.33 $[10e^{-4000t} \sin 3000t]u(t)$ V

Chapter 9

9.3

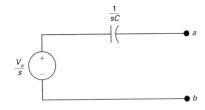

9.5 (a) $Z = 5(s^2 + 2000s + 10^7)/s$;

(b) Zeros at $-1000 + j3000$ rad/s and $-1000 - j3000$ rad/s, pole at 0

9.7 $Z = \frac{2}{s+1}$, no zeros, one pole at -1 rad/s

9.11 $[-45te^{-5t} + 24e^{-5t}]u(t)$ V

9.15 (a)

(b) $I_o = \dfrac{-300}{(s + 5000)(s + 20,000)}$

$i_o = -(20e^{-5000t} - 20e^{-20,000t})u(t)$ mA

(c) $V_o = \dfrac{-80}{(s + 5000)} + \dfrac{20}{(s + 20,000)}$

$v_o = -(80e^{-5000t} - 20e^{-20,000t})u(t)$ V

9.19 $10e^{-7000t} \cos(24,000t - 16.26°)u(t)$ mA

9.23 (a) $-6 \times 10^5/(s^2 + 10,000s + 16 \times 10^6)$;

(b) $[100e^{-8000t} - 100e^{-2000t}]u(t)$ V

9.29 (a) $I_a = \left(\dfrac{8}{3}\right)\left(\dfrac{1}{s} - \dfrac{1}{s+6}\right)$,

$I_b = \left(\dfrac{4}{3}\right)\left(\dfrac{1}{s} + \dfrac{2}{s+6}\right)$;

(b) $i_a = \left(\dfrac{8}{3}\right)\left(1 - e^{-6t}\right)u(t)$ A,

$i_b = \left(\dfrac{4}{3}\right)\left(1 + 2e^{-6t}\right)u(t)$ A;

(c) $V_a = \dfrac{(400/3)}{s^2} + \dfrac{(400/9)}{s} - \dfrac{(400/9)}{s+6}$,

$V_b = \dfrac{(400/3)}{s^2} - \dfrac{(800/9)}{s} + \dfrac{(800/9)}{s+6}$,

$V_c = \dfrac{(400/3)}{s^2} + \dfrac{(400/9)}{s} - \dfrac{(400/9)}{s+6}$;

(d) $v_a = \dfrac{400}{9}(3t + 1 - e^{-6t})u(t)$ V,

$v_b = \dfrac{400}{9}(3t - 2 + 2e^{-6t})u(t)$ V,

$v_c = \dfrac{400}{9}(3t + 1 - e^{-6t})u(t)$ V;

(e) $\approx 7.17s$

9.35 $[5 - 5000te^{-1000t} - 5e^{-1000t}]u(t)$ mA

9.39 246.28 ms

9.43 (a) $H(s) = \dfrac{50}{s + 50}$, pole at -50 rad/s;

(b) $H(s) = \dfrac{s}{s + 50}$, zero at 0, pole at -50 rad/s;

(c) $H(s) = \dfrac{s}{s + 3000}$, zero at 0, pole at -3000 rad/s;

(d) $H(s) = \dfrac{3000}{s + 3000}$, pole at -3000 rad/s;

(e) $H(s) = \dfrac{100}{s + 125}$, pole at -125 rad/s;

9.45 (a) $H(s) = \dfrac{-15,625s}{(s + 5000 - j10,000)(s + 5000 + j10,000)}$;

(b) zero at 0, pole at $-5000 + j10,000$ rad/s, pole at $-5000 - j10,000$ rad/s

9.51 (a) $v_o = t^2$ V $0 \le t \le 10s$,

$v_o = 40t - 200 - t^2$ V $10 \le t \le 20s$,

$v_o = 200$ V $20 \le t \le \infty s$;

(b)

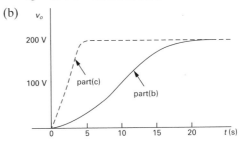

(c) $v_o = 16t^2$ V $0 \le t \le 2.5s$,

   $v_o = 160t - 200 - 16t^2$ V $2.5 \le t \le 5s$,

   $v_o = 200$ V $5 \le t \le \infty s$,

(d) The waveform in part (c) is closer to replicating the input waveform. In part (c) $h(\lambda)$ is closer to being an ideal impulse response.

9.55 (a) $i_o = 8(1 - e^{-1000t})$ μA $0 \le t \le 1$ ms

   $i_o = (12e^{-1000(t-0.001)} - 8e^{-1000t} - 4)$ μA

      1 ms $\le t \le 4$ ms

   $i_o = (12e^{-1000(t-0.001)} - 4e^{-1000(t-0.004)} - 8e^{-1000t})$ μA

      4 ms $\le t \le \infty$

(b) $v_o = 200e^{-1000t}$ μV $0 < t < 1$ ms

   $v_o = 200e^{-1000t} - 300e^{-1000(t-0.001)}$ μV

      1 ms $< t < 4$ ms

   $v_o = (200e^{-1000t} - 300e^{-1000(t-0.001)} + 100e^{-1000(t-0.004)})$ μV

      4 ms $< t < \infty$

(c) solutions are consistent with circuit behavior at $t = 1^-$ ms, 1^+ ms, 4^- ms, and 4^+ ms.

9.59 -27.43 V

9.61 $60e^{-5t}$ V $0 < t < \infty$

9.63 $50\cos(8000t + 36.87°)$ V

9.67 (a) $-16 \times 10^4 s/(s + 8000)(s + 16,000)$;

(b) $4\cos(8000t - 161.57°)$ V

9.74 (a) 0.5 J;

(b) 0 J;

(c) $1118.03e^{-3000t}\cos(4000t - 26.57°)u(t)$ V;

(d) show alternative approach leads to the same answer

9.75 (a) $100e^{-500t}u(t)$ V;

(b) At $t = 0^+$ impulsive current establishes 100 V on 0.5 μF capacitor which agrees with the solution. The equivalent capacitance is $(1/27)$ μF. Therefore the time constant is 2 ms. Thus $1/\tau$ equals 500, which agrees with the solution.

Index

Ollscoil na hÉireann, Gaillimh

3 1111 40072 5709

An Abbreviated List of Laplace Transform Pairs

TYPE	$f(t)(t > 0^-)$	$F(s)$
(impulse)	$\delta(t)$	1
(step)	$u(t)$	$\dfrac{1}{s}$
(ramp)	t	$\dfrac{1}{s^2}$
(exponential)	e^{-at}	$\dfrac{1}{s+a}$
(sine)	$\sin \omega t$	$\dfrac{\omega}{s^2 + \omega^2}$
(cosine)	$\cos \omega t$	$\dfrac{s}{s^2 + \omega^2}$
(damped ramp)	te^{-at}	$\dfrac{1}{(s+a)^2}$
(damped sine)	$e^{-at} \sin \omega t$	$\dfrac{\omega}{(s+a)^2 + \omega^2}$
(damped cosine)	$e^{-at} \cos \omega t$	$\dfrac{s+a}{(s+a)^2 + \omega^2}$

An Abbreviated List of Operational Transforms

OPERATION	$f(t)$	$F(s)$
Multiplication by a constant	$Kf(t)$	$KF(s)$
Addition/subtraction	$f_1(t) + f_2(t) - f_3(t) + \cdots$	$F_1(s) + F_2(s) - F_3(s) + \cdots$
First derivative (time)	$\dfrac{df(t)}{dt}$	$sF(s) - f(0^-)$
Second derivative (time)	$\dfrac{d^2 f(t)}{dt^2}$	$s^2 F(s) - sf(0^-) - \dfrac{df(0^-)}{dt}$
nth derivative (time)	$\dfrac{d^n f(t)}{dt^n}$	$s^n F(s) - s^{n-1} f(0^-) - s^{n-2}\dfrac{df(0^-)}{dt}$ $- s^{n-3}\dfrac{df^2(0^-)}{dt^2} - \cdots - \dfrac{d^{n-1} f(0^-)}{dt^{n-1}}$
Time integral	$\displaystyle\int_0^t f(x)\, dx$	$\dfrac{F(s)}{s}$
Translation in time	$f(t-a)u(t-a),\, a > 0$	$e^{-as} F(s)$
Translation in frequency	$e^{-at} f(t)$	$F(s+a)$
Scale changing	$f(at),\, a > 0$	$\dfrac{1}{a} F\left(\dfrac{s}{a}\right)$
First derivative (s)	$tf(t)$	$-\dfrac{dF(s)}{ds}$
nth derivative (s)	$t^n f(t)$	$(-1)^n \dfrac{d^n F(s)}{ds^n}$
s integral	$\dfrac{f(t)}{t}$	$\displaystyle\int_s^\infty F(u)\, du$